Nanotechnology to Overcome World's Most Critical Health Issues: Liposomes and beyond—a Themed Issue Dedicated to Professor Yechezkel Barenholz

Nanotechnology to Overcome World's Most Critical Health Issues: Liposomes and beyond—a Themed Issue Dedicated to Professor Yechezkel Barenholz

Editors

Marina A. Dobrovolskaia
Kirill A. Afonin

MDPI • Basel • Beijing • Wuhan • Barcelona • Belgrade • Manchester • Tokyo • Cluj • Tianjin

Editors
Marina A. Dobrovolskaia
Frederick National Laboratory for Cancer Research
Frederick, MD, USA

Kirill A. Afonin
University of North Carolina Charlotte
Charlotte, NC, USA

Editorial Office
MDPI
St. Alban-Anlage 66
4052 Basel, Switzerland

This is a reprint of articles from the Special Issue published online in the open access journal *Molecules* (ISSN 1420-3049) (available at: https://www.mdpi.com/journal/molecules/special_issues/Dedicated_Barenholz).

For citation purposes, cite each article independently as indicated on the article page online and as indicated below:

LastName, A.A.; LastName, B.B.; LastName, C.C. Article Title. *Journal Name* **Year**, *Volume Number*, Page Range.

ISBN 978-3-0365-8166-8 (Hbk)
ISBN 978-3-0365-8167-5 (PDF)

Contents

About the Editors

Marina A. Dobrovolskaia

Dr. Dobrovolskaia is the Laboratory Co-Director, Director of Operations, and the Head of the Immunology Section at the Nanotechnology Characterization Laboratory (NCL). She leads the NCL to provide preclinical nanoparticle characterization services to the nanotechnology research community, advance the translation of promising nanotechnology concepts from bench to clinic, and contribute to the education of the next generation of scientists in the field of preclinical development of nanotechnology-based products. In her role as the Head of the Immunology section, Dr. Dobrovolskaia leads a team conducting preclinical studies to monitor nanoparticles' toxicity to the immune system both in vitro and in vivo using a variety of immune function and animal models.

Kirill A. Afonin

Dr. Afonin is a full professor at the Department of Chemistry, the University of North Carolina at Charlotte. His expertise is in computational and experimental RNA biology. Dr. Afonin's research focuses on biopolymers, including both DNA and RNA, that can be programmed to fold into nucleic acid nanoparticles (NANPs) for beneficial uses in nanotechnology with broad applications. His team is using this promising technology to pursue several objectives in various fields, including nanodesign, immunology, delivery, biosensing, therapeutics, and dynamic structures. Dr. Afonin also serves as a founding council member and the vice-president of the International Society of RNA Nanotechnology and Nanomedicine. Among other awards, he is the recipient of two NIH Fellows Awards for Research Excellence (FARE), a prestigious NIH Maximizing Investigators' Research Award (MIRA R35), and two NIH R01s.

Preface to "Nanotechnology to Overcome World's Most Critical Health Issues: Liposomes and beyond—a Themed Issue Dedicated to Professor Yechezkel Barenholz"

The Special Issue "Nanotechnology to Overcome World's Most Critical Health Issues: Liposomes and beyond—a Themed Issue Dedicated to Professor Yechezkel Barenholz" was written by experts in the field of nanomedicine to honor Professor Barenholz of the Hebrew University, Jerusalem, whose distinguished career influenced many young scientists and set up high scientific quality standards for all researchers worldwide. Throughout his career, Professor Barenholz taught at leading universities in Israel and the United States of America. He is known worldwide for his work on the development of a PEGylated liposomal doxorubicin formulation, known as Doxil®, which has completely transformed care for patients worldwide. This Special Issue includes both original research papers and review articles describing a variety of nanotechnology platforms, including (but not limited to) liposomes for drug delivery, which are geared toward overcoming the world's most critical health issues and supporting the sustainability of the environment. We are immensely grateful to Professor Jindrich Kopecek of the University of Utah, Professor Phillip Low of Purdue University, Professor Kwangmeyung Kim of the Korea Institute of Science and Technology, Professors Szilard Pal and Aleksandar Széchenyi of the University of Pécs, Professor Kausik Chakrabarti of the University of North Carolina at Charlotte, and Dr. Stephan Stern of the Nanotechnology Characterization Lab and their team members for contributing their knowledge and sharing their research in this Special Issue. We also want to acknowledge all members of the Immunology Section at the Nanotechnology Characterization Lab and Professor Afonin's Laboratory at the University of North Carolina at Charlotte, for their dedication and tireless work in helping to complete this Special Issue. We cordially thank Kylie Xiao and Marci Mao of the MDPI Editorial team for their help with finalizing the issue and preparing a book. We enjoyed working with everyone and wish them continued success in their future endeavors. On behalf of all authors and editors of this Special Issue, we wish Professor Barenholz a lifetime of wellness and happiness to cherish; thank you for everything you have taught us, and we will continue to learn from your wisdom.

Marina A. Dobrovolskaia and Kirill A. Afonin
Editors

molecules

Editorial

Special Issue "Nanotechnology to Overcome the World's Most Critical Health Issues: Liposomes and Beyond—A Themed Issue Dedicated to Professor Yechezkel Barenholz"

Marina A. Dobrovolskaia [1,*] and Kirill A. Afonin [2,*]

1 Nanotechnology Characterization Laboratory, Cancer Research Technology Program, Frederick National Laboratory for Cancer Research Sponsored by the National Cancer Institute, Frederick, MD 21701, USA
2 Nanoscale Science Program, Department of Chemistry, University of North Carolina Charlotte, Charlotte, NC 28223, USA
* Correspondence: marina@mail.nih.gov (M.A.D.); kafonin@uncc.edu (K.A.A.)

This Special Issue is intended to celebrate Professor Yechezkel Barenholz's distinguished achievements. Professor Barenholz is Professor Emeritus of the Hebrew University in Jerusalem, Israel. He joined the University's faculty in 1968, received his PhD in 1971, and became a Professor in 1981. Throughout his career, Prof. Barenholz has taught young scientists at leading universities worldwide. One of the prominent achievements in Prof. Barenholz's career was the development of a PEGylated liposomal doxorubicin formulation, known as Doxil®, which completely transformed care for cancer patients worldwide. Professor Barenholz's research focuses on the biochemistry of lipids and membranes in addition to the biophysics laws underlying the fluidity of cellular membranes. Another focus area is the development of liposomes and lipid-based nanocarriers to overcome the shortcomings of current therapeutics by improving drug delivery. Professor Barenholz has authored more than 400 papers and has an h-index of 94; he is also the inventor of over 55 patents and an awardee of many prestigious national and international awards in the biomedical field. Professor Barenholz is highly regarded by his peers and students. One of the examples of his continuous contributions to the education of the next generation of scientists is the "Barenholz Prize", which supports Israeli PhD students in applied sciences and encourages their professional growth and innovation. Professor Barenholz has founded and is currently leading the steering committee of the Hebrew University School of Business Administration BioMed-MBA program, through which he organized an online platform that enables the Israeli BioMed ecosystem.

This Special Issue comprises a collection of ten research and review articles prepared by international leaders in the fields of biomedical nanotechnology and drug delivery. The research articles describe several innovative technologies that span from the formulation of new nanomedicines and imaging agents to the in vitro assessment of their immunological properties.

The work presented by Professor Kim's team from the Korea Institute of Science and Technology introduces new nanoparticles for combinational photochemotherapy of pancreatic cancer [1]. The formulation was based on light-activated monomethyl auristatin E prodrug linked to a photosensitizer (Ce6) through a caspase-3-specific cleavable peptide. Under irradiation with visible light, Ce6-generated reactive oxygen species induced the overexpression of caspase-3 in cancer cells, which, in turn, released the drug. The resulting formulations were extensively characterized, and in vivo data confirmed significant delays in tumor progression.

Professor Széchenyi, from the University of Pécs, and colleagues have developed a nanotechnology-based platform with which to treat onychomycosis [2]. The tested antifungal nanoformulations were based on silica-nanoparticle-stabilized Pickering emulsions, which were designed for the site-specific delivery of tioconazole and Melaleuca alternifolia

Citation: Dobrovolskaia, M.A.; Afonin, K.A. Special Issue "Nanotechnology to Overcome the World's Most Critical Health Issues: Liposomes and Beyond—A Themed Issue Dedicated to Professor Yechezkel Barenholz". *Molecules* **2023**, *28*, 4788. https://doi.org/10.3390/molecules28124788

Received: 1 June 2023
Accepted: 13 June 2023
Published: 15 June 2023

essential oil. Microbiological in vitro experiments with relevant pathogens confirmed the significant antifungal effect and indicated promise for the topical treatment of onuchomycosis.

In order to overcome some of the commonly accepted hurdles associated with the clinical use of NIR dyes, the group of Professor Low from Purdue University has synthesized and tested a series of novel PEGylated UreterGlow derivatives [3]. The team identified promising bioimaging candidates with prolonged kidney retention times and unique emission profiles, not overlapping with other commonly used NIR probes.

Professors Yang and Kopeček, with colleagues from the University of Utah, have expanded the recently introduced notion of drug-free macromolecular therapeutics, or DFMT [4]. In their approach, the antibodies conjugated to short oligonucleotides could bind specific cell surface receptors, characteristic to diseased cells, and then become crosslinked via human serum albumin decorated with complementary oligonucleotides, which in turn induces apoptosis. The work published in this Special Issue demonstrated a scenario where DFMT was designed to crosslink CD38 receptors on lymphoma and multiple myeloma cells.

The immunology team of the Nanotechnology Characterization Laboratory at the Frederick National Laboratory for Cancer Research and Dr. Pang of the U.S. Food and Drug administration have evaluated the suitability of several in vitro assays that use peripheral blood mononuclear cells (PBMCs) as model systems to detect the innate immune responses induced by ten common immune-modulating impurities, as well as by a peptide drug product [5]. Based on the results of this comprehensive study, the sets of signature cytokines have been identified for further use in multiplex assays. In addition, the authors have demonstrated that the logistics of blood storage and handling must be taken into consideration and further evaluated, since they may influence the measured immunostimulatory responses.

Lastly, a collaborative effort between the Nanotechnology Characterization Laboratory at the Frederick National Laboratory for Cancer Research and Professor Afonin from the University of North Carolina at Charlotte have demonstrated how compositional variations in commercially available lipid-like carriers influence the immunostimulatory properties of nucleic acid nanoparticles that have different architectural characteristics [6].

The reviews cover several important topics that deal with the immunotoxicity of nanomedicines, drug formulations, and biomedical applications of nucleic-acid-based nanomaterials.

Dr. Stern and colleagues of the Nanotechnology Characterization Laboratory at the Frederick National Laboratory for Cancer Research provide a comprehensive review on nanomedicine reformulations of chloroquine and hydroxychloroquine to improve their therapeutic performance and broaden clinical applications [7]. Several reformulation nanomedicine approaches, ranging from liposomes to metal nanoparticles, have been discussed, as have the gaps in the current understanding of new nanoformulations; future perspectives and recommendations that may help to overcome the current limitations have been suggested.

Professor Chakrabarti and his team from the University of North Carolina at Charlotte discussed the ways how functional biological interactions can be studied using RNA nanotechnology [8], and Professor Afonin's group from the same university elaborated on the immunorecognition of nucleic-acid-based nanoparticles designed for therapeutic applications [9].

Finally, the Nanotechnology Characterization Laboratory at the Frederick National Laboratory for Cancer Research reviewed the cellular and molecular mechanisms of inflammation caused by innate immunity-modulating impurities, with an emphasis on the safety and efficacy of pharmaceutical products [10].

Acknowledgments: The study was funded in part (M.A.D.) by federal funds from the National Cancer Institute, National Institutes of Health, under contract 75N91019D00024. The content of this publication does not necessarily reflect the views or policies of the Department of Health and Human Services, nor does mention of trade names, commercial products, or organizations imply endorsement by the U.S. Government. Research reported in this publication was supported by the National Institute of General Medical Sciences of the National Institutes of Health under award number R35GM139587 (to K.A.A.). The content is solely the responsibility of the authors and does not necessarily represent the official views of the National Institutes of Health.

Conflicts of Interest: The authors declare no conflict of interest.

References

1. Cho, I.K.; Shim, M.K.; Um, W.; Kim, J.H.; Kim, K. Light-Activated Monomethyl Auristatin E Prodrug Nanoparticles for Combinational Photo-Chemotherapy of Pancreatic Cancer. *Molecules* **2022**, *27*, 2529. [CrossRef] [PubMed]
2. Voros-Horvath, B.; Das, S.; Salem, A.; Nagy, S.; Boszormenyi, A.; Koszegi, T.; Pal, S.; Szechenyi, A. Formulation of Tioconazole and Melaleuca alternifolia Essential Oil Pickering Emulsions for Onychomycosis Topical Treatment. *Molecules* **2020**, *25*, 5544. [CrossRef] [PubMed]
3. Mahalingam, S.M.; Putt, K.S.; Srinivasarao, M.; Low, P.S. Design of a Near Infrared Fluorescent Ureter Imaging Agent for Prevention of Ureter Damage during Abdominal Surgeries. *Molecules* **2021**, *26*, 3739. [CrossRef] [PubMed]
4. Gambles, M.T.; Li, J.; Wang, J.; Sborov, D.; Yang, J.; Kopecek, J. Crosslinking of CD38 Receptors Triggers Apoptosis of Malignant B Cells. *Molecules* **2021**, *26*, 4658. [CrossRef] [PubMed]
5. Holley, C.K.; Cedrone, E.; Donohue, D.; Neun, B.W.; Verthelyi, D.; Pang, E.S.; Dobrovolskaia, M.A. An In Vitro Assessment of Immunostimulatory Responses to Ten Model Innate Immune Response Modulating Impurities (IIRMIs) and Peptide Drug Product, Teriparatide. *Molecules* **2021**, *26*, 7461. [CrossRef] [PubMed]
6. Newton, H.S.; Radwan, Y.; Xu, J.; Clogston, J.D.; Dobrovolskaia, M.A.; Afonin, K.A. Change in Lipofectamine Carrier as a Tool to Fine-Tune Immunostimulation of Nucleic Acid Nanoparticles. *Molecules* **2023**, *28*, 4484. [CrossRef] [PubMed]
7. Stevens, D.M.; Crist, R.M.; Stern, S.T. Nanomedicine Reformulation of Chloroquine and Hydroxychloroquine. *Molecules* **2020**, *26*, 175. [CrossRef] [PubMed]
8. Klotz, K.; Radwan, Y.; Chakrabarti, K. Dissecting Functional Biological Interactions Using Modular RNA Nanoparticles. *Molecules* **2022**, *28*, 228. [CrossRef] [PubMed]
9. Bila, D.; Radwan, Y.; Dobrovolskaia, M.A.; Panigaj, M.; Afonin, K.A. The Recognition of and Reactions to Nucleic Acid Nanoparticles by Human Immune Cells. *Molecules* **2021**, *26*, 4231. [CrossRef] [PubMed]
10. Holley, C.K.; Dobrovolskaia, M.A. Innate Immunity Modulating Impurities and the Immunotoxicity of Nanobiotechnology-Based Drug Products. *Molecules* **2021**, *26*, 7308. [CrossRef] [PubMed]

molecules

MDPI

Article

Light-Activated Monomethyl Auristatin E Prodrug Nanoparticles for Combinational Photo-Chemotherapy of Pancreatic Cancer

In Kyung Cho [1,2,3,†], Man Kyu Shim [2,†], Wooram Um [4], Jong-Ho Kim [3] and Kwangmeyung Kim [2,5,*]

1 Laboratory Animal Resource Center, Korea Research Institute of Bioscience and Biotechnology, Cheongju 28116, Korea; cik@kribb.re.kr
2 Biomedical Research Institute, Korea Institute of Science and Technology, Seoul 02792, Korea; mks@kist.re.kr
3 Department of Pharmaceutical Science, College of Pharmacy, Kyung Hee University, 26 Kyungheedae-ro, Dongdaemun-gu, Seoul 02447, Korea; jonghokim@khu.ac.kr
4 School of Chemical Engineering, College of Engineering, Sungkyunkwan University, Suwon 16419, Korea; tings0609@nate.com
5 KU-KIST Graduate School of Converging Science and Technology, Korea University, Seoul 02841, Korea
* Correspondence: kim@kist.re.kr
† These authors contributed equally to this work.

Abstract: Pancreatic cancer is a highly fatal disease that is becoming an increasingly leading cause of cancer-related deaths. In clinic, the most effective approach to treat pancreatic cancers is the combination treatment of several chemotherapeutic drugs, including fluorouracil, leucovorin, irinotecan, and oxaliplatin (FOLFIRINOX), but this approach is not adequate to manage patients due to their severe toxic side effects. Herein, we proposed light-activated monomethyl auristatin E (MMAE) prodrug nanoparticles for combinational photo-chemotherapy and optimized its applications for pancreatic cancer treatment. The photosensitizer (Ce6) and chemotherapeutic drug (MMAE) were conjugated through caspase-3-specific cleavable peptide (KGDEVD). The resulting CDM efficiently promoted the reactive oxygen species (ROS) under visible light irradiation and thereby induced caspase-3 overexpression in pacreatic cancers, which subsequently released the MMAE from the system. Importantly, MMAE released from CDM further amplified the activation of CDM into MMAE by inducing extensive apoptotic cell death in tumor microenvironment for treatment of tumor cells in deep in the tumor tissues as far visible light cannot reach. In addition, CDM formed prodrug nanoparticles via intermolecular π-π stacking and hydrophobic interactions, allowing durable and reliable treatment by preventing fast leakage from the pancreatic cancers via the lymphatic vessels. The CDM directly (intratumoral) injected into pancreatic cancers in orthotopic models through an invasive approach significantly delayed the tumor progression by combinational photo-chemotherapy with less toxic side effects. This study offers a promising and alternative approach for safe and more effective pancreatic cancer treatment via prodrug nanoparticles that combine photodynamic therapy and chemotherapy.

Keywords: pancreatic cancer; prodrug nanoparticles; combination treatment; synergistic effect

Citation: Cho, I.K.; Shim, M.K.; Um, W.; Kim, J.-H.; Kim, K. Light-Activated Monomethyl Auristatin E Prodrug Nanoparticles for Combinational Photo-Chemotherapy of Pancreatic Cancer. *Molecules* **2022**, *27*, 2529. https://doi.org/10.3390/molecules27082529

Academic Editors: Marina A. Dobrovolskaia and Kirill A. Afonin

Received: 22 March 2022
Accepted: 12 April 2022
Published: 14 April 2022

1. Introduction

Pancreatic cancer, with a 5-year survival rate of less than 5%, is a highly fatal disease that is becoming an increasingly leading cause of cancer-related deaths [1]. Most patients are diagnosed with unresectable or metastatic disease; thus, their treatment options are limited [2,3]. In addition, even for the small subset of patients who have resectable pancreatic cancers, the prognosis remains very poor with only 20% survival rate of 5 years after surgery [4]. In the past decades, significant advances have been made in diagnostic methods, perioperative management, and local or systemic therapies for advanced malignancies, but there is only modest incremental progress in patient survival [5]. The most effective approach to treat advanced pancreatic cancers in patients is the combination treatment of several chemotherapeutic drugs, including fluorouracil, leucovorin, irinotecan, and oxaliplatin (FOLFIRINOX), which led to significantly prolonged survival of patients [6,7]. However, systemic administration of such drugs has risk of the severe toxic side effects due to their low cancer specificity; thus, there is still a desperate need for new treatment options that have a potent antitumor therapeutic potential with less side effects for safe and more effective pancreatic cancer treatment [8].

Photodynamic therapy (PDT), a mature localized phoththerapy with approval for clinical use in 1993, is an alternative treatment option for advanced cancers [9]. Under visible light irradiation, the photosensitizers absorb a photon of the light and released ions and energy is directly transferred to the molecular oxygen to produce a reactive oxygen species (ROS), such as singlet oxygen ($^1O_2$), superoxide (O_2^-), and peroxide anions (O_2^{2-}), which damage the cancer tissues [10]. However, limited penetration depth of visible light to biological tissues and oxygen depletion during treatment greatly reduce the antitumor therapeutic potential of PDT [11,12]. Therefore, it is hard to expect great outcomes of patients for complete regression of pancreatic cancer by monotherapy of PDT. To overcome these obstacles, PDT can be combined with chemotherapeutic drugs to expect synergistic effect for improved therapeutic efficacy in pancreatic cancer; thus, developing a promising drug to combine PDT with a chemotherapeutic drug for combinational photo-chemotherapy and optimizing its treatment methods for pancreatic cancer are formidable challenges [13,14].

Herein, we proposed light-activated monomethyl auristatin E (MMAE) prodrug nanoparticles, constructed with photosensitizer chorin e6 (Ce6), caspase-3-specific cleavable KGDEVD peptide, and a chemotherapeutic drug (MMAE) for combinational photo-chemotherapy of pancreatic cancer. The MMAE was chemically conjugated to C-terminus of KGDEVD peptide via the self-immolative linker (PABC; *p*-aminobenzylcarbamate), and Ce6 was further introduced to N-terminus of peptide (Figure 1a). The resulting Ce6-KGDEVD-PABC-MMAE prodrug (CDM) promotes a significant ROS by visible light-induced PDT, and overexpressed caspase-3 in tumor tissues by apoptosis, subsequently trigger release of MMAE by enzymatic cleavage of KGDEVD peptide (Figure 1b). Importantly, MMAE released from CDM further amplified the activation of CDM into MMAE by inducing extensive apoptotic cell death in the tumor microenvironment for treatment of tumor cells deep in the tissues as far visible light cannot reach. Compared with the efficacy of monotherapy of PDT that is restricted by depletion of oxygen during treatment and depth limitation of light, CDM induced considerable cytotoxic effects by releasing additionally MMAE that has 100–1000 times higher potency than doxorubicin [8]. In contrast, the release of MMAE from CDM is greatly minimized in the absence of visible light due to lack of caspase-3 expression, thereby reducing the risk of the severe toxic side effects during treatment. Interestingly, CDM form stable nanoparticles via the intermolecular π-π stacking and hydrophobic interactions, which prevent fast leakage from the tumors via the lymphatic vessels to allow durable and reliable treatment [15–18]. In this study, we established the optimal treatment methods for pancreatic cancers using the CDM-mediated photo-chemotherapy. The CDM is directly (intratumoral) injected into pancreatic cancers in orthotopic models through invasive approach, and cancer tissue is locally irradiated by visible light to promote the caspase-3 overexpression by inducing apoptosis, resulting in subsequent release of MMAE (Figure 1c). This study demonstrated a promising therapeutic potential of combinational photo-chemotherapy by CDM in orthotopic pancreatic cancer models compared with monotherapy of photodynamic therapy or chemotherapy.

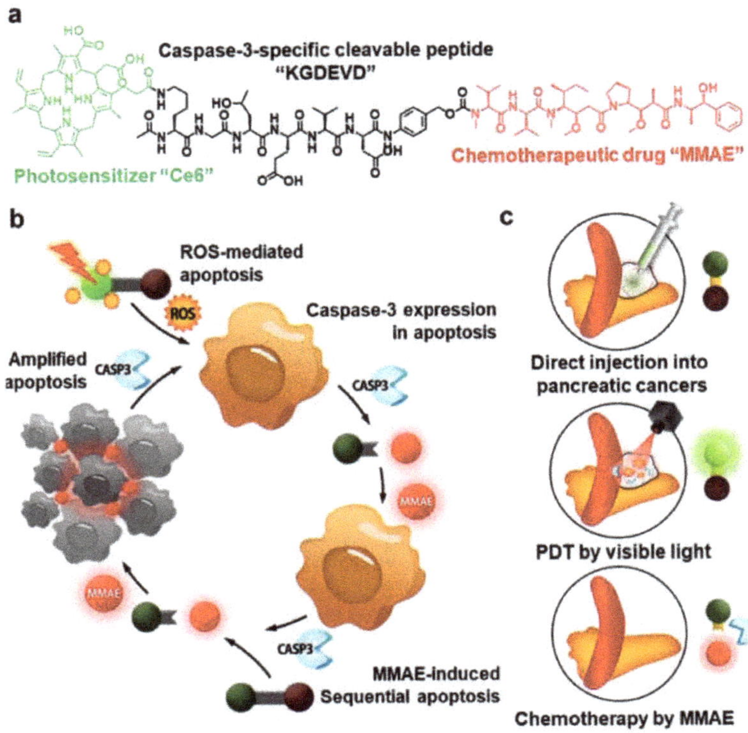

Figure 1. Light-activated monomethyl auristatin E (MMAE) prodrug nanoparticles and its application for pancreatic cancer treatment. (**a**) Photosensitizer (Ce6) and chemotherapeutic drug (MMAE) are chemically conjugated through the caspase-3-specific cleavable KGDEVD peptide, resulting in CDM. (**b**) The CDM promotes a significant reactive oxygen species (ROS) in cancer cells and thereby induces caspase-3 overexpression by apoptosis, which subsequently triggers release of MMAE. (**c**) Directly injected CDM in pancreatic cancers causes a potent antitumor therapeutic potential by combinational photo-chemotherapy under visible light irradiation, with less toxic side effects.

2. Results and Discussion

2.1. Preparation and Characterization of CDM

As an alternative therapeutic approach to combine PDT and chemotherapy for pancreatic cancer treatment, light-activated monomethyl auristatin E (MMAE) prodrug was prepared by conjugating photosensitizer (Ce6), caspase-3-specific cleavable peptide (KGDEVD), and a chemotherapeutic drug (MMAE). The MMAE was conjugated to the C-terminus of KGDEVD peptide through a self-immolative linker (PABC), and Ce6 was further introduced to the N-terminus of peptide via EDC/NHS reaction, resulting in CDM (Figure S1). The target enzyme of caspase-3 is the most widely studied as a biomarker for cancer-specific diagnosis and therapy, as it plays a key role in both the intrinsic and extrinsic death receptor pathways [19]. The KGDEVD peptide was used as an apoptosis-specific cleavable linker because it has great responsiveness to active caspase-3 at an early stage of apoptosis [20]. In addition, the self-immolative linker, PABC, was selected to design the prodrug owing to its favorable electronic and steric characteristics for enzymatic activation [21]. Finally, the MMAE that has 100–1000 times higher potency than doxorubicin was conjugated to amplify the antitumor therapeutic potential by subsequent release after visible light irradiation for PDT [8]. After the reaction, 99% of CDM was purified with high-performance liquid chromatography (HPLC; Figure S2). The molecular weight of CDM was also measured using a MALDI-TOF mass spectrometer, wherein the exact molecular weight was calculated to be 2131.1 Da and then confirmed to be 2131.1 m/z (Figure S3). Interestingly, CDM efficiently self-assembled into prodrug nanoparticles via intermolecular π-π stacking and hydrophobic interactions, showing a size distribution of 50–110 nm with an average size of 75.1 $\pm$ 4.4 nm in aqueous condition (Figure 2a) [15–17]. A TEM image further showed the spherical structure of CDM nanoparticles in saline (Figure 2b). In addition, CDM nanoparticles showed great stability in saline; significant changes of particle size were not observed for 48 h of incubation (Figure 2c). These nano-sized particles of CDM can prevent the rapid clearance through lymphatic vessels from the pancreatic cancers, which allow more durable and reliable treatment compared with small molecular drugs [18]. Next, we assessed the target enzyme-specific cleavage of CDM nanoparticles. When the CDM nanoparticles were incubated with caspase-3 for 24 h, approximately 99% of CDM was cleaved, resulting in release of MMAE (Figure 2d). In contrast, CDM nanoparticles were not cleaved after incubation with caspase-9, caspase-8, cathepsin B, cathepsin L, cathepsin K, and cathepsin D for 24 h, indicating high target enzyme-specificity. Upon visible light irradiation, the ROS production from CDM was nearly similar with Ce6 in same experimental conditions (Figure 2e). The photophysical property of CDM was further confirmed by the Singlet Oxygen Sensor Green (SOSG) assays, wherein efficiency for singlet oxygen generation of CDM was similar with Ce6 (Figure S4). This result demonstrated that chemical modification with KGDEVD-MMAE did not affect the ability to promote ROS under visible light irradiation of photosensitizer, Ce6. Taken together, CDM efficiently formed prodrug nanoparticles with photosensitizer and chemotherapeutic drug, and it was expected that they promote a significant ROS in pancreatic cancers under visible light irradiation to induce caspase-3 overexpression, which subsequently release the MMAE.

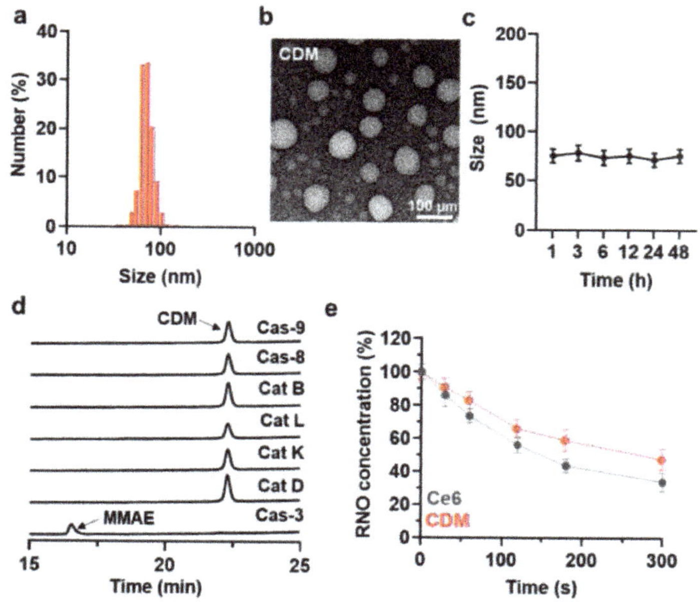

Figure 2. Physicochemical characterization of CDM. (**a**) Size distribution of CDM in aqueous condition (saline; 1 mg/kg). (**b**) TEM image showing spherical morphology of CDM. (**c**) Stability of CDM nanoparticles in saline. (**d**) Target enzyme-specific cleavage of CDM was assessed after incubation with caspase-3 and various other enzymes. (**e**) ROS generation of CDM or Ce6 in the presence of visible light irradiation.

2.2. In Vitro Cytotoxicity and Caspase-3 Overexpression by CDM

We assessed in vitro cytotoxicity and caspase-3 overexpression by CDM under visible light irradiation. First, we found the optimal irradiation timing by evaluating cellular uptake of CDM over time after incubation in human pancreatic cancer cell line KPC960. When the CDM (500 nM) was incubated with KPC960, the fluorescence (Ex/Em:633/660) of CDM was gradually increased in an incubation time-dependent manner, and that was saturated after 12 h of incubation (Figure 3a). Then, the caspase-3 expression levels in the KPC960 cells were assessed after treatment of MMAE, Ce6 with visible light, or CDM with visible light (Figure 3b). The KPC960 cells were incubated with an equivalent dose (500 nM) of MMAE, Ce6, or CDM for 12 h, and Ce6- or CDM-treated KPC960 cells were exposed to visible light with power of 40 mW for 5 min. As expected, MMAE- or Ce6 (with light)-treated KPC960 cells showed significantly elevated amounts of caspase-3 due to apoptosis induction of cells compared with the control group. Notably, CDM with light-treated KPC960 expressed the highest amount of caspase-3, which was significantly upregulated compared with a single treatment of Ce6 (with light) and MMAE. These results suggested that caspase-3 overexpression during PDT by CDM subsequently promotes the release of MMAE, thus amplifying apoptosis of pancreatic cancer cells. As a control, caspase-3 expression in KPC960 was not upregulated after treatment with CDM in the absence of visible light irradiation; thus, we expected that CDM can reduce toxic side effects during treatment by restricting the drug activation in off-target sites. Finally, combinational photo-chemotherapy of CDM resulted in a potent cytotoxicity in KPC960 cells. The viability of KPC960 cells were assessed after treatment with MMAE, Ce6 with light, or CDM with or without light for 12 h (Figure 3c). As expected, CDM in the presence of light showed significantly higher cytotoxicity in KPC960 cells compared with MMAE and Ce6 with light, which further confirmed the enhanced antitumor therapeutic potential of combinational photo-chemotherapy.

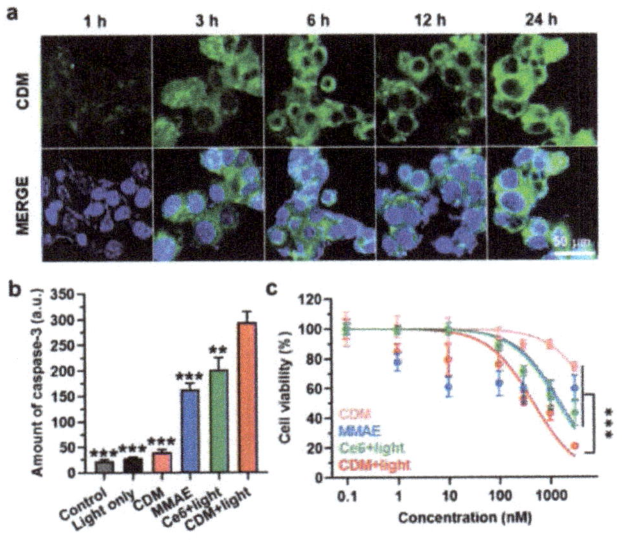

Figure 3. In vitro cytotoxicity and caspase-3 overexpression by CDM. (**a**) Time-dependent cellular uptake of CDM in human pancreatic cancer cell, KPC960. (**b**) The caspase-3 expression levels in KPC960 cells after treatment of CDM, MMAE, or Ce6. CDM- or Ce6-treated KPC960 cells were exposed to visible light with power of 40 mW for 5 min. The asterisks in Figure indicate the comparison to the CDM+light group. (**c**) The viability of KPC960 cells after treatment of CDM, MMAE, or Ce6. CDM- or Ce6-treated KPC960 cells were exposed to visible light with power of 40 mW for 5 min.

As a control, the cytotoxicity of CDM in cardiomyocytes (H9C2) and human dermal fibroblasts (HDF) was also assessed in absence of visible light to confirm whether CDM can prevent non-specific cleavage in the off-target tissues for minimizing severe toxic side effects. As expected, significant cytotoxicity was not observed in both cell lines (Figure S5). These in vitro results clearly demonstrated the mode of action (MOA) of CDM that promotes caspase-3 overexpression in the pancreatic cancer cells by Ce6-mediated PDT and the subsequently release of the MMAE for synergistic effects.

2.3. *In Vivo Therapeutic Efficacy of CDM in Orthotopic Pancreatic Cancer Models*

We next evaluated a therapeutic efficacy of combinational photo-chemotherapy by CDM in orthotopic pancreatic cancer models. The mice models were prepared by direct inoculation of KPC960 cells (1×10^5) into the pancreas tissue after incision at the left abdominal side. After 14 days of inoculation, the mice were randomly divided into five groups of saline, MMAE, Ce6 with light, or CDM with or without light, and each drug (0.1 mg/kg of Ce6 and MMAE, 0.3 mg/kg of CDM as equivalent dose of 0.1 mg/kg of MMAE) was directly injected into pancreatic cancers via the invasive approach. In case of Ce6 or CDM groups, pancreatic cancer tissues were locally irradiated with power of 160 mW for 10 min. The visible light irradiation was performed after minimum incision, as shown in Figure S6. Importantly, CDM with light (299.31 ± 22.1 mm^3) significantly delayed the growth of pancreatic cancer compared with saline (2105.11 ± 205.1 mm^3), Ce6 with light (631.51 ± 60.22 mm^3), and CDM without light (1261.21 ± 188.31 mm^3) groups on day 15 after treatments (Figure 4a). The mice treated with MMAE showed delayed progression of pancreatic cancer (211.21 ± 20.51 mm^3 on day 8), but they were all dead within 8 days of treatment due to severe toxic side effects of MMAE. The pancreatic cancer tissues stained with H&E or TUNEL exhibited the elevated damaged areas in CDM with the light group compared with other treatments, which further confirmed the enhanced therapeutic efficacy of CDM-mediated combinational photo-chemotherapy (Figure 4b). Taken together, these in vivo results demonstrated that the combination of Ce6-mediated PDT and MMAE-mediated chemotherapy by CDM induces a potent therapeutic efficacy in the pancreatic cancers.

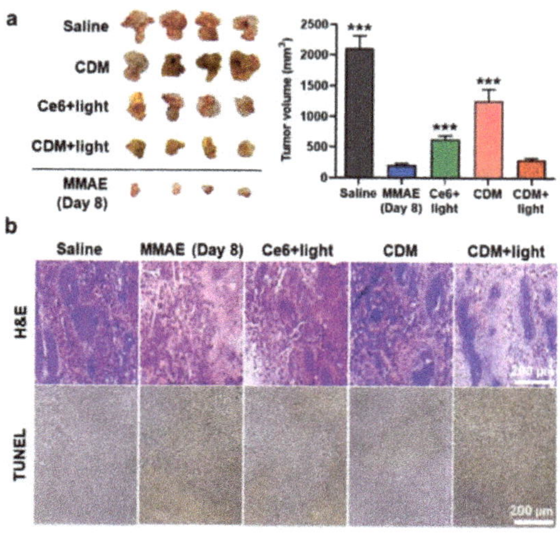

Figure 4. In vivo therapeutic efficacy of CDM in orthotopic pancreatic cancer models. (**a**) Optical images and volume of cancer tissues from orthotopic pancreatic cancer models after treatment with saline, MMAE, Ce6, or CDM. The tumors of Ce6- or CDM-treated mice were locally irradiated with visible light with power of 160 mW for 10 min. The asterisks in Figure indicate the comparison to the CDM+light group. (**b**) Pancreatic cancer tissues stained with H&E or TUNEL.

2.4. The Safety of CDM Treatment in Orthotopic Pancreatic Cancer Models

The safety of CDM treatment was assessed in orthotopic pancreatic cancer models, which were treated as same protocol in Figure 4. First, mice treated with saline, Ce6 with light, or CDM with or without light did not show significant body weight changes during treatments (Figure 5a). In contrast, MMAE treatment resulted in significant body weight loss of mice due to their severe toxicity, and mice were eventually dead within 8 days after treatment (Figure 5b). The median survival of the mice treated with saline, Ce6 with light, and CDM without light was determined to be 15 days, 24 days, and 18 days, wherein the mice were dead owing to the tumor progression. On the contrary, mice in the CDM with light group survived over 30 days by delaying tumor growth and reducing toxic side effects. The safety of CDM treatment was further evaluated by analyzing normal organ tissues on day 8 after treatment. As expected, since the spleen is located nearby pancreatic cancer, MMAE-treated mice revealed a significant reduction in spleen weight (Figure 5c). However, CDM greatly reduced the MMAE-mediated toxicity during treatment by releasing MMAE specifically under visible light irradiation that promotes caspase-3 expression and minimizing the drug leakage from the pancreatic cancers via the nano-sized particle.

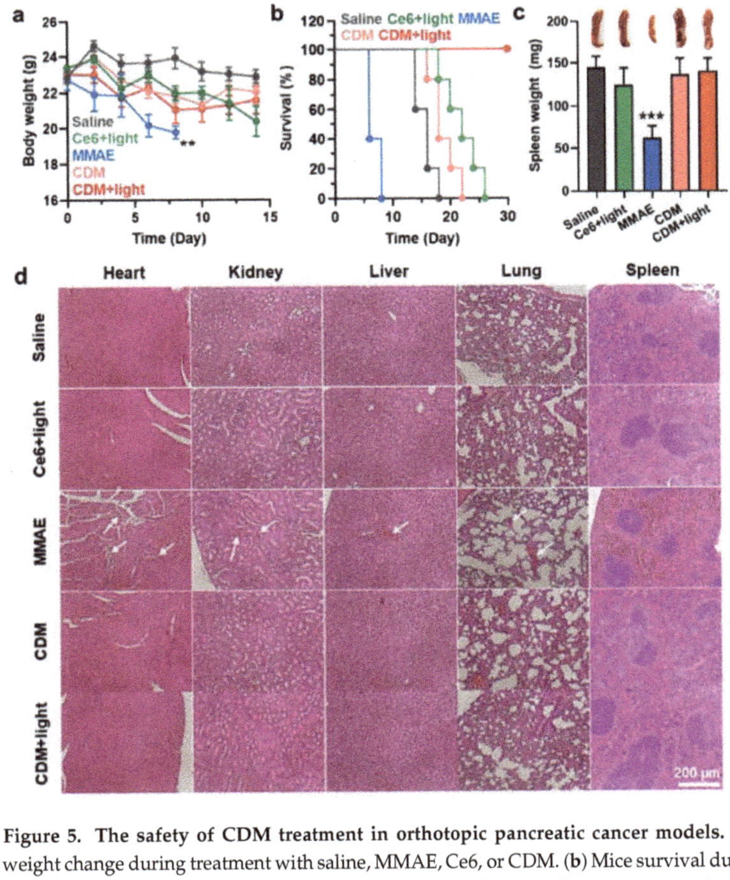

Figure 5. The safety of CDM treatment in orthotopic pancreatic cancer models. (**a**) The body weight change during treatment with saline, MMAE, Ce6, or CDM. (**b**) Mice survival during treatment with saline, MMAE, Ce6, or CDM. (**c**) The spleen weights of orthotopic pancreatic cancer models after 8 days of each treatment. (**d**) The organ tissues stained with H&E after 8 days of each treatment. The asterisks in Figure indicate the comparison to the CDM+light group.

As a result, the spleen weight of mice treated with CDM was nearly similar to that of saline-treated mice. In addition to spleen tissue, other tissues were also stained with H&E to observe damaged areas on day 8 after treatments (Figure 5d). Importantly, MMAE treatment resulted in extensive toxicity in every organ (white arrows), but CDM-treated mice showed only negligible structural abnormalities in the organ tissues. These results showed that even with the 100–1000 times more potent efficacy than doxorubicin, the clinical use of MMAE is strictly hindered owing to the severe toxicity, but CDM efficiently mitigated MMAE-related side effects based on prodrug nanoparticle formulation that can be specifically activated by overexpressed caspase-3 at defined tumor tissues under visible light irradiation.

3. Materials and Methods

3.1. Materials

Ac-K(Alloc)GD(All)E(All)VD(All)-OH (alloc-KGDEVD) was purchased from Peptron (Daejeon, Korea). Chlorin e6 (Ce6) was purchased from Frontier Scientific Inc. (Logan, UT, USA). 2-ethoxy-1-ethoxycarbonyl-1,2-dihydroquinoline (EEDQ), Bis(p-nitrophenyl) carbonate, 1-ethyl-3(3-dimethylaminopropyl) carbodiimide (EDC), Dimethyl sulfoxide (DMSO), *N,N*-diisopropylethylamine (DIPEA), hydroxybenzotriazole (HOBt), p-aminobenzyl alcohol, tetrakis(triphenylphosphine)palladium, and N-hydroxysuccinimide (NHS) were purchased from Sigma Chemical Co. (St. Louis, MO, USA). Anhydrous dimethylformamide (DMF) and dimethylsulfoxide (DMSO) were purchased from Merck (Darmstadt, Germany). Tributyltin hydride (Bu3SnH) and glacial acetic acid were purchased from Acros (St. Louis, MO, USA). DMEM medium, penicillin–streptomycin, and fetal bovine serum (FBS) were purchased from WelGENE Inc. (Daegu, Korea). Cathepsin B, D, K, L, caspase-9, caspase-3, and Caspase-3 ELISA Kit (cat# SMIF00) were purchased from R&D system (Minneapolis, MN, USA). KPC960 (Human pancreatic cancer) cell lines were purchased from American Type Culture Collection (ATCC; Manassas, VA, USA). Tem grid (Carbon Film 200 Mesh copper) was purchased from Electron Microscopy Sciences (Hatfield, PA, USA). Cell counting kit-8 (CCK-8) was purchased from Vitascientific (Beltsville, MD, USA).

3.2. Preparation of CDM Nanoparticles

Briefly, alloc-KGDEVD (0.5 g, 0.55 mmol), 4-aminobenzyl alcohol (0.34 g, 1.1 mmol, 1 eq.) and 2-Ethoxy-1-ethoxycarbonyl-1,2-digydroquinoline (0.135 g, 1.1 mmol, 1 eq.) were reacted in anhydrous *N,N*-Dimethylformamide (15 mL) overnight in room temperature. The solution was added into diethyl ether to from dried powder. Then, the obtained chemicals were further incubated in *N,N*-Dimethylformamide (25 mL) with bis(p-nitrophenyl)carbonate (5 eq.) and *N,N*-Diisopropylethylamine (3 eq.) at room temperature for 1 h. The mixed solution was precipitated in diethyl ether. The dried sediment (300 mg), MMAE (250 mg, 0.6 eq.), and 1-Hydroxybenzotriazole hydrate (30 mg, 0.75 eq.) were mixed in anhydrous DMF (20 mL). After 1 h of reaction, azabenzene (5 mL) and *N,N*-Diisopropylethylamine (100 μL, 1 eq.) were further added and mixed for 72 h at room temperature. For the removal of protection groups, Alloc-KGDEVD-PABC-MMAE was reacted with tributyltin hydride (10 eq.), tetrakis(triphenylphosphine)palladium (1 eq.), and acetic acid (20 eq.) under nitrogen gas atmosphere for 2 h. Then, the KGDEVD-PABC-MMAE (1 eq.) was reacted with Ce6 (1 eq.) in the presence of EDC (3 eq.) and NHS (3 eq.). After 24 h of reaction, the resulting Ce6-(Ac)KGDEVD-PABC-MMAE was purified using the C18 reverse analytical column. After preparation, the purity was analyzed by the reverse-phase HPLC system.

3.3. Characterization of CDM Nanoparticles

The size distribution of CDM in saline (1 mg/mL) was characterized by using dynamic light scattering (DLS; SZ-100, Horiba. Ltd., Kyoto, Japan, 532 nm, 10 mW). The morphology of CDM nanoparticles in saline (1 mg/mL) was observed using the transmission electron microscope (TEM; Talos F200X; FEI Compant, Hillsboro, OR, USA) after negative staining with 1% uranyl acetate. The ROS generation was assessed by bleaching test using p-nitroso-N,N'dimethylaniline (RNO). The Ce6 and Ce6-DEVD-MMAE (1 μM) was dissolved in saline containing 1.2 mM of L-histidine and 1% DMSO with 10 μM of RNO. Then, the samples were irradiated with visible light (Shanghai Dream Laser Technology Co., Ltd., Shanghai, China) with power of 40 mW for 5 min. Finally, the RNO absorbance was measured via UV-vis spectrometer (Agilent Spectroscopy System, Agilent Technology, Santa Clara, CA, USA) at 405 nm of each sample (n = 5). In the same conditions, the photophysical property of CDM was further confirmed by the Singlet Oxygen Sensor Green (SOSG) assays. The Ce6 and Ce6-DEVD-MMAE (1 μM) were mixed with SOSG (0.25 μM). Then, solutions were irradiated with visible light with power of 40 mW for different time periods, and the fluorescence intensity was measured using a fluorescence spectrometer (Hitachi, Tokyo, Japan). The caspase-3-specific cleavage of CDM nanoparticles was analyzed by reverse-phase HPLC system (Agilent Technologies 1200 series, Agilent Technologies, USA). The CDM nanoparticles (10 μM) were incubated with caspase-3, caspase-9, caspase-8, cathepsin B, cathepsin L, cathepsin K, or cathepsin D (10 μg) in buffer (0.1% CHAPS, 0.9% NaCL, 50 mM HEPES, 1 mM EDTA, 10 mM DTT, 10% glycerol, pH 7.6) for 24 h at 37 °C.

3.4. In Vitro Cellular Uptake of CDM Nanoparticles

To observe the cellular uptake of CDM nanoparticles, 3×10^5 KPC960 cells were seeded into 35-mm confocal dishes with DMEM media. Then, the cells were treated with 10 μM of CDM nanoparticles for 1, 3, 6, 12, and 24 h, respectively. After treatment, all cells were fixed with 2% paraformaldehyde solution for 10 min and stained with 4,6-diamidino-2-phenylindole (DAPI) for 15 min. The fluorescence imaging was performed using a confocal laser microscope (Leica TCS SP8, Leica Microsystems GmbH, Wetzlar, Germany).

3.5. Quantitative Assay of Caspase-3 Expression

To assess the caspase-3 expression after each treatment, the KPC960 cells were incubated with an equivalent amount (500 nM) of MMAE, Ce6 or CDM. In case of Ce6 or CDM groups, the cells were exposed to visible light with power of 40 mW for 5 min. After treatment, amount of caspase-3 in the cell lysates were measured using the caspase-3 assay kit.

3.6. In Vitro Cytotoxicity of CDM Nanoparticles

The cytotoxicity of CDM nanoparticles was evaluated via the cell counting kit-8. The KPC960 cells (5×10^4) were seeded in a 96-well cell culture plates. After 24 h of stabilization, the cells were treated with various concentrations of Ce6, MMAE, or CDM nanoparticles for 24 h. The cells treated with Ce6 or CDM were irradiated with visible light (40 mW for 5 min) after 6 h of drug treatment. Then, 200 μL of medium containing 10% CCK-8 solution was added in each well, and cell viability was measured by a microplate reader (VERSAmac™; Molecular Devices Corp., San Jose, CA, USA) (n = 5). As a control, cell viability of H9C2 and HDF after treatment with CDM nanoparticles for 24 h in absence of visible light irradiation was assessed.

3.7. Preparation of Orthotopic Pancreatic Cancer Models and Treatment Protocol

Mice were bred and maintained under specific pathogen-free conditions at the Korea Institute of Science and Technology (KIST). All experiments were conducted using protocols approved by the Association for Assessment and Accreditation of Laboratory Animal Care at the KIST. The BALB/C nude mice were anesthetized with Zoletil (8%), Ketamine (2%), and the left abdominal side was incised. Then, KPC960 cells (1×10^5) suspended in 10 μL saline were directly injected into the pancreas. The ventral wound was sutured in one layer with 6-0 non-absorbable sutures (Ailee Co., Busan, Korea). After 14 days of inoculation, mice were randomly divided into five groups (n = 4); (i) Saline (10 μL); (ii) Ce6 (0.1 mg/kg in 10 μL of saline) with visible light; (iii) MMAE (0.1 mg/kg in 10 μL of saline); (iv) CDM

(0.3 mg/kg in 10 μL of saline); or (v) CDM (0.3 mg/kg in 10 μL of saline) with visible light. Each drug was directly injected into pancreatic cancers after incision at the left abdominal side, and tumors of Ce6 and CDM nanoparticle groups were irradiated with visible light with power of 160 mW for 10 min. The weight was measured every 2 days until 2 weeks after post-injection and the animals ware sacrificed. The tumor was fixed with 4% paraformaldehyde solution for paraffin embedding.

3.8. Statistics

The statistical significance between two groups was analyzed using Student's t-test. One-way analysis of variance (ANOVA) was performed for comparisons of more than two groups, and multiple comparisons were analyzed using Tukey–Kramer post-hoc test. Statistical significance is indicated with an asterisk (* $p < 0.05$, ** $p < 0.01$, and *** $p < 0.001$) in the figures.

4. Conclusions

In this study, we proposed light-activated monomethyl auristatin E (MMAE) prodrug nanoparticles and optimized their application for pancreatic cancer treatment. The prodrug nanoparticles (CDM) were prepared by self-assembly of photosensitizer (Ce6), caspase-3-specific cleavable peptide (KGDEVD), and MMAE conjugates. The CDM could promote ROS under visible light irradiation and thereby induce caspase-3 overexpression in cancer cells, which subsequently triggered the release of MMAE. When the CDM was directly injected into pancreatic cancers of orthotopic models with visible light irradiation, the progression of cancers was significantly delayed by a potent therapeutic efficacy of combinational Ce6-mediated PDT and MMAE-mediated chemotherapy, compared with a single treatment of PDT or MMAE. In particular, CDM also greatly reduced the MMAE-related severe toxic side effects by restricting the drug activation in off-target tissues. Overall, these results suggested that combinational photo-chemotherapy by light-activated MMAE prodrug nanoparticles provides a promising and alternative therapeutic approach for pancreatic cancer.

Supplementary Materials: The following supporting information can be downloaded at: https://www.mdpi.com/article/10.3390/molecules27082529/s1, Figure S1: Synthetic route to prepare light-activated MMAE prodrugs (CDM). Figure S2: The purity of CDM after preparation, confirmed via the HPLC. Figure S3: The exact molecular weight of CDM was calculated to be 2131.1 Da and measured to be 2131.1 m/z [M], which was confirmed via the MALDI-TOF mass spectrometer. Figure S4: Singlet oxygen generation from Ce6 and CDM under visible light irradiation, as con-firmed via the Singlet Oxygen Sensor Green (SOSG) assays. Figure S5: The viability of H9C2 and HDF cells after treatment of CDM in absence of visible light. Figure S6: The photos to show the procedure of visible light irradiation for in vivo experiments.

Author Contributions: Conceptualization, K.K.; methodology, W.U. and J.-H.K.; validation, I.K.C.; formal analysis, I.K.C. and M.K.S.; investigation, I.K.C. and M.K.S.; resources, K.K.; data curation, I.K.C. and M.K.S.; writing—original draft preparation, M.K.S. and K.K.; visualization, I.K.C.; supervision, K.K.; project administration, K.K.; funding acquisition, M.K.S. and K.K. All authors have read and agreed to the published version of the manuscript.

Funding: This work was supported by the National Research Foundation of Korea (NRF) grant funded by the Korea government (NRF-2019R1A2C3006283 and NRF-2021R1C1C2005460), the KU-KIST Graduate School of Converging Science and Technology (Korea University & KIST), and the Intramural Research Program of KIST.

Institutional Review Board Statement: The study was conducted according to the guidelines of the Declaration of Helsinki and approved by the Institutional Review Board (or Ethics Committee) of Korea Institute of Science and Technology (approved number of 2020-123).

Informed Consent Statement: Not applicable.

Data Availability Statement: Not applicable.

Conflicts of Interest: The authors declare no conflict of interest.

Sample Availability: Samples of the compounds CDM are available from the authors.

References

1. Hidalgo, M. Pancreatic Cancer. *N. Engl. J. Med.* **2010**, *362*, 1605–1617. [CrossRef] [PubMed]
2. Kamisawa, T.; Wood, L.D.; Itoi, T.; Takaori, K. Pancreatic cancer. *Lancet* **2016**, *388*, 73–85. [CrossRef]
3. Liu, L.; Kshirsagar, P.G.; Gautam, S.K.; Gulati, M.; Wafa, E.I.; Christiansen, J.C.; White, B.M.; Mallapragada, S.K.; Wannemuehler, M.J.; Kumar, S.; et al. Nanocarriers for pancreatic cancer imaging, treatments, and immunotherapies. *Theranostics* **2022**, *12*, 1030–1060. [CrossRef] [PubMed]
4. Bardeesy, N.; DePinho, R.A. Pancreatic cancer biology and genetics. *Nat. Rev. Cancer* **2002**, *2*, 897–909. [CrossRef] [PubMed]
5. Zhang, L.; Sanagapalli, S.; Stoita, A. Challenges in diagnosis of pancreatic cancer. *World J. Gastroenterol.* **2018**, *24*, 2047–2060. [CrossRef] [PubMed]
6. Conroy, T.; Desseigne, F.; Ychou, M.; Bouché, O.; Guimbaud, R.; Bécouarn, Y.; Adenis, A.; Raoul, J.-L.; Gourgou-Bourgade, S.; de la Fouchardière, C.; et al. FOLFIRINOX versus Gemcitabine for Metastatic Pancreatic Cancer. *N. Engl. J. Med.* **2011**, *364*, 1817–1825. [CrossRef]
7. Janssen, Q.P.; van Dam, J.L.; Doppenberg, D.; Prakash, L.R.; van Eijck, C.H.J.; Jarnagin, W.R.; O'Reilly, E.M.; Paniccia, A.; Besselink, M.G.; Katz, M.H.G.; et al. FOLFIRINOX as Initial Treatment for Localized Pancreatic Adenocarcinoma: A Retrospective Analysis by the Trans-Atlantic Pancreatic Surgery Consortium. *JNCI J. Natl. Cancer Inst.* **2022**, djac018. [CrossRef]
8. Kyu Shim, M.; Yang, S.; Sun, I.-C.; Kim, K. Tumor-activated carrier-free prodrug nanoparticles for targeted cancer Immunotherapy: Preclinical evidence for safe and effective drug delivery. *Adv. Drug Deliv. Rev.* **2022**, *183*, 114177. [CrossRef]
9. Yang, S.; Sun, I.-C.; Hwang, H.S.; Shim, M.K.; Yoon, H.Y.; Kim, K. Rediscovery of nanoparticle-based therapeutics: Boosting immunogenic cell death for potential application in cancer immunotherapy. *J. Mater. Chem. B* **2021**, *9*, 3983–4001. [CrossRef]
10. Dolmans, D.E.J.G.J.; Fukumura, D.; Jain, R.K. Photodynamic therapy for cancer. *Nat. Rev. Cancer* **2003**, *3*, 380–387. [CrossRef]
11. Bansal, A.; Yang, F.; Xi, T.; Zhang, Y.; Ho John, S. In vivo wireless photonic photodynamic therapy. *Proc. Natl. Acad. Sci. USA* **2018**, *115*, 1469–1474. [CrossRef] [PubMed]
12. Lee, D.; Kwon, S.; Jang, S.-y.; Park, E.; Lee, Y.; Koo, H. Overcoming the obstacles of current photodynamic therapy in tumors using nanoparticles. *Bioact. Mater.* **2022**, *8*, 20–34. [CrossRef] [PubMed]
13. Um, W.; Park, J.; Ko, H.; Lim, S.; Yoon, H.Y.; Shim, M.K.; Lee, S.; Ko, Y.J.; Kim, M.J.; Park, J.H.; et al. Visible light-induced apoptosis activatable nanoparticles of photosensitizer-DEVD-anticancer drug conjugate for targeted cancer therapy. *Biomaterials* **2019**, *224*, 119494. [CrossRef] [PubMed]
14. Choi, J.; Shim, M.K.; Yang, S.; Hwang, H.S.; Cho, H.; Kim, J.; Yun, W.S.; Moon, Y.; Kim, J.; Yoon, H.Y.; et al. Visible-Light-Triggered Prodrug Nanoparticles Combine Chemotherapy and Photodynamic Therapy to Potentiate Checkpoint Blockade Cancer Immunotherapy. *ACS Nano* **2021**, *15*, 12086–12098. [CrossRef] [PubMed]
15. Shim, M.K.; Park, J.; Yoon, H.Y.; Lee, S.; Um, W.; Kim, J.-H.; Kang, S.-W.; Seo, J.-W.; Hyun, S.-W.; Park, J.H.; et al. Carrier-free nanoparticles of cathepsin B-cleavable peptide-conjugated doxorubicin prodrug for cancer targeting therapy. *J. Control Release* **2019**, *294*, 376–389. [CrossRef]
16. Shim, M.K.; Moon, Y.; Yang, S.; Kim, J.; Cho, H.; Lim, S.; Yoon, H.Y.; Seong, J.-K.; Kim, K. Cancer-specific drug-drug nanoparticles of pro-apoptotic and cathepsin B-cleavable peptide-conjugated doxorubicin for drug-resistant cancer therapy. *Biomaterials* **2020**, *261*, 120347. [CrossRef]
17. Moon, Y.; Shim, M.K.; Choi, J.; Yang, S.; Kim, J.; Yun, W.S.; Cho, H.; Park, J.Y.; Kim, Y.; Seong, J.-K.; et al. Anti-PD-L1 peptide-conjugated prodrug nanoparticles for targeted cancer immunotherapy combining PD-L1 blockade with immunogenic cell death. *Theranostics* **2022**, *12*, 1999–2014. [CrossRef]
18. Lim, S.; Park, J.; Shim, M.K.; Um, W.; Yoon, H.Y.; Ryu, J.H.; Lim, D.-K.; Kim, K. Recent advances and challenges of repurposing nanoparticle-based drug delivery systems to enhance cancer immunotherapy. *Theranostics* **2019**, *9*, 7906–7923. [CrossRef]
19. Shim, M.K.; Yoon, H.Y.; Lee, S.; Jo, M.K.; Park, J.; Kim, J.-H.; Jeong, S.Y.; Kwon, I.C.; Kim, K. Caspase-3/-7-Specific Metabolic Precursor for Bioorthogonal Tracking of Tumor Apoptosis. *Sci. Rep.* **2017**, *7*, 16635. [CrossRef]
20. Cho, Y.S.; Kim, G.C.; Lee, H.M.; Kim, B.; Kim, H.R.; Chung, S.W.; Chang, H.W.; Ko, Y.G.; Lee, Y.S.; Kim, S.W.; et al. Albumin metabolism targeted peptide-drug conjugate strategy for targeting pan-KRAS mutant cancer. *J. Control Release* **2022**, *344*, 26–38. [CrossRef]
21. Shim, M.K.; Yoon, H.Y.; Ryu, J.H.; Koo, H.; Lee, S.; Park, J.H.; Kim, J.-H.; Lee, S.; Pomper, M.G.; Kwon, I.C.; et al. Cathepsin B-Specific Metabolic Precursor for In Vivo Tumor-Specific Fluorescence Imaging. *Angew. Chem. Int. Ed.* **2016**, *55*, 14698–14703. [CrossRef] [PubMed]

molecules

MDPI

Article

Formulation of Tioconazole and *Melaleuca alternifolia* Essential Oil Pickering Emulsions for Onychomycosis Topical Treatment

Barbara Vörös-Horváth [1], Sourav Das [2,3], Ala' Salem [1], Sándor Nagy [1], Andrea Böszörményi [4], Tamás Kőszegi [2,3], Szilárd Pál [1,*,†] and Aleksandar Széchenyi [1,*,†]

1 Institute of Pharmaceutical Technology and Biopharmacy, Faculty of Pharmacy, University of Pécs, H-7624 Pécs, Hungary; barbara.horvath@aok.pte.hu (B.V.-H.); ala.salem@aok.pte.hu (A.S.); sandor.nagy@aok.pte.hu (S.N.)
2 Department of Laboratory Medicine, Medical School, University of Pécs, H-7624 Pécs, Hungary; pharma.souravdas@gmail.com (S.D.); koszegi.tamas@pte.hu (T.K.)
3 János Szentágothai Research Center, University of Pécs, H-7624 Pécs, Hungary
4 Department of Pharmacognosy, Faculty of Pharmacy, Semmelweis University, H-1085 Budapest, Hungary; aboszormenyi@gmail.com
* Correspondence: szilard.pal@aok.pte.hu (S.P.); szechenyi.aleksandar@gytk.pte.hu (A.S.); Tel.: +36-72-501-500 (ext. 23554) (S.P.); +36-72-501-500 (ext. 28801) (A.S.)
† These authors contributed equally to this work.

Academic Editors: Marina A. Dobrovolskaia and Francesca Mancianti
Received: 13 October 2020; Accepted: 23 November 2020; Published: 26 November 2020

Abstract: Onychomycosis is a disease that affects many adults, whose treatment includes both oral and topical therapies with low cure rates. The topical therapy is less effective but causes fewer side effects. This is why the development of an effective, easy to apply formulation for topical treatment is of high importance. We have used a nanotechnological approach to formulate Pickering emulsions (PEs) with well-defined properties to achieve site-specific delivery for antifungal drug combination of tioconazole and *Melaleuca alternifolia* essential oil. Silica nanoparticles with tailored size and partially hydrophobic surface have been synthesized and used for the stabilization of PEs. In vitro diffusion studies have been performed to evaluate the drug delivery properties of PEs. Ethanolic solution (ES) and conventional emulsions (CE) have been used as reference drug formulations. The examination of the antifungal effect of PEs has been performed on *Candida albicans* and *Trichophyton rubrum* as main pathogens. In vitro microbiological experimental results suggest that PEs are better candidates for onychomycosis topical treatment than CE or ES of the examined drugs. The used drugs have shown a significant synergistic effect, and the combination with an effective drug delivery system can result in a promising drug form for the topical treatment of onychomycosis.

Keywords: pickering emulsions; onychomycosis topical treatment; tioconazole; tea tree essential oil; antifungal activity

1. Introduction

Onychomycosis is a fungal infection of nails and nail bed and occurs on both finger and toenails. This fungal infection affects about 11.4% [1] of the adult population and is responsible for more than 50% of nail diseases [2]. For the treatment of onychomycosis, oral, topical, mechanical, and chemical therapies or a combination of these methods are used in practice [3]. The therapy is a long process (10–12 months) and has a poor cure rate [4]. Oral therapy is the most effective, but in the case of prolonged use, orally administered drugs can cause severe side effects because of their high toxicity [5]. Drug interactions can also occur, which is the main reason for the contraindication of oral therapy.

In such cases, topical therapy is recommended, and it is a more attractive alternative for patients. Unfortunately, the topical treatment of onychomycosis is limited because the nail plate acts as a barrier to drug diffusion. Its hydrophilic nature and keratinized structure reduce the diffusion of high molecular weight or lipophilic antifungal drugs [6]. In order to enhance the penetration of drugs, diffusion enhancers (e.g., mechanical pretreatment [7], phosphoric acid [8], keratolytes [9]) or an appropriate formulation should be used [3].

The azole antifungal agents have been used since 1980 for topical and oral therapy of fungal infections, among others for onychomycosis; in clinical treatment, they are used for superficial and systemic fungal infections with safety [10]. The water solubility of generally applied antifungal azole derivatives in the treatment of fungal nail infections is very low (<0.01 mg/mL) [10]; therefore, their formulation contains organic solvents in most cases. Commercially available nail lacquers contain organic solvents to increase the solubility of antifungal drugs, but these solvents act unfavorably to the drug permeability [11]. Their restricted drug delivery ability is caused by the rapid evaporation of organic solvents, so some drugs remain on the surface of the nail [12]. Several researches have proved that drugs with aqueous-based formulations have higher nail permeability than non-aqueous ones [13,14]. Another problem with azole derivatives is that fungi can become resistant to the drug in a long-lasting treatment [15]. A combination of azoles with antifungal essential oil (EO) could solve this problem because fungi cannot easily acquire resistance to multiple antifungal components of EOs [16]. Moreover, the azole derivatives could show synergistic antifungal activity with some essential oils (EOs) [17], presumably because of their different mode of action. The azole derivatives inhibit the action of cytochrome P-450 enzyme, lanosterol demethylase of fungi [18], thereby preventing the synthesis of ergosterol, while the EOs damage the cell membranes and organelles of fungi [19]. Because of the lipophilic character of azole derivatives [10], it is likely that it can be dissolved in lipophilic EO, and their solution can be used for drug formulation.

The oil in water type emulsions as water-based drug formulations can provide a possible way to overcome the water solubility problems. Conventionally, the emulsions are stabilized with surfactants. In a long-lasting topical treatment, the use of surfactants should be avoided because they can cause irritation and side effects, or in some cases, they can get into the blood circulation [20]. Using particle-stabilized emulsions, i.e., Pickering emulsions (PEs) [21] instead of a conventional emulsion, has several advantages. The solid particles spontaneously adsorb on the oil-water interface and form a shell-like structure on the PE droplet surface [22]. This adsorption can be considered as irreversible adsorption because the solid particles have higher adsorption energy than surfactants on oil-water interfaces [23], so the stability of PEs can be the same or better than conventional emulsions. Another important parameter for emulsion-based drug formulation and therapy is the size of emulsion droplets. The droplet size of PE can be influenced by several parameters, such as emulsification time and energy, oil to solid particle volume ratio, and concentration of oil phase [24]. The fungal hypha damages the nail structure, creating pores in μm size range [25], with porosity in the 5–20% range depending on pretreatment of the nail [26]. An emulsion droplet with the appropriate size could penetrate into the porous nail structure and retain the antifungal drug on the nail bed for a longer period, which is the main site of reinfection [27], and thereby a targeted drug delivery can be achieved. Inert and biocompatible particles should be chosen as stabilizing particles in PEs drug formulation.

Silica nanoparticles (SNPs) are widespread in pharmaceutical technology in topical treatments because of their favorable chemical and surface properties, thermal stability, mechanical resistance, and biocompatibility [28,29]. The effects of topically applied SNPs have been examined in detail [30], and it has been found that they have no toxic effect even after prolonged usage. Because of the above-mentioned advantageous properties of SNPs, they can be suitable for PEs stabilization. The PEs are most stable when the partial wetting conditions of stabilizing particles are the same for the oil and water phase [31]. The native SNPs are hydrophilic because of the high number of free silanol groups at their surface [32], which has to be modified with organic ligands to achieve appropriate wettability and strong stabilizing effect.

In the present study, our aim was to formulate PE of an azole derivative and antifungal essential oil as an alternative formulation for onychomycosis topical treatment. Tioconazole (TIO) has a broad antifungal activity for common dermatophytes, which has proved to be efficient for the topical treatment of fungal infections [33]. Nenoff et al. determined that *Melaleuca alternifolia* (MA) EO (tea tree EO) inhibited the growth of several clinical fungal isolates, so they suggested its use in the topical treatment of fungal infections [34]. For PE formulation, we used the solution of TIO in MA EO. Synthesis and surface modification of SNPs with different sizes were performed, and they were used as stabilizing agents of PEs. We characterized the stability, droplet size, and emulsion type of PEs. In vitro diffusion studies were also performed through artificial membranes. The aim of the in vitro diffusion test was to compare drug delivery characteristics of different formulations in the membranes that have similar porous structure and surface properties as the nail plate and nail bed. The antifungal activity against *Candida albicans* and *Trichophyton rubrum*—the species mainly responsible for fungal nail infections—has been investigated [3].

2. Results and Discussion

2.1. Characterization and In Vitro Diffusion Study of PEs

2.1.1. Characterization of SNPs

The size distribution and PDI values for synthesized and surface-modified SNPs were determined by DLS and TEM. Data for mean diameter and PDI values are presented in Table 1. The TEM images showed that SNPs were monodispersed, spherical, and had a smooth surface. It can be clearly seen that the size and morphology did not change significantly during the surface modification (Table 1 and Figure 1). The surface modification of SNPs was confirmed by FTIR spectroscopy; the results were previously published [35].

Table 1. Properties of SNPs. HS: hydrophilic particle. ET: with ethyl functional group modified particle. Size represented as mean ± SD of three parallel syntheses. The numbers refer to the particle sizes.

Samples	d_{DLS} (nm)	PDI_{DLS}	d_{TEM} (nm)	PDI_{TEM}
20HS	20.1 ± 0.2	0.008	20.0	0.011
50HS	52.7 ± 0.9	0.017	53.0	0.037
100HS	105.2 ± 3.6	0.021	103.0	0.083
20ET	20.1 ± 0.8	0.158	22.0	0.210
50ET	54.2 ± 2.7	0.178	55.0	0.337
100ET	110.7 ± 4.1	0.231	112.0	0.349

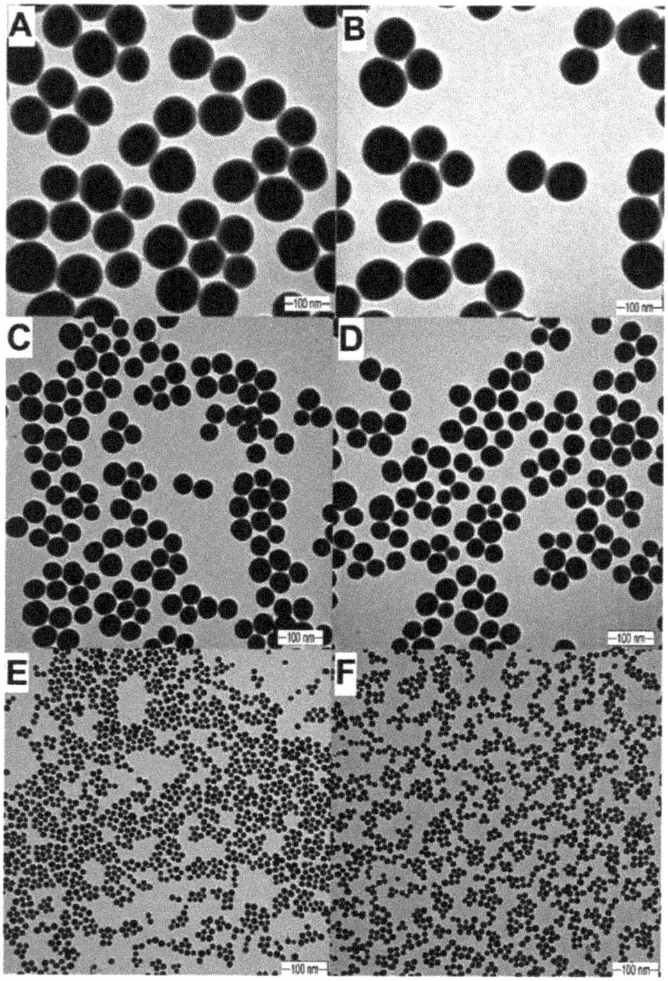

Figure 1. TEM images of silica nanoparticles (SNPs). (**A**): HS100, d = 103.0 nm. (**B**): ET100, d = 110.7 nm. (**C**): HS50, d = 53.0 nm. (**D**): ET50, d = 55.0 nm. (**E**): HS20, d = 20.0 nm. (**F**): ET20, d = 20.0 nm.

2.1.2. GC Analysis of *Melaleuca Alternifolia* EO

The composition of MA was determined by gas chromatography. The components were identified by comparing their retention times and relative retention factors with standards and oils of known composition. Two parallel measurements were performed. The main compounds were *p*-cymene 35.2% and terpinene-4-ol 32.5%. A detailed composition is presented in Table 2.

Table 2. Composition of *Melaleuca alternifolia* EO. The results of GC analysis showed the average of the two parallel measurements in the percentage of volatile compounds. The main components of MA have been highlighted.

Compounds	Retention Time t_R (min)	Percentage Ratio of Compounds (%)
α-thujene	5	1.7
α-pinene	5.2	4.6
β-phellandrene	5.2	0.6
β-pinene	6.2	1.2
β-mycrene	6.4	0.8
α-terpinene	7.0	1.4
***p*-cymene**	**7.3**	**35.2**
terpinyl-acetate	7.3	2.1
cineole	7.4	5.8
γ-terpinene	8	7.6
terpinolene	8.6	1.7
terpinene-4-ol	**10.7**	**32.5**
α-terpineol	11	2.6
aromadendrene	15.5	0.7
epiglobulol	16.4	1.2

2.1.3. Characterization of PEs

Properties of PEs are influenced by many parameters, like the interfacial surface tension of the phases, size, wettability, and concentration of stabilizing particles, o/w phase ratio, and emulsification energy. In this study, we examined the influence of the o/w phase ratio and size of the stabilizing SNPs on the droplet size and stability of formulated PEs, while other parameters were kept constant.

The results can be seen in Table 3, including data for composition, droplet size, and appearance of PEs. For the microbiological experiments and in vitro diffusion studies, we used emulsions that were stable for at least one week, which means that their droplet size did not change within this period, and creaming, phase separation, aggregation, or sedimentation of SNPs did not occur.

As shown by Binks and Horozov [31], the size of stabilizing particles influences the emulsion droplet size at the same o/w phase ratio. We found that the increment of oil phase concentration caused an increment of emulsion droplet size at all examined oil phase concentrations. When 20 nm SNPs were used, the droplet size increased until reaching a droplet size of 1.8 µm. Further increment of oil phase concentration did not cause significant droplet size increment, and the stability of emulsions was much higher in the 11.16–16.12 mg/mL concentration range (for oil phase) (see Table 3). We observed a similar effect when 50 nm SNPs were used in the 4.48–11.19 mg/mL concentration range and 1.6 µm droplet size (Table 3). We could not observe such behavior for PE 100ET. In this case, the droplet size continuously increased as the oil phase concentration increased, and the stability of emulsions was much lower (less than one week).

The stability of PE was also influenced by the size of the stabilizing SNPs. We found that with the increasing SNPs size, the stability of PE decreased, at the same o/w ratio. The PE stability was 20 weeks using 20ET, 8 weeks for 50ET, and 1 week for 100ET SNPs, at 11.19 mg/mL oil phase concentration. The zeta potential values of PEs could also provide their colloidal stability, whose values did not differ significantly from the SNPs suspensions. The zeta potential of 20ET, 50ET, and 100ET SNPs was −28.3, −25.2, and −25.0 mV, while the values of PE 20ET, PE 50ET, and PE100Et in the 0.90–17.91 mg/mL oil concentration range were from −28.0 to −19.3 mV, −24.9 to −19.0 mV, and −24.3 to −17.6 mV, respectively.

The type of emulsions was determined by conductivity measurements. The conductivity values of stabilizing SNPs suspended in distilled water were 215.0, 211.4, and 268.3 $\mu S \cdot cm^{-1}$ for 20ET, 50ET, and 100ET, respectively, while conductivity for the oil phase was 0.058 $\mu S \cdot cm^{-1}$. The conductivity values of PEs were in the 157.33–257.50 $\mu S \cdot cm^{-1}$ range, which means that all the PEs were o/w type emulsions.

Table 3. Parameters of Pickering emulsions of *Melaleuca alternifolia* EO and tioconazole stabilized with 20ET, 50ET, and 100ET SNPs. * These emulsions were creaming, but after 30 s shaking, their droplet size recovered to the original value, and they retained again their stability for 1 week.

	PE 20ET		PE 50ET		PE 100ET	
Coil Phase (mg/mL)	**Droplet Size (nm)**	**Stability**	**Droplet Size (nm)**	**Stability**	**Droplet Size (nm)**	**Stability**
0.90	4306 ± 39.6	10 min	1280 ± 62.8	1 day	1070 ± 438.5	30 min
1.79	615 ± 22.2	30 min	1320 ± 95.9	1 day	1350 ± 531.8	30 min
2.69	890 ± 103.3	30 min	1440 ± 83.5	1 day	1630 ± 464.5	30 min
3.58	1250 ± 94.5	1 day	1650 ± 51.5	2 day	1730 ± 514.5	10 min
4.48	1320 ± 32.5	8 weeks	1670 ± 216.8	2 day	1850 ± 107.9	10 min
5.37	1440 ± 100.2	8 weeks	1620 ± 79.7	8 weeks	1890 ± 333.8	10 min
6.27	1470 ± 35.2	8 weeks	1610 ± 34.4	8 weeks	1950 ± 95.0	10 min
7.16	1470 ± 62.5	8 weeks	1670 ± 62.8	8 weeks	1940 ± 20.1	1 week *
8.96	1660 ± 56.7	8 weeks	1690 ± 70.4	8 weeks	2070 ± 51.2	1 week *
11.19	1890 ± 41.2	20 weeks	2200 ± 188.9	8 weeks	2200 ± 59.5	1 week *
13.43	1840 ± 141.0	20 weeks	2250 ± 170.8	2 weeks *	2800.0 ± 85.7	1 week
16.12	1820.0 ± 99.6	20 weeks	2080 ± 160.1	2 weeks *	2850 ± 184.3	1 week
17.91	1850 ± 496.6	8 weeks	2380 ± 157.0	2 weeks *	3090 ± 116.6	1 week
Coil Phase (mg/mL)	**Appearance PET 20ET**		**Appearance PET 50ET**		**Appearance PET 100ET**	
0.90	creaming		sedimentation		creaming	
1.79	creaming		sedimentation		creaming	
2.69	creaming		sedimentation		creaming	
3.58	opalescent		opalescent		creaming	
4.48	opalescent		opalescent		creaming	
5.37	opalescent		milky		creaming	
6.27	opalescent		milky		creaming	
7.16	opalescent		milky		milky	
8.96	opalescent		milky		milky	
11.19	milky		milky		milky	
13.43	milky		milky		aggregation, sedimentation	
16.12	milky		milky		aggregation, sedimentation	
17.91	opalescent		milky		aggregation, sedimentation	

2.1.4. In Vitro Diffusion Studies through Artificial Membranes

Our goal was to formulate an emulsion that is capable of delivering the antifungal drugs through the nail plate and retain the drugs in the site of the infection (nail bed) for a prolonged time to provide a sustained drug release. The diffusion studies of PEs on the artificial membranes were performed in Franz diffusion vertical cells in order to examine the drug delivery ability of the formulated PEs. For diffusion studies, we applied PEs, CE, and ES of the same concentration, 17.91 mg/mL, as an antifungal drug combination. Because of the droplet size similarity between the CE and PEs, we can assume that only the dosage form determined the diffusion properties of the drug.

We found that PEs possessed better drug delivery properties through agar gel membrane compared to CE and ES (Table 4 and Figure 2). We examined the diffusion properties of PEs with different droplet sizes and found that the PEs with smaller droplet size (1.85 μm) could deliver as much as 89.9% of TIO through the agar membrane. In the experiment where the composite membrane was used, we found that the ES had diffused through the composite membrane structure, and only a small portion (2.4%) of the drug remained in the composite membrane (Table 4 and Figure 3). The PE 20ET delivered 89.9% of TIO through the agar gel membrane, and only 5.7% had diffused through composite membranes, suggesting that 84.2% of the applied drug remained in the targeted area. This amount was 61.1% at PE 50ET and 45.13% at PE 100ET. These in vitro experimental results suggested that PEs had better on-site drug delivery properties.

Table 4. Results of in vitro diffusion studies. ES: ethanolic solution, CE: conventional emulsion, PE: Pickering emulsion, CA: cumulative TIO amount after 2 h. The concentration of TIO was 3.58 mg/mL in each formulation.

Samples	Stabilizing Agent	Droplet Size (nm)	CA Agar Gel (%)	CA Agar Gel (mg/cm^2)	CA Composite Membrane (%)	CA Composite Membrane (mg/cm^2)
ES	-	-	18.33	0.26	15.90	0.22
CE	Tween80	2470.0 ± 89.1	35.02	0.49	11.01	0.15
PE 20ET	20ET SNPs	1850 ± 496.6	89.88	1.26	5.70	0.08
PE 50ET	50ET SNPs	2380 ± 157.0	67.18	0.95	6.06	0.05
PE 100ET	100ET SNPs	3090 ± 116.6	45.22	0.63	0.09	0.001

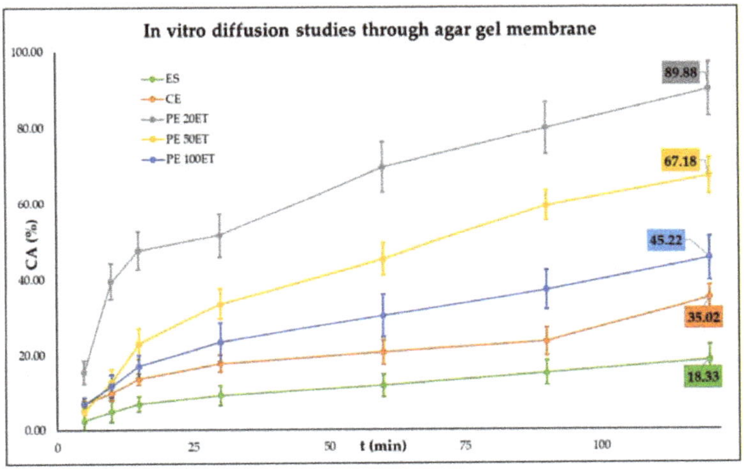

Figure 2. In vitro diffusion studies through agar gel membrane. ES: ethanolic solution, CE: conventional emulsion, PE: Pickering emulsion, CA: cumulative TIO amount after 2 h. C_{TIO} = 3.58 mg/mL.

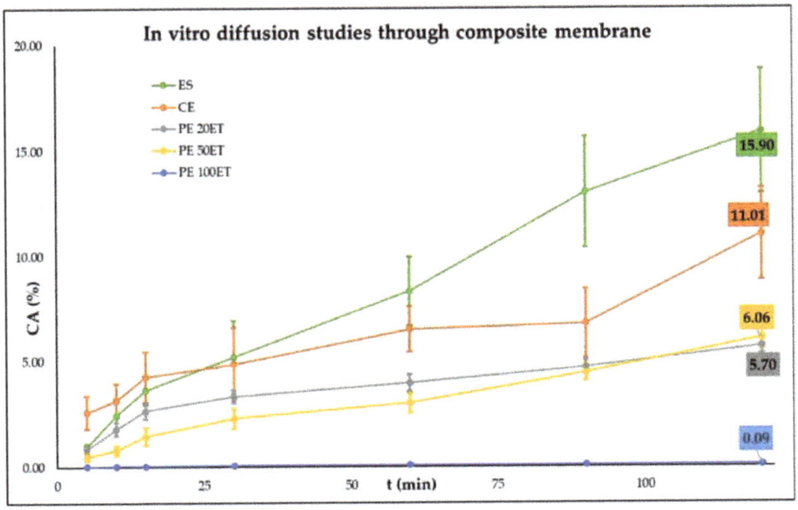

Figure 3. In vitro diffusion studies through the composite membrane. ES: ethanolic solution, CE: conventional emulsion, PE: Pickering emulsion, CA: cumulative TIO amount after 2 h. C_{TIO} = 3.58 mg/mL.

2.2. Microbiological Tests Using Candida albicans and Trichophyton rubrum

Data obtained for the minimum inhibitory concentration (MIC) and minimum fungicidal concentration (MFC) on *T. rubrum* and on *C. albicans* are shown in Table 5 and Figures 4 and 5 for the ethanolic solution of TIO (ES-TIO) and ethanolic solution of MA (ES-MA) and their combinations. The TIO and MA combination showed a significant synergistic effect. When *T. rubrum* and *C. albicans* were treated with the combination of TIO and MA, both the MIC and MFC values decreased significantly compared to the separately used drugs.

Analyzing the antimicrobial data of the different formulations of TIO and MA clearly showed that the PEs were more effective than CE or ES against the two pathogens. The PE 100ET showed the most effective growth inhibition against both *T. rubrum* and *C. albicans*, and this formulation had the highest fungicidal activity.

Table 5. Minimum inhibitory concentration (MIC) and minimum fungicidal concentration (MFC) of the test samples in combinations on *T. rubrum* and on *C. albicans*.

Sample	*T. rubrum*		*C. albicans*	
	MIC (μg/mL)	MFC (μg/mL)	MIC (μg/mL)	MFC (μg/mL)
ES-TIO	4.68	37.5	18.75	74.88
ES-MA	74.88	149.92	74.88	149.92
ES-TIO-MA	10.43	83.47	11.24	89.95
CE	5.21	83.47	11.24	89.95
PE 20ET	5.21	41.73	5.61	22.48
PE 50ET	2.6	20.86	2.8	11.24
PE 100ET	1.29	10.43	2.8	11.24

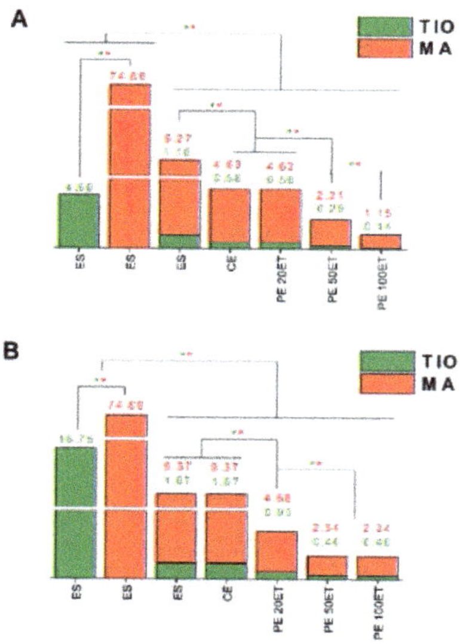

Figure 4. Minimum inhibitory concentration (MIC_{90}) of ES-TIO, ES-MA, ES-TIO-MA, CE-TIO-MA, PE 20ET-TIO-MA, PE 50ET-TIO-MA, and PE 100ET-TIO-MA in μg/mL on *T. rubrum* (**A**) and *C. albicans* (**B**). Six independent experiments with three technical replicates were performed. The green (*) and red (*) asterisks represent a significance value of $p < 0.01$ for the MIC_{90}, respectively.

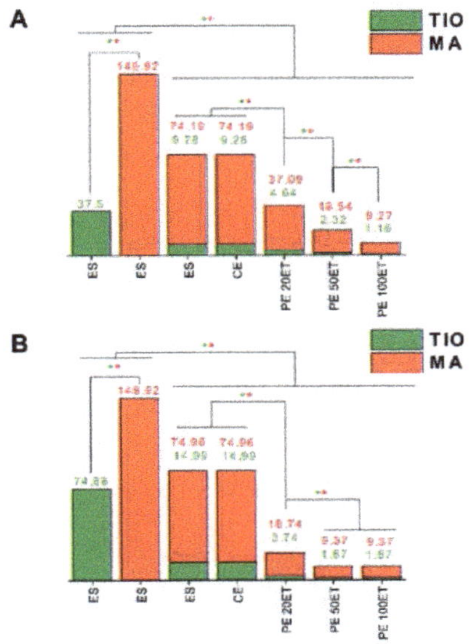

Figure 5. Minimum fungicidal concentration (MFC) of ES-TIO, ES-MA, ES-TIO-MA, CE-TIO-MA, PE 20ET-TIO-MA, PE 50ET-TIO-MA, and PE 100ET-TIO-MA in μg/mL on *T. rubrum* (**A**) and *C. albicans* (**B**). Six independent experiments with three technical replicates were performed. The green (*) and red (*) asterisks represent a significance value of $p < 0.01$ for the MFC, respectively.

3. Materials and Methods

3.1. Preparation and Characterization of Silica Nanoparticle-Stabilized Pickering Emulsions

3.1.1. Synthesis, Surface Modification, and Characterization of Silica Nanoparticles

Synthesis of hydrophilic SNPs (HS) was performed based on the work of Stöber et al. [36]. Size-selective synthesis parameters were set based on our previous work [37], as well as the reaction circumstances for surface modification. The synthesis route and details can be found in the Supplementary Materials. We previously reported that SNPs that have theoretical surface coverage of 20% with ethyl groups could stabilize the MA droplets to give a stable PE [24]. For the synthesis and surface modification of silica nanoparticles, the following chemicals were used: tetraethoxysilane ([TEOS] (Alfa Aesar GmbH, Karlsruhe Germany, purity 98%), ethyltriethoxysilane ([ETES] Alfa Aesar Karlsruhe Germany, purity 96%), absolute ethanol (VWR Chemicals Ltd., Debrecen Hungary, AnalaR NORMAPUR® ≥99.8%), and 28 m/m% ammonium solution (VWR Chemicals Ltd., Debrecen Hungary, AnalaR NORMAPUR®, analytical reagent).

The size distribution was determined by dynamic light scattering (DLS) (Malvern Zetasizer Nano S, Malvern Panalytical Ltd., Great Malvern, Worcestershire, UK). The size distribution was confirmed, and the morphology of silica nanoparticles was studied by transmission electron microscopy (TEM) (JEOL-1400 electron microscopy, JEOL Ltd., Tokyo, Japan). For TEM experiments, 400 mesh copper grids coated with carbon were used (Micro to Nano Ltd., Haarlem, Netherlands).

3.1.2. Gas Chromatography Analysis of *Melaleuca Alternifolia* Essential Oil

Solid-Phase Microextraction (SPME) Conditions

Samples were loaded into vials (20 mL headspace) sealed with a silicon/PTFE septum prior to SPME-GC/MS analysis. Sample preparation using the static headspace solid-phase microextraction (sHS-SPME) technique was carried out with a CTC Combi PAL (CTC Analytics AG, Zwingen, Switzerland) automatic multipurpose sampler using a 65 μM StableFlex polydimethyl siloxane/carboxene/divinyl benzene (CAR/PDMS/DVB) SPME fiber (Supelco, Bellefonte, PA, USA). After an incubation period of 5 min at 100 °C, extraction was performed by exposing the fiber to the headspace of a 20 mL vial containing the sample for 10 min at 100 °C. The fiber was then immediately transferred to the injector port of the GC/MS and desorbed for 1 min at 250 °C, in split mode, and the split ratio was 1:90. The SPME fiber was cleaned and conditioned in a Fiber Bakeout Station in a pure nitrogen atmosphere at 250 °C for 15 min.

GC-MS Conditions

The analyses were carried out with an Agilent 6890N/5973N GC-MSD (Agilent Technologies, Santa Clara, CA, USA) system equipped with Supelco (Sigma-Aldrich Ltd., Budapest, Hungary) SLB-5MS capillary column (30 m × 250 μm × 0.25 μm). The GC oven temperature was programmed to increase from 60 °C (3 min isothermal) to 250 °C at 8°C/min (1 min isothermal). High purity helium (6.0) was used as carrier gas at 1.0 mL/min (37 cm/s) in constant flow mode. The mass selective detector (MSD) was equipped with a quadrupole mass analyzer and was operated in electron ionization mode at 70 eV in full scan mode (41–500 amu at 3.2 scan/s). The data were evaluated using MSD ChemStation D.02.00.275 software (Agilent Technologies, Santa Clara, CA, USA). The identification of the compounds was carried out by comparing retention data and the recorded spectra with the data of the NIST 2.0 library. The percentage evaluation was carried out by area normalization.

3.1.3. Determination of Solubility of Tioconazole in *Melaleuca Alternifolia* Essential Oil

Solubility Calculations by Hansen Solubility Parameters (HSPs)

As a preliminary study, the calculations of solubility parameters were performed using the Hansen Solubility Parameters in Practice (HSPiP) software version 5.0.11 using the simplified molecular-input line-entry system (SMILES), obtained from PubChem. HSPs (Equation (1)) use group contribution to split the total cohesion energy of a solvent into contributions from atomic dispersion (δ_d), polar interactions (δ_p), and hydrogen bonding (δ_h) [38].

$$\delta = \left(\delta_d^2 + \delta_p^2 + \delta_h^2\right)^{0.5} \tag{1}$$

Differences in solubility parameters were calculated with the HSP difference (Equation (2)). A value below that of the reported cut-off value 7 $\mathrm{Mpa}^{0.5}$ indicates miscibility [39].

$$\Delta\delta = |\delta_{solvent} - \delta_{tioconazole}| \tag{2}$$

For the calculation, the three main components of MA were used (*p*-cymene, terpinene-4-ol, γ-terpinene), and it could be established that TIO can be dissolved in the EO. The results of the calculation can be found in Supplementary Materials Tables S2–S4. In order to determine the exact solubility of TIO in MA, the solvent addition method was performed (Section 3.1.3 Determination of Kinetic Solubility).

Determination of Kinetic Solubility

The kinetic solubility of TIO (tioconazole, purity ≥98%, Alfa Aesar, Karlsruhe, Germany) in water-saturated MA (*Melaleuca alternifolia* essential oil, Tebamol®, BIO-DIÄT-BERLIN GmbH, Berlin, Germany) was determined by the solvent addition method [40]. The examination was performed at ambient temperature (25 °C). The initial suspension was prepared by weighing the exact amount of 1.0 mg TIO and the addition of 500 μL of MA. The volume of MA was increased until the suspension turned into a clear solution. The light scattering of suspension was determined visually and with instrumental measurement of scattered light intensity (DLS, Malvern Zetasizer Nano S). The kinetic solubility of TIO in MA was found to be 0.213 mg/mL (23.8 *m/m*%). The concentration of TIO in MA was set to 20 *m/m*% for the PE preparations.

3.1.4. Preparation and Characterization of Pickering Emulsions

The concentration of emulsifiers in the water phase was set to 1 mg/mL and was kept constant for all experiments. Three different sizes of SNPs were used for PE formulation (20ET, 50ET, 100ET) and Tween80® surfactant (Tw80) (Polysorbate80 Acros Organics, Thermo Fisher Scientific, Waltham, MA, USA) for conventional emulsions (CE). The concentration of oil phase varied between 0.90 and 17.91 mg/mL, and the ratio of TIO and MA was always constant (20 *m/m*%).

The emulsification was performed in two steps. The coarse emulsions were prepared by sonication for 2 min (Bandelin Sonorex RK 52H, BANDELIN electronic GmbH & Co. KG, Berlin, Germany). The final emulsification was performed with UltraTurrax (IKA Werke T-25 basic, IKA Werke GmbH, Staufen im Breisgau, Germany) for 2 min at 13,500 rpm. The emulsions' droplet size was determined with DLS using a Malvern Zetaziser Nano S instrument (Malvern Panalytical Ltd., Great Malvern, Worcestershire, UK). The stability of the emulsions was determined from periodical droplet size determination. The emulsions were stored at room temperature (25 °C).

The type of emulsions was determined with conductivity test using Mettler Toledo Seven2Go S3 conductivity meter (Mettler Toledo GmbH, Giessen, Germany) and InLab® 738-ISM sensor (Mettler Toledo GmbH, Giessen, Germany).

All experiments, measurements, and standard deviation calculations were performed from 3 parallel sample preparations.

3.1.5. In Vitro Diffusion Studies—Static Franz Diffusion Cell Method

Accepted models for testing drugs and their formulations for onychomycosis treatments include penetration tests through cadaver nails [41], nail clippings, bovine hoof slices, or keratin films [42] made from human keratin source. The non-uniformity of natural membranes causes huge inhomogeneity in the results [43–45], which makes the comparison of different formulations impossible. The aim of our study was to examine the diffusion properties of applied drugs in complex colloidal systems; therefore, in our opinion, the similarity in hydrophilicity and surface charge between the nail plate or nail bed and artificial membranes was of the highest importance for testing and comparison of the formulations. The nail plate acts as a negatively charged aqueous hydrogel, as it is described in the literature [46], and it has properties similar to that of the agar gel [47]. Based on the literature data obtained from independent researches, we compared the diffusion coefficient and flux of well-studied antibiotic chloramphenicol (5 mg/mL in phosphate-buffered saline) with different membranes, namely agar gel [48], bovine hoof slice, and cadaver nail plate [49]. We found that the diffusion coefficients and flux values were very similar for agar gel and bovine hoof slice membranes. Flux for bovine hoof was $4.07 \pm 1.18 \cdot 10^{-6}$ mg/cm^2·s, for agar gel $1.96 \pm 0.47\ 10^{-6}$ mg/cm^2·s, and $8.21 \pm 2.11\ 10^{-7}$ mg/cm^2·s for the cadaver nail plate. The flux values for agar gel were closer to the value for the cadaver nail plate, which might suggest that agar gel is a good model membrane for water-based formulations.

The agar gel membrane was used to model the nail plate. The composite membrane, consisting of the agar gel layer on top of the cellulose acetate membrane, was used to simulate the complex structure of nail plate and nail bed since nail bed has similar properties as skin [50,51], and the cellulose membrane has been commonly used as a model for skin permeability [52]. The main aim of the study on two types of membranes was to examine whether the examined formulations could deliver the applied lipophilic drugs through agar gel as a model for nail plate, and the composite membrane was used to examine if the formulation could retain the drugs on the main site of the infection, namely nail bed. The amount of the drug transported through the membrane was calculated based on the amount introduced to the membrane. In the case of agar membrane, the goal was to prepare the drug delivery system that could deliver the highest drug amount through that membrane. The composite membrane was used to test the on-site retention of drugs in different formulations. The amount of retained drug was calculated as a difference between the drug amount passed through the agar gel membrane and the amount passed through the composite membrane

For in vitro testing, the 2.1 mm thick 6 *m*/*m*% agar gel membrane (Agar powder, purity >95%, VWR Chemicals Ltd., Debrecen, Hungary) and the same agar gel membrane combined with 0.8 mm thick cellulose acetate with effective penetration area of 2.54 cm^2 (Membranfilter Porafil, Macherey-Nagel GmbH&Co. KG, Düren, Germany, pore size 0.2 µm) were used. Before each measurement, the agar gel was always freshly prepared. The agar powder was dispersed in demineralized water, and the mixture was boiled in a closed vial for 3 min until all agar was completely dissolved. Exactly 10 mL of agar gel was poured into a plastic vessel (i.d. 70.8 mm), then left to cool (25 °C) and gelate. After the gelation, the agar gel was soaked in PBS buffer for 12 h. Finally, the agar membrane was cut out with a sharp home-made tool and placed on the Franz cell. The cellulose acetate membranes were also freshly soaked in PBS buffer before the experiments.

The examination of diffusion properties was performed at 32 °C in static vertical Franz diffusion cells (Hanson Microette Plus, Hanson Research 60-301-106, Hanson Research Corporation, Chatsworth, CA, USA); six parallel cells with effective penetration area 2.54 cm^2 were used, and each experiment was made in triplicates. The volume of the receiver chamber was 7 mL; the receiver solution was PBS buffer. For PBS preparation, the following salts were used: NaCl (high purity, VWR Chemicals Ltd., Debrecen Hungary), KCl (purity 99–100.5%, VWR Chemicals Ltd., Debrecen Hungary), $Na_2HPO_4{\cdot}2H_2O$ (AnalaR NORMAPUR®, purity ≥99.0%, VWR Chemicals Ltd., Debrecen Hungary), and KH_2PO_4 (purity ≥99.0%, VWR Chemicals Ltd., Debrecen Hungary). The 600 µL volume of emulsion or solution sample was placed into the donor chamber, and the diffusion was examined for 2 h; the stirring rate was 700 min^{-1}, and the samples were collected after 5, 10, 15, 30, 60, 90, and 120 min. The withdrawn sample volume was replaced with a fresh PBS buffer.

The TIO content was determined with HPLC measurements using UV-Vis detector (SPD 10-A, Shimadzu Europa GmbH, Duisburg, Germany); the method is based on Bagary et al. [53]. Separations were carried out using a monolithic silica type column (ODS-AM302, S-5µm, 120A, YMC Co., Kyoto, Japan). The mobile phase consisted of methanol/0.02 M K_2HPO_4 = 85/15 *V*/*V*% and 0.2 *V*/*V*% trimethylamine (methanol dehydrated, ultrapure ≥99.8%, VWR Chemicals Ltd., Debrecen Hungary; trimethylamine: HiPerSolv CHROMANORM®, VWR Chemicals Ltd., Debrecen Hungary), pH = 7.0. The mobile phase was freshly filtered through Millipore Nylon membrane (pore size: 0.2 µm, Merck KGaA, Darmstadt, Germany) before the analysis. Isocratic elution was programmed with a 1.5 mL/min flow rate; the temperature of measurement was 32 °C. The detection wavelength of tioconazole was 254 nm, and its retention time was 3.5 min.

3.2. *Microbiological Tests against Candida albicans and Trichophyton rubrum*

3.2.1. Instruments Used in the Microbiological Experiments

UV-Vis spectrophotometer (Hitachi U-3900, Hitachi High-Tech Corporation, Japan), microbiological incubator (Thermo Scientific Heraeus B12, Thermo Fischer Scientific, Waltham, MA, USA), Bürker cell counting chamber (Hirschmann Laborgeräte GmbH & Co., Germany), Multiskan EX 355 (Thermo Fischer Scientific, Waltham, MA, USA) spectrophotometer were used throughout the experiments.

3.2.2. Materials Used in the Microbiological Experiments

For the microbiological experiments, the following materials were used: sterile 96-well microtiter plates (Greiner Bio-One, Kremsmunster, Austria), potato dextrose agar (PDA) (BioLab, Budapest, Hungary), sterile filter inserts (pore size 10 µm) from PluriSelect (pluriSelect Life Science, Leipzig, Germany), dextrose, adenine, bacteriological peptone and agar-agar (Reanal Labor, Budapest, Hungary), sterile centrifuge tubes (TPP Techno Plastic Products, Trasadingen, Switzerland), homemade Sabouraud dextrose agar or SDA (containing 4% dextrose, 1% bacteriological peptone, and 1.5% agar-agar in double-distilled water), yeast extract peptone dextrose agar (containing 2% bacteriological peptone, 1% yeast extract, 2% dextrose, and 1.5% agar-agar in double-distilled water), 3-(*N*-Morpholino)-propanesulfonic acid (MOPS) from Serva Electrophoresis GmBH (Heidelberg, Germany), and RPMI 1640 medium (containing 3.4% MOPS, 1.8% dextrose, and 0.002% adenine) from Sigma-Aldrich Chemie GmBH (Steinheim, Germany). Highly purified water (<1.0 μS) was applied throughout the experiments.

3.2.3. Fungal Cultures and Inoculum Preparation

Trichophyton rubrum (*T. rubrum*) DSM 21146 and *Candida albicans* (*C. albicans*) ATCC 001 were obtained from Leibniz Institute DSMZ GmbH (Braunschweig, Germany) and from Department of General and Environmental Microbiology (Institute of Biology, University of Pécs, Hungary), respectively.

We followed the methods described previously [54–57] for *T. rubrum* and *C. albicans* culture preparation. In brief, *T. rubrum* stock inoculum suspensions were prepared from 7-day old cultures grown on PDA at 28 °C for sporulation. Ten days later, the observed fungal colonies were flooded with 10 mL distilled water, followed by scraping the surface using a sterile loop. Conidia and hyphal mixed suspensions were withdrawn and were transferred to a sterile centrifuge tube through sterile filter inserts (10 μm, pluriSelect) to remove hyphae, leaving a filtered inoculum containing spores only. The inoculum cell population was adjusted to 0.5 to 5×10^6 spores/mL visually using a Bürker cell counting chamber, followed by further turbidity calibration with a UV-Vis spectrophotometer (Hitachi U-3900) at 520 nm. The spores were further diluted to the desired population according to the experimental requirements.

C. albicans stock inoculum was prepared from 48 h old culture grown on YEPD agar plates at 30 °C. After 18 h of incubation at 30 °C in a microbiological incubator, on YEPD agar slant, the cells were looped out, diluted with 0.9% sterile saline, and were counted by a Bürker cell counting chamber, followed by turbidity calibration with a UV-Vis spectrophotometer (Hitachi U-3900) at 595 nm. The fungal cell population was set to $\sim 1 \times 10^6$ cells/mL and was diluted later according to the experimental designs.

3.2.4. Determination of Antifungal Activities

For the evaluation of the minimum inhibitory concentration (MIC) of *T. rubrum* and *C. albicans*, we followed previously published methods [56–59]. The ethanolic solutions of TIO and MA in a wide concentration range (0.5–300 μg/mL) were used for the assay. CE and PEs formulations were also tested; an initial concentration of the oil phase was 160 μg/mL for *T. rubrum*, whereas 180 μg/mL for *C. albicans* treatment was applied. The treating mixtures were further diluted up to 256 times in a serial half-dilution format.

One hundred microliters of fungal cell suspensions (see Sections 3.2.5 and 3.2.6) with equal fungal contents were applied thereafter to the microplate wells containing 100 μL of the different samples. Detailed information on the assay conditions can be found in Sections 3.2.5 and 3.2.6. As a blank, suspensions of 20ET, 50ET, 100ET SNPs, pure ethanol, Tw80 solution were used.

3.2.5. Determination of Minimum Inhibitory Concentration of *T. rubrum*

The *T. rubrum*, inoculum size of $\sim 2.5 \times 10^4$ spores/mL, containing the test drugs in half-dilution format, was incubated in RPMI media for 7 days in a microbiological incubator at 28 °C. The microplates containing *T. rubrum* incubated for 7 days with the test drugs were evaluated following the protocol as recommended by the Clinical and Laboratory Standards Institute (CLSI) M38-A2. The untreated cell samples and the medium without cells were considered as the growth control and blank, respectively. The endpoint determination readings for the minimum inhibitory concentrations (MIC) were performed visually based on the comparison of the growth in the wells containing the test drugs with that of the growth control [60]. All evaluations were performed in triplicates in six independent experiments.

3.2.6. Determination of Minimum Inhibitory Concentration of *C. albicans*

A population size of ~2×10^3 CFU/mL was incubated in RPMI media with the above-mentioned test drug concentration range at 30 °C for 48 h in the case of *C. albicans*. A Multiskan EX 355 spectrophotometer was used to measure the absorbance (at 595 nm) of the samples in the microtiter plate in the case of *C. albicans*. The absorbance values of the respective treatments were converted to a percentage and were compared to growth control (100%). The untreated fungal samples and the medium without cells were considered as the growth control and blank, respectively. All evaluations were performed in triplicates in six independent experiments.

3.2.7. Determination of the Minimum Fungicidal Concentration (MFC)

Determination of MFC was performed using the methods as described earlier with modifications [56]. After performing the MIC, 10 μL of the content from each well (not visibly turbid) was inoculated onto sterile SDA plates. The plates were incubated at 30 °C for 48 h. MFC was evaluated as the lowest drug concentration, resulting in no growth (≥99.9% growth inhibition). Measurements were performed by applying three technical replicates in six independent experiments.

3.2.8. Statistical Analyses

The statistical analyses were conducted using a one-way ANOVA test (Origin 2016, OriginLab Corp., Northampton, MA, USA), and the significance was set at $p \leq 0.05$.

4. Conclusions

The choice of drugs used in this research was based on careful consideration. The TIO was chosen as a drug with high antifungal activity but low water solubility and permeability through the nail plate. MA EO was selected because of the antifungal activity and because it is a liquid and can be used as a solvent for TIO. The combination of the drugs applied in this study showed a significant synergistic effect. The solution of TIO in MA EO was successfully formulated into stable Pickering emulsions. In vitro studies have demonstrated that PEs are effective drug formulations that can provide site-specific and effective drug delivery through artificial membranes. 20ET PE achieved the highest drug delivery efficiency as it could deliver 40% of the drug introduced to the artificial membrane within 10 min. The amount delivered at this time was 572 μg of TIO through the agar model membrane, while the MFC of the TIO in this formulation was 4.69 μg/mL. To prove the real applicability of the suggested drug combination and PE formulation, we have to perform experiments on the natural nail model. Still, from the presented data, we can conclude that the application of both site-specific drug delivery and synergistic antifungal drug combinations is a promising route for the development of effective onychomycosis topical treatment formulation.

Supplementary Materials: The following are available online: Table S1. Parameters of hydrophilic and surface-modified silica nanoparticle synthesis. Table S2. HSPs of tioconazole and the three main components of *Melaleuca alternifolia* essential oil. Table S3. Calculated solubility parameters of tioconazole compared to the three main components of *Melaleuca alternifolia* essential oil. Table S4. Calculated solubility parameters.

Author Contributions: Conceptualization A.S. (Aleksandar Széchenyi), B.V.-H., S.P.; methodology B.V.-H., S.N., S.D., T.K.; formal analysis, B.V.-H.; investigation, B.V.-H., S.D., S.N., A.S. (Ala' Salem), A.B.; resources A.S. (Ala' Salem), T.K.; data curation A.S. (Ala' Salem); writing—original draft preparation, B.V.-H., A.S. (Aleksandar Széchenyi); writing—review and editing A.S. (Aleksandar Széchenyi); visualization, B.V.-H.; supervision, A.S. (Ala' Salem); funding acquisition A.S. (Ala' Salem), S.P. All authors have read and agreed to the published version of the manuscript.

Funding: The project has been supported by the European Union, co-financed by the European Social Fund grant No. EFOP-3.6.1.-16-2016-00004 (Comprehensive Development for Implementing Smart Specialization Strategies at the University of Pécs). Transmission electron microscopy studies were performed using the JEOL-1400 TEM electron microscope, which was funded by the grant GINOP-2.3.3-15-2016-0002 (New generation electron microscope: 3D ultrastructure). The work was also funded by the grant KA-2018-17, University of Pécs, Medical School.

Conflicts of Interest: The authors declare no conflict of interest.

References

1. Sigurgeirsson, B.; Baran, R. The prevalence of onychomycosis in the global population—A literature study. *J. Eur. Acad. Dermatol. Venereol.* **2013**, *28*, 1480–1491. [CrossRef] [PubMed]
2. Scher, R. Onychomycosis is more than a cosmetic problem. *Br. J. Dermatol.* **1994**, *130*, 15. [CrossRef] [PubMed]
3. Gupta, A.K.; Tu, L.Q. Therapies for Onychomycosis: A Review. *Dermatol. Clin.* **2006**, *24*, 375–379. [CrossRef] [PubMed]
4. Ameen, M.; Lear, J.T.; Madan, V.; Mohd, M.F.; Richardson, M. British Association of Dermatologists' guidelines for the management of onychomycosis. *Br. J. Dermatol.* **2014**, *171*, 937–958. [CrossRef] [PubMed]
5. Gauwerky, K.; Borelli, C.; Korting, H.C. Targeting virulence: A new paradigm for antifungals. *Drug Discov. Today* **2009**, *14*, 214–222. [CrossRef] [PubMed]
6. Stuttgen, G.; Bauer, E. Bioavailability, skin- and nail-penetration of topically applied antimycotics. *Mykosen* **1982**, *25*, 74–80. [CrossRef] [PubMed]
7. Di Chiacchio, N.; Kaduc, B.V.; De Almeida, A.R.T.; Madeira, C.L. Nail abrasion. *J. Cosmet. Dermatol.* **2003**, *2*, 150–152. [CrossRef]
8. Repka, M.A.; Mididoddi, P.K.; Stodghill, S.P. Influence of human nail etching for the assessment of topical onychomycosis therapies. *Int. J. Pharm.* **2004**, *282*, 95–106. [CrossRef]
9. Mohorčič, M.; Torkar, A.; Friedrich, J.; Kristl, J.; Murdan, S. An investigation into keratinolytic enzymes to enhance ungual drug delivery. *Int. J. Pharm.* **2007**, *332*, 196–201. [CrossRef]
10. Fromtling, R.A. Overview of medically important antifungal azole derivatives. *Clin. Microbiol. Rev.* **1988**, *1*, 187–217. [CrossRef]
11. Smith, K.A.; Hao, J.; Li, S.K. Effects of Organic Solvents on the Barrier Properties of Human Nail. *J. Pharm. Sci.* **2011**, *100*, 4244–4257. [CrossRef] [PubMed]
12. Bohn, M.; Kraemer, K.T. Dermatopharmacology of ciclopirox nail lacquer topical solution 8% in the treatment of onychomycosis. *J. Am. Acad. Dermatol.* **2000**, *43*, S57–S69. [CrossRef]
13. Walters, K.A.; Flynn, G.L.; Marvel, J.R. Physicochemical characterization of the human nail: Permeation pattern for water and the homologous alcohols and differences with respect to the stratum corneum*. *J. Pharm. Pharmacol.* **1983**, *35*, 28–33. [CrossRef] [PubMed]
14. Monti, D.; Tampucci, S.; Chetoni, P.; Burgalassi, S.; Mailland, F. Ciclopirox vs amorolfine: In vitro penetration into and permeation through human healthy nails of commercial nail lacquers. *J. Drugs Dermatol.* **2014**, *13*, 143–147. [PubMed]
15. Canuto, M.M.; Gutiérrez, F. Antifungal drug resistance to azoles and polyenes. *Lancet Infect. Dis.* **2002**, *2*, 550–563. [CrossRef]
16. Nazzaro, F.; Fratianni, F.; Coppola, R.; De Feo, V. Essential Oils and Antifungal Activity. *Pharmaceuticals* **2017**, *10*, 86. [CrossRef] [PubMed]
17. Rosato, A.; Vitali, C.; Piarulli, M.; Mazzotta, E.; Argentieri, M.P.; Mallamaci, R. In vitro synergic efficacy of the combination of Nystatin with the essential oils of Origanum vulgare and Pelargonium graveolens against some Candida species. *Phytomedicine* **2009**, *16*, 972–975. [CrossRef] [PubMed]
18. Yang, W.; Wiederhold, N.P.; O Williams, R. Drug delivery strategies for improved azole antifungal action. *Expert Opin. Drug Deliv.* **2008**, *5*, 1199–1216. [CrossRef]
19. Yu, D.; Wang, J.; Shao, X.; Xu, F.; Wang, H. Antifungal modes of action of tea tree oil and its two characteristic components against Botrytis cinerea. *J. Appl. Microbiol.* **2015**, *119*, 1253–1262. [CrossRef]
20. Pape, W.J.; Pfannenbecker, U.; Hoppe, U. Validation of the red blood cell test system as in vitro assay for the rapid screening of irritation potential of surfactants. *Mol. Toxicol.* **1987**, *1*, 525–536.
21. Yang, Y.; Fang, Z.; Chen, X.; Zhang, W.; Xie, Y.; Chen, Y.; Liu, Z.G.; Yuan, W. An Overview of Pickering Emulsions: Solid-Particle Materials, Classification, Morphology, and Applications. *Front. Pharmacol.* **2017**, *8*, 287. [CrossRef] [PubMed]
22. Chevalier, Y.; Bolzinger, M.-A. Emulsions stabilized with solid nanoparticles: Pickering emulsions. *Colloids Surfaces A Physicochem. Eng. Asp.* **2013**, *439*, 23–34. [CrossRef]

23. Binks, B.P. Particles as surfactants–similarities and differences. *Curr. Opin. Colloid. Interface Sci.* **2002**, *7*, 21–41. [CrossRef]
24. Horváth, B.; Pál, S.; Széchenyi, A. Preparation and in vitro diffusion study of essential oil Pickering emulsions stabilized by silica nanoparticles. *Flavour Fragr. J.* **2018**, *33*, 385–396. [CrossRef]
25. Scherer, W.P.; Scherer, M.D. Scanning Electron Microscope Imaging of Onychomycosis. *J. Am. Podiatr. Med. Assoc.* **2004**, *94*, 356–362. [CrossRef] [PubMed]
26. Baraldi, A.; Jones, S.A.; Guesné, S.; Traynor, M.J.; McAuley, W.J.; Brown, M.B.; Murdan, S. Human Nail Plate Modifications Induced by Onychomycosis: Implications for Topical Therapy. *Pharm. Res.* **2015**, *32*, 1626–1633. [CrossRef]
27. Elewski, B.E. Onychomycosis: Pathogenesis, Diagnosis, and Management. *Clin. Microbiol. Rev.* **1998**, *11*, 415–429. [CrossRef]
28. Arriagada, F.; Nonell, S.; Morales, J. Silica-based nanosystems for therapeutic applications in the skin. *Nanomedicine* **2019**, *14*, 2243–2267. [CrossRef]
29. Frelichowska, J.; Bolzinger, M.-A.; Pelletier, J.; Valour, J.-P.; Chevalier, Y. Topical delivery of lipophilic drugs from o/w Pickering emulsions. *Int. J. Pharm.* **2009**, *371*, 56–63. [CrossRef]
30. An, S.S.A.; Ryu, H.J.; Seong, N.-W.; So, B.J.; Seo, H.-S.; Kim, J.-H.; Hong, J.-S.; Park, M.-K.; Kim, M.-S.; Kim, Y.-R.; et al. Evaluation of silica nanoparticle toxicity after topical exposure for 90 days. *Int. J. Nanomed.* **2014**, *9*, 127–136. [CrossRef]
31. Aveyard, R.; Binks, B.P.; Clint, J.H. Emulsions stabilised solely by colloidal particles. *Adv. Colloid Interface Sci.* **2003**, *100*, 503–546. [CrossRef]
32. Leung, K.; Nielsen, I.M.B.; Criscenti, L.J. Elucidating the Bimodal Acid−Base Behavior of the Water−Silica Interface from First Principles. *J. Am. Chem. Soc.* **2009**, *131*, 18358–18365. [CrossRef] [PubMed]
33. Jevons, S.; Gymer, G.E.; Brammer, K.W.; Cox, D.A.; Leeming, M.R.G. Antifungal Activity of Tioconazole (UK-20,349), a New Imidazole Derivative. *Antimicrob. Agents Chemother.* **1979**, *15*, 597–602. [CrossRef] [PubMed]
34. Nenoff, P.; Haustein, U.-F.; Brandt, W. Antifungal Activity of the Essential Oil of Melaleuca alternifolia (Tea Tree Oil) against Pathogenic Fungi in vitro. *Ski. Pharmacol. Physiol.* **1996**, *9*, 388–394. [CrossRef]
35. Horváth, B.; Balázs, V.L.; Varga, A.; Böszörményi, A.; Kocsis, B.; Horváth, G.; Széchenyi, A. Preparation, characterisation and microbiological examination of Pickering nano-emulsions containing essential oils, and their effect on *Streptococcus mutans* biofilm treatment. *Sci. Rep.* **2019**, *9*, 16611.
36. Stöber, W.; Fink, A.; Bohn, E. Controlled growth of monodisperse silica spheres in the micron size range. *J. Colloid Interface Sci.* **1968**, *26*, 62–69. [CrossRef]
37. Horváth, B.; Salem, A.; Šafarik, T.; Pál, S.; Széchenyi, A. Systematic study of reaction conditions for size controlled synthesis of silica nanoparticles. In Proceedings of the 6th Nano Today Conference, Lissbon, Portugal, 16–20 June 2019; p. 4.12.
38. Hansen, C.M. *The Three Dimensional Solubility Parameter and Solvent Diffusion Coefficient*; Danish Technical Press: Copenhagen, Denmark, 1967.
39. Mohammad, M.A.; Alhalaweh, A.; Velaga, S. Hansen solubility parameter as a tool to predict cocrystal formation. *Int. J. Pharm.* **2011**, *407*, 63–71. [CrossRef]
40. Reus, M.A.; Van Der Heijden, A.E.D.M.; Ter Horst, J.H. Solubility Determination from Clear Points upon Solvent Addition. *Org. Process. Res. Dev.* **2015**, *19*, 1004–1011. [CrossRef]
41. Shivakumar, H.N.; Juluri, A.; Desai, B.G.; Murthy, S.N. Ungual and Transungual drug delivery. *Drug Dev. Ind. Pharm.* **2011**, *38*, 901–911. [CrossRef]
42. Lusiana; Reichl, S.; Müller-Goymann, C.C. Keratin film made of human hair as a nail plate model for studying drug permeation. *Eur. J. Pharm. Biopharm.* **2011**, *78*, 432–440. [CrossRef]
43. Van Hoogdalem, E.; Hoven, W.V.D.; Terpstra, I.; Van Zijtveld, J.; Verschoor, J.; Visser, J. Nail penetration of the antifungal agent oxiconazole after repeated topical application in healthy volunteers, and the effect of acetylcysteine. *Eur. J. Pharm. Sci.* **1997**, *5*, 119–127. [CrossRef]
44. Hui, X.; Baker, S.J.; Wester, R.C.; Barbadillo, S.; Cashmore, A.K.; Sanders, V.; Hold, K.M.; Akama, T.; Zhang, Y.; Plattner, J.J.; et al. In Vitro Penetration of a Novel Oxaborole Antifungal (AN2690) Into the Human Nail Plate. *J. Pharm. Sci.* **2007**, *96*, 2622–2631. [CrossRef] [PubMed]
45. Tiwary, A.K.; Sapra, B. High failure rate of transungal drug delivery: Need for new strategies. *Ther. Deliv.* **2017**, *8*, 239–242. [CrossRef] [PubMed]

46. Mercer, D.; Stewart, C.S.; Miller, L.; Robertson, J.; Duncan, V.M.S.; O'Neil, D. Improved Methods for Assessing Therapeutic Potential of Antifungal Agents against Dermatophytes and Their Application in the Development of NP213, a Novel Onychomycosis Therapy Candidate. *Antimicrob. Agents Chemother.* **2019**, *63*, e02117-18. [CrossRef]
47. Lahaye, M.; Rochas, C. Chemical structure and physico-chemical properties of agar. *Hydrobiologia* **1991**, *221*, 137–148. [CrossRef]
48. Meulemans, A.; Paycha, F.; Hannoun, P.; Vulpillat, M. Measurement and clinical and pharmacokinetic implications of diffusion coefficients of antibiotics in tissues. *Antimicrob. Agents Chemother.* **1989**, *33*, 1286–1290. [CrossRef]
49. Mertin, D.; Lippold, B.C. In-vitro Permeability of the Human Nail and of a Keratin Membrane from Bovine Hooves: Penetration of Chloramphenicol from Lipophilic Vehicles and a Nail Lacquer. *J. Pharm. Pharmacol.* **1997**, *49*, 241–245. [CrossRef]
50. Baran, R. *Agache's Measuring the Skin*; Springer: Cham, Switzerlands, 2017.
51. Berker, D. Nail Anatomy. *Clin. Dermatol.* **2013**, *31*, 509–515. [CrossRef]
52. Haigh, J.M.; Smith, E.W. The selection and use of natural and synthetic membranes for in vitro diffusion experiments. *Eur. J. Pharm. Sci.* **1994**, *2*, 311–330. [CrossRef]
53. Bagary, R.I.E.; Elkady, E.F.; Tammam, M.H.; El-maaty, A.A. Simultaneous HPLC and derivative spectrophotometry detrmination of tioconazole and benzyl alcohol in bulk and cream with tioconazole forced degradation study. *Anal. Chem. Ind. J.* **2014**, *14*, 462–473.
54. Hammer, K.A.; Carson, C.F.; Riley, T.V. Antifungal effects of Melaleuca alternifolia (tea tree) oil and its components on Candida albicans, Candida glabrata and Saccharomyces cerevisiae. *J. Antimicrob. Chemother.* **2004**, *53*, 1081–1085. [CrossRef] [PubMed]
55. Barros, L.; Calhelha, R.C.; Vaz, J.A.; Ferreira, I.; Baptista, P.; Estevinho, L.M. Antimicrobial activity and bioactive compounds of Portuguese wild edible mushrooms methanolic extracts. *Eur. Food Res. Technol.* **2006**, *225*, 151–156. [CrossRef]
56. Araújo, C.R.; Miranda, K.C.; Fernandes, O.D.F.L.; Soares, A.J.; Silva, M.D.R.R. In vitro susceptibility testing of dermatophytes isolated in Goiania, Brazil, against five antifungal agents by broth microdilution method. *Revista do Instituto de Medicina Tropical de São Paulo* **2009**, *51*, 9–12. [CrossRef] [PubMed]
57. Cantón, E.; Pemán, J.; Viudes, A.; Quindós, G.; Gobernado, M.; Espinel-Ingroff, A. Minimum fungicidal concentrations of amphotericin B for bloodstream Candida species. *Diagn. Microbiol. Infect. Dis.* **2003**, *45*, 203–206. [CrossRef]
58. Das, S.; Czuni, L.; Báló, V.; Papp, G.; Gazdag, Z.; Papp, N.; Kőszegi, T. Papp Cytotoxic Action of Artemisinin and Scopoletin on Planktonic Forms and on Biofilms of Candida Species. *Molecules* **2020**, *25*, 476. [CrossRef]
59. Santos, D.; Hamdan, J.S. In vitro activities of four antifungal drugs against Trichophyton rubrum isolates exhibiting resistance to fluconazole. *Mycoses* **2007**, *50*, 286–289. [CrossRef]
60. Majoros, L.; Kardos, G.; Feiszt, P.; Szabó, B. Efficacy of amphotericin B and flucytosine against fluconazole-resistant Candida inconspicua clinical isolates. *J. Antimicrob. Chemother.* **2005**, *56*, 253–254. [CrossRef]

Sample Availability: Samples of the Pickering emulsions are available from the authors.

molecules

MDPI

Article

Design of a Near Infrared Fluorescent Ureter Imaging Agent for Prevention of Ureter Damage during Abdominal Surgeries

Sakkarapalayam M. Mahalingam [1], Karson S. Putt [2], Madduri Srinivasarao [3] and Philip S. Low [2,3,*]

1 Department of Chemistry, PSG College of Arts and Science and PSG Institute of Advanced Studies, Coimbatore 641004, Tamil Nadu, India; mahalpurdue@gmail.com
2 Institute for Drug Discovery, Purdue University, West Lafayette, IN 47907, USA; puttk@purdue.edu
3 Department of Chemistry, Purdue University, West Lafayette, IN 47907, USA; msriniva@purdue.edu
* Correspondence: plow@purdue.edu; Tel.: +1-765-494-5272

Abstract: The inadvertent severing of a ureter during surgery occurs in as many as 4.5% of colorectal surgeries. To help prevent this issue, several near-infrared (NIR) dyes have been developed to assist surgeons with identifying ureter location. However, the majority of these dyes exhibit at least some issue that precludes their widespread usage such as high levels of uptake in other tissues, overlapping emission wavelengths with other NIR dyes used for other fluorescence-guided surgeries, and/or rapid excretion times through the ureters. To overcome these limitations, we have synthesized and characterized the spectral properties and biodistribution of a new series of PEGylated UreterGlow derivatives. The most promising dye, UreterGlow-11 was shown to almost exclusively excrete through the kidneys/ureters with detectable fluorescence observed for at least 12 h. Additionally, while the excitation wavelength is similar to that of other NIR dyes used for cancer resections, the emission is shifted by ~30 nm allowing for discrimination between the different fluorescence-guided surgery probes. In conclusion, these new UreterGlow dyes show promising optical and biodistribution characteristics and are good candidates for translation into the clinic.

Keywords: ureter imaging; fluorescence-guided surgery; near-infrared dye; PEG pharmacokinetics

Citation: Mahalingam, S.M.; Putt, K.S.; Srinivasarao, M.; Low, P.S. Design of a Near Infrared Fluorescent Ureter Imaging Agent for Prevention of Ureter Damage during Abdominal Surgeries. *Molecules* **2021**, *26*, 3739. https://doi.org/10.3390/molecules26123739

Academic Editors: Kirill A. Afonin

Received: 29 May 2021
Accepted: 18 June 2021
Published: 19 June 2021

1. Introduction

Because ureters are not commonly visible on visceral surfaces, their localization during abdominal surgery can be problematic, leading to accidental severance of the ureter in as many as 4.5% of colorectal surgeries [1] and 0.3% of all gynecological procedures [2,3]. To prevent the resulting leakage of urine into the peritoneum and the ensuing long-term complications [1,4,5], any cleaved ureter must be immediately religated by a time-consuming, complicated, and expensive procedure, thereby dramatically increasing the cost and complexity of the surgery. Not surprisingly, considerable effort has been focused on the development of methods to prevent ureter injuries during surgery.

One of the earliest approaches to avoid accidental ureter cleavage was to insert a stent into the ureter that would rigidify the duct and render it detectable by palpation [6,7]. However, because the process of stent insertion was found to cause occasional injury [1,8,9] and since physical palpation was not possible during robotic or endoscopic surgeries, the stent insertion strategy never attracted significant usage. Systemically administered near-infrared (NIR) fluorescent dyes such as indocyanine green (ICG) [10] and UreterGlow [11] were then explored for similar intraoperative ureter visualization, but these initial fluorescent dyes were found to clear primarily through the liver, bile duct, and intestines [10,11], creating high background fluorescence that could mask the location of proximal ureters. While much brighter fluorescent signals have been achieved by intraureter dye injection [12–14], the injection process has been considered by many surgeons to be too involved for routine ureter localization, leading to similar problems with widespread adoption [15]. Finally, although a zwitterionic near-infrared (NIR) fluorescent dye has been recently designed to excrete predominantly through the ureter [16], its transit time in ureters has been found to be very brief, and its fluorescence excitation and emission unnecessarily overlap with many tumor-targeted NIR dyes (e.g., IR800CW [17,18], LS288 [19,20], ICG [10,12–15], and OTL38 [11], Table 1), creating a potential discrimination problem when malignant lesions reside near a ureter.

Table 1. Photo-physical properties of NIR dyes used for fluorescence-guided surgeries and new PEGylated probes.

NIR Dyes	Excitation Maximum (nm)	Emission Maximum (nm)	Stokes Shift (nm)
IR800CW	780	795	15
LS288	770	785	15
ZW800-1	770	788	18
ICG	780	802	22
OTL38	776	796	20
UreterGlow	800	830	30
UreterGlow-0	800	830	30
UreterGlow-3	800	830	30
UreterGlow-11	800	830	30
UreterGlow-45	800	830	30

In an effort to create a ureter imaging agent with (1) reduced fluorescence in healthy tissues (except ureters), (2) prolonged transit through the ureters, and (3) an emission spectrum distinct from that of commonly used tumor-targeted fluorescent dyes, we have conjugated a longer wavelength NIR dye to a series of polyethylene glycol (PEG) oligomers of different lengths and examined their clearance following intravenous injection into mice. We report that conjugation of the NIR dye S0456 via a thioether bridge to PEG oligomers of 11–45 oxyethylene units yields ureter imaging agents with little uptake in healthy tissues, prolonged excretion almost exclusively through the ureters, and facile ureter visualization at an emission maximum that is easily distinguished from the common tumor-targeted NIR dyes.

2. Methods

2.1. Materials

S0456 was purchased from Few Chemicals (Bitterfeld-Wolfen, Germany). 2-(1*H*-7-Azabenzotriazol-1-yl)-1,1,3,3-tetramethyl uronium hexafluorophosphate methanaminium (HATU) was obtained from Genscript Inc. (Piscataway, NJ, USA). The PEG3, PEG11, and PEG2000 oligomers were purchased from TCI America (Portland, OR, USA), BroadPharm (San Diego, CA, USA), and Laysan Bio (Arab, AL, USA), respectively. Diisopropylethylamine (DIPEA), dimethyl sulfoxide (DMSO), and all other chemical reagents were purchased from Sigma-Aldrich (St. Louis, MO, USA). Tubes, pipette tips, microtiter plates, and all other consumables were purchased from Fisher Scientific (Waltham, MA, USA).

2.2. Synthesis of PEGylated S0456 NIR Dyes

The S0456 near-infrared dye (1 equiv) was reacted with 2-(4-mercaptophenyl) acetic acid (1.1 equiv) in 2 mL DMSO for 12 h (90% yield, 95% purity). The resulting compound (1 equiv) was coupled to one of three different PEG linkers (1.1 equiv) of sizes 3, 11, and ~45 (PEG2000) in the presence of HATU (1.1 equiv) and DIPEA (5 equiv) in 5 mL DMSO for 6 h to yield the final PEGylated S0456 dyes.

The crude product was purified by preparative reverse-phase high-performance liquid chromatography using a mobile phase of 20 mM ammonium acetate buffer and gradient of 5 to 80% acetonitrile over 30 min (xTerra C18; Waters; 10 µm; 19 × 250 mm). Elution of the conjugate was monitored at 280 nm, and identities of eluted compounds were analyzed by liquid chromatography-mass spectrometry (LC-MS) (see Supplementary Materials). The molecular weight of UreterGlow-0 was calculated (calcd) for $[M + H]^+$

($C_{46}H_{54}N_2O_{14}S_5$): 1019.2, found 1019.3. UreterGlow-3 calcd. ($C_{54}H_{72}N_4O_{16}S_5$): 1193.4, found 1193.6. UreterGlow-11 calcd ($C_{54}H_{72}N_4O_{16}S5$): 1545.9, found 1545.9. UreterGlow-45 was synthesized using commercially available PEG2000 which is comprised of PEG oligomers of various lengths that should be centered around a PEG chain length of 45. Calcd. for PEG45 length ($C_{138}H_{240}N_4O_{58}S_5$): 3043.7, found 3001.3, which corresponds to the most prominent PEG chain length centered around 44.

2.3. *Characterization of Spectral Properties*

The excitation and emission wavelengths of the various dyes (1 μM in PBS) were scanned using a Cary Eclipse fluorimeter (Agilent, Santa Clara, CA, USA). The wavelength which resulted in the maximum excitation and emission values for each dye was determined.

2.4. *Animal Husbandry*

ND4 Swiss Webster mice (Harlan Laboratories, Indianapolis, IN, USA) were maintained on normal rodent chow and housed in a sterile environment on a standard 12 h light and dark cycle for the duration of the study. All animal procedures were approved by the Purdue Animal Care and Use Committee in accordance with NIH guidelines (protocol #1111000316 approved 2 February 2021).

2.5. *In Vivo Biodistribution*

Mice were injected via tail vein with 10 nmol of a fluorescent dye conjugate. Mice were sacrificed at 2, 4, and 6 h post-injection, and organs were removed ($n = 1$ per timepoint per conjugate for the initial UreterGlow-0, -3, -11, and -45 biodistribution studies and $n = 3$ per timepoint per conjugate for the UreterGlow-11, IR800CW, IR800BK, and ZW800-1 biodistribution studies). For urine analysis, mice were administered 10 nmol UreterGlow-11 via tail vein injection, sacrificed at various timepoints, and urine was removed from the bladder via syringe. The organs and urine were imaged using a Caliper IVIS Lumina II Imaging Station (PerkinElmer, Waltham, MA, USA) coupled with an ISOON5160 Andor Nikon camera equipped with Living Image Software Version 4.0 (PerkinElmer, Waltham, MA, USA). The settings were as follows: lamp level, high; excitation, 745 nm; emission, ICG; epi illumination; binning (M) 4; FOV, 12.5; f-stop, 4; acquisition time, 1 s.

2.6. *Effect of pH on UreterGlow-11 Emission Spectra*

UreterGlow-11 (1 μM) was added to PBS (pH 7.4), freshly collected human urine (pH 5.5) or sodium carbonate buffered saline with pH values ranging from 2.5 to 10. The dye was excited using 800 nm light and the emission spectrum was obtained using a Cary Eclipse fluorimeter (Agilent, Santa Clara, CA, USA).

3. Results and Discussion

3.1. *Design and Synthesis of the Ureter Probes*

In an effort to remedy the deficiencies of current ureter imaging agents, we undertook to design a water-soluble NIR dye that would (i) excrete for several hours primarily through the ureters, (ii) avoid uptake by normal tissues, (iii) excite with the same light source used for visualization of tumor-targeted fluorescent dyes, and (iv) emit at a longer wavelength than the tumor-targeted fluorescent dyes; i.e., to allow discrimination of the tumor from ureter fluorescence. Because PEGylation can prolong circulation times and reduce nonspecific uptake by healthy tissues [21–23], we synthesized a series of optical probes comprised of PEG oligomers of different lengths linked to the cyanine dye, S0456, via a thioether bond to 4-mercaptophenylacetic acid (see Methods). As summarized in Scheme 1, PEG oligomers containing 0, 3, 11, and ~45 oxyethylene units were conjugated to the modified S0456 dye and designated as UreterGlow-0, -3, -11, -45. All four conjugates were purified using preparative reverse-phase HPLC and then characterized by LC-MS (see Supplementary Materials). With sufficient quantities synthesized and purities of >95%

achieved for all conjugates, characterization of their physical and biological properties could commence.

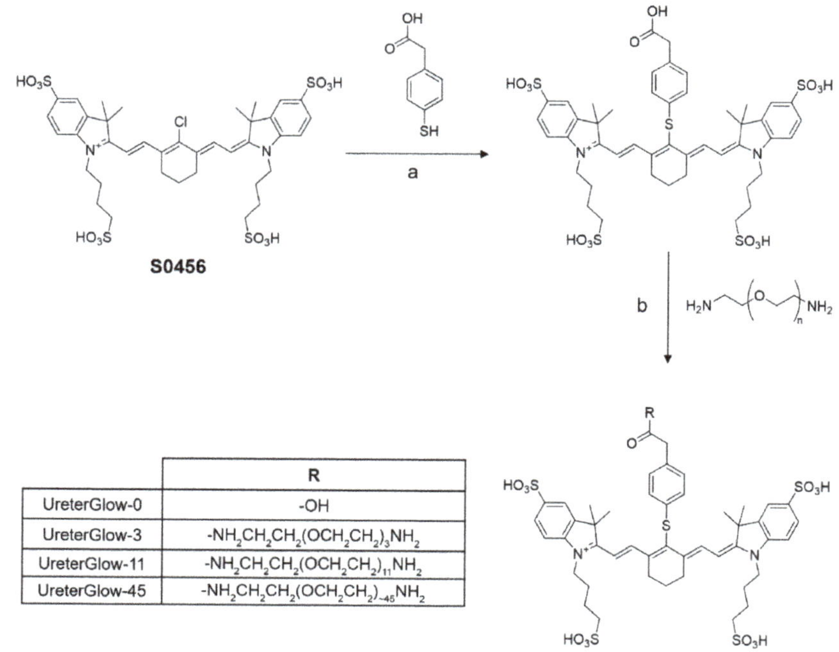

	R
UreterGlow-0	-OH
UreterGlow-3	$-NH_2CH_2CH_2(OCH_2CH_2)_3NH_2$
UreterGlow-11	$-NH_2CH_2CH_2(OCH_2CH_2)_{11}NH_2$
UreterGlow-45	$-NH_2CH_2CH_2(OCH_2CH_2)_{\sim 45}NH_2$

Scheme 1. Synthesis of PEGylated NIR dyes for improving renal clearance. (**a**) DMSO, RT, 12 h. (**b**) HATU (1.1 equiv), DIPEA (5 equiv), DMSO, RT, 6 h.

3.2. Characterization of Physical Properties

Following synthesis of the desired conjugates, molecular weights were confirmed by mass spectrometry, and excitation and emission spectra were obtained using a fluorescence spectrophotometer. As shown in Table 1 and Figure 1, use of a thioether in these UreterGlow conjugates instead of an oxoether bridge connecting the S0456 dye to a phenyl ring shifted the excitation maximum of the conjugate from 776 nm (i.e., similar to OTL38 and many other tumor-targeted fluorescent dyes) to 800 nm. The corresponding emission maxima also shifted from 796 nm to 830 nm, respectively. Because the excitation spectra of both the oxo- and thioether bridged S0456 dyes, as well as the other major NIR dyes used for tumor imaging (IR800CW, LS288, ZW800-1, ICG, and OTL38) overlap over most of their excitation spectra (Figure 1), all of the above NIR dyes should be excitable with the same light source; i.e., avoiding the need to change light sources or cameras to image cancer tissues and ureters simultaneously. Moreover, because the emission spectra of the thioether dyes are shifted ~30–40 nm to longer wavelengths from the major tumor-imaging dyes (Table 1 and Figure 1), it should be possible to display tumor tissue and ureters in different colors on any imaging monitor [24]. Based on these considerations, we expect our ureter probes to function well in combination with most tumor-targeted fluorescent dyes to help prevent accidental ureter damage during abdominal surgeries.

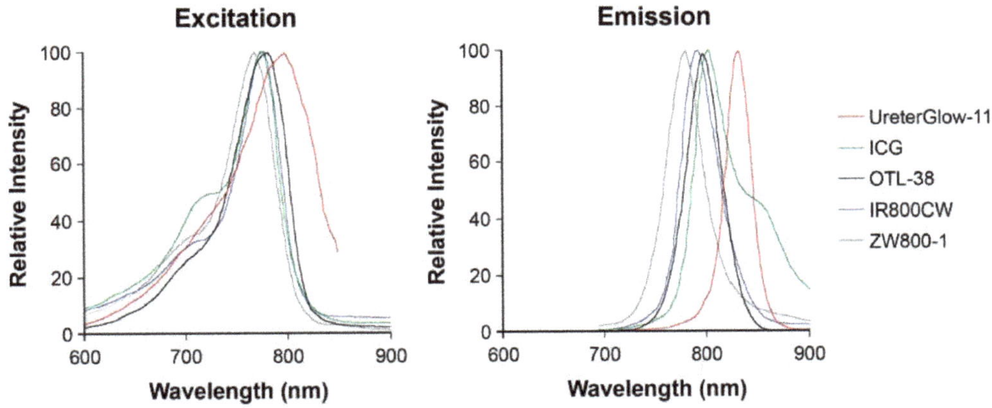

Figure 1. Excitation and emission spectra of various NIR dyes for fluorescence-guided surgery applications. Dyes (1 μM in PBS, pH 7.4) were excited at 680 nm and their emissions were scanned from 700 to 900 nm. Alternatively, excitation wavelengths were scanned from 600 nm to 850 or 900 nm while the emission wavelength was set at 900 nm followed by normalization of their intensities.

3.3. *In Vivo Imaging and Biodistribution*

To test the hypothesis that a PEG linker of the appropriate length can reduce healthy tissue uptake while prolonging passage of the conjugate through the ureter, the aforementioned PEGylated probes were injected via tail vein into live mice and allowed to circulate for different lengths of time (i.e., 2, 4, and 6 h) before euthanasia and analysis of tissue fluorescence ($n = 1$ for each timepoint and each conjugate). As shown in Figures 2 and 3, UreterGlow-0 showed significant uptake in all major organs except the heart, spleen, and lungs, demonstrating that UreterGlow-0 would not function well for ureter imaging. However, as the length of the appended PEG chain was increased, healthy tissue retention decreased at all time points, with minimal if any healthy tissue fluorescence of UreterGlow-11 and -45 detected at the 2 h time point and no significant healthy tissue fluorescence observed at any subsequent time points. These data demonstrate that the longer PEG chains suppress uptake of the UreterGlow conjugates by healthy tissues, and that their capture by the liver and subsequent excretion via the bile duct into the intestines is also suppressed by longer PEGylation. Because compounds not excreted via the liver/bile duct must excrete through the kidneys, this redirection of UreterGlow-11 and -45 to clearance through the kidneys should enhance and prolong their flow through the ureters. However, because both UreterGlow-11 and -45 performed similarly, UreterGlow-11 was employed in all further studies because it could be synthesized as a homogeneous molecular species.

To compare the properties of UreterGlow-11 with other NIR dyes previously examined for ureter imaging, mice ($n = 3$ per time point per conjugate) were intravenously injected with UreterGlow-11, IR800CW, IR800BK, or ZW800-1) [16,25] and sacrificed 2, 4, or 6 h after injection prior to analysis of tissue-retained fluorescence. As shown in Figure 4, UreterGlow-11 showed little or no uptake in any tissues except the kidneys at all time points examined, suggesting its signal to background contrast along the urinary tract should be very high. In contrast, all other dyes investigated displayed significant accumulation in healthy organs, likely due to their partial excretion through the liver, bile duct and intestines and/or nonspecific retention by an unknown process in these tissues. These nonspecific uptake properties could be troublesome during fluorescence-guided surgeries of metastatic cancers since the latter dyes are also commonly used in fluorescent probes for imaging malignant lesions.

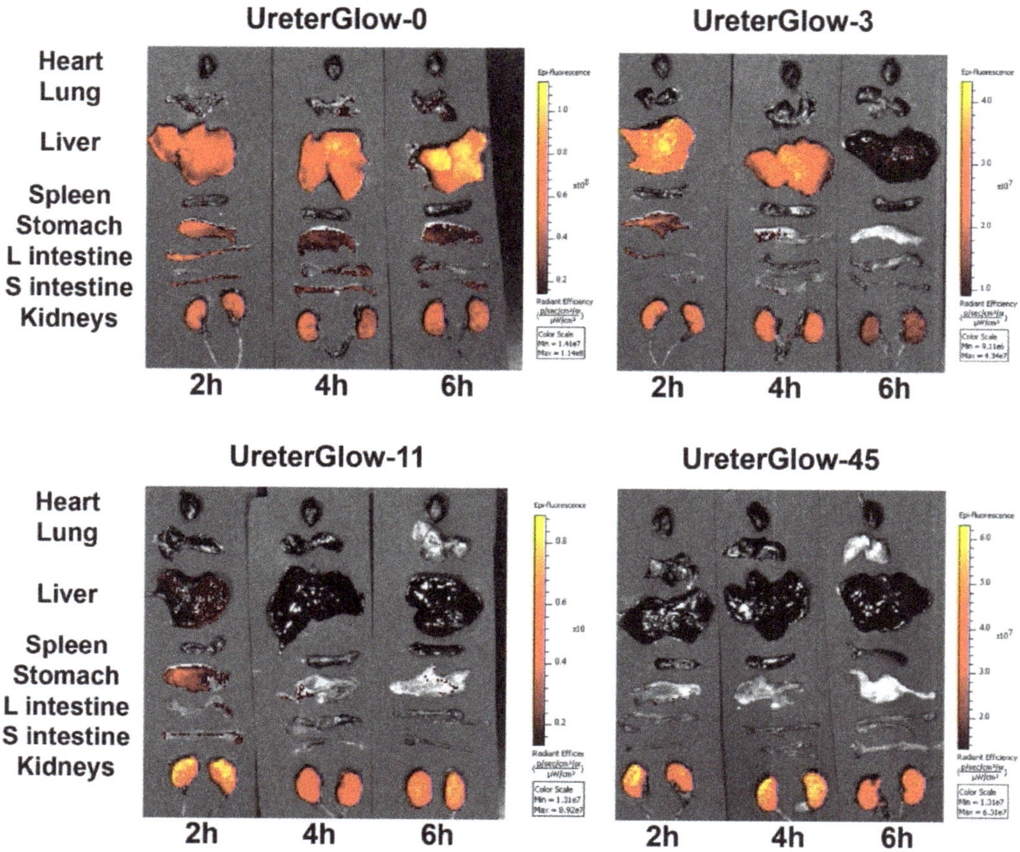

Figure 2. In vivo biodistribution of UreterGlow conjugates. Mice were administered 10 nmol of various conjugates ($n = 1$ per time point per conjugate) via tail vein injection. After varying times, mice were euthanized and their organs were removed. Organs were imaged and the fluorescence intensity recorded.

Although the small size of murine ureters rendered them difficult to image, because any dye that appears in the urine will have recently passed through the ureters, we collected urine samples at different times, post-intravenous injection, and measured their fluorescence intensities in order to confirm that UreterGlow-11 could provide strong ureter fluorescence for prolonged periods following administration. As shown in Figure 5, urine fluorescence remained high for at least 12 h after UreterGlow-11 infusion and then only gradually declined over the subsequent 12 h. These data suggest that UreterGlow-11 should illuminate ureters well, even during protracted abdominal surgeries.

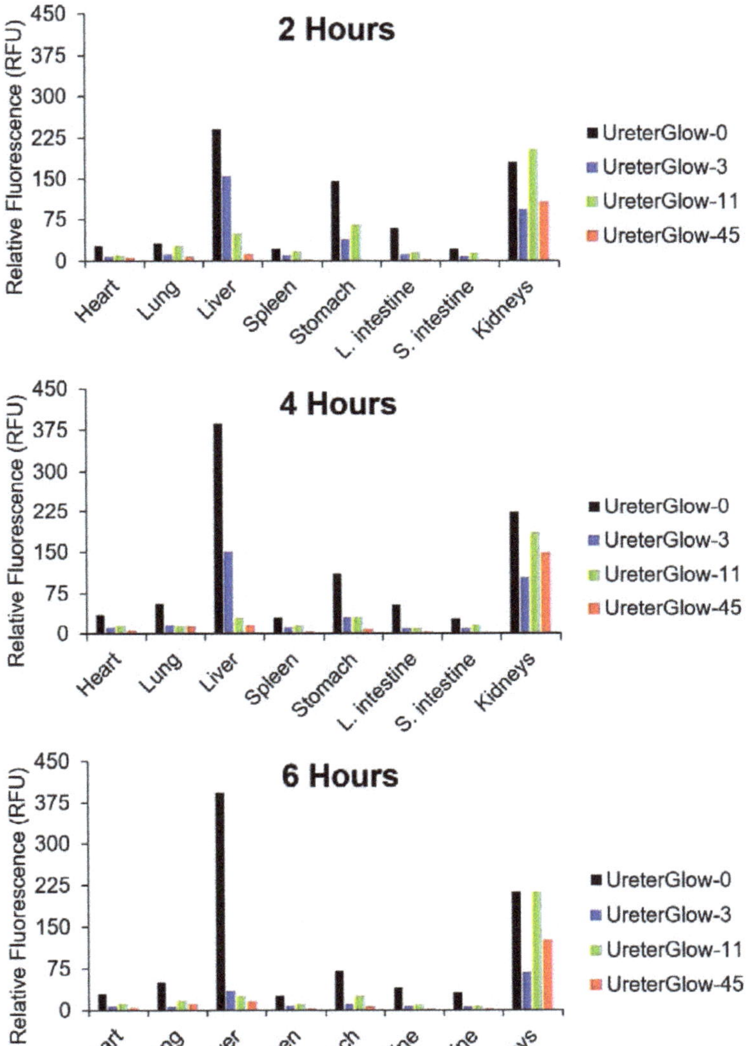

Figure 3. Quantitation of in vivo biodistribution of UreterGlow conjugates. Mice were administered 10 nmol of various conjugates (n = 1 per time point per conjugate) via tail vein injection. After varying times, mice were euthanized and their organs were removed. Organs were imaged, fluorescence intensity was recorded, and relative fluorescence was plotted.

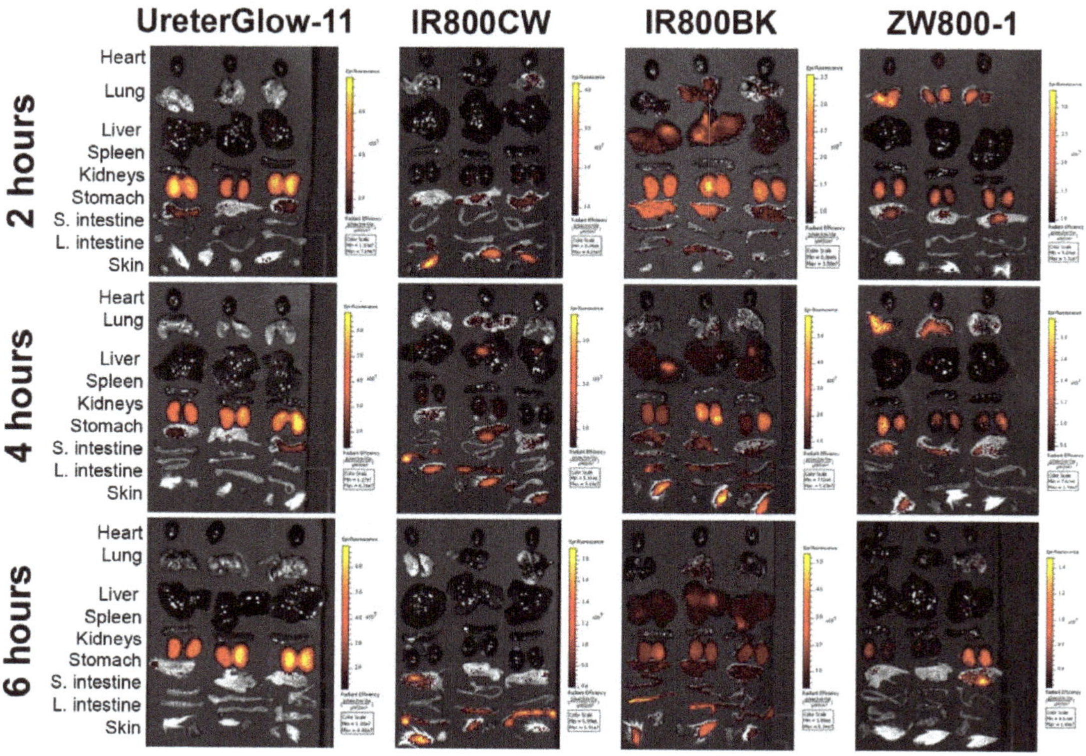

Figure 4. Comparison of in vivo biodistribution of UreterGlow-11 and other NIR ureter imaging dyes/conjugates. Mice were administered 10 nmol of various conjugates (n = 3 per time point per conjugate) via tail vein injection. After varying times, mice were euthanized, their organs were removed, fluorescence intensity was imaged.

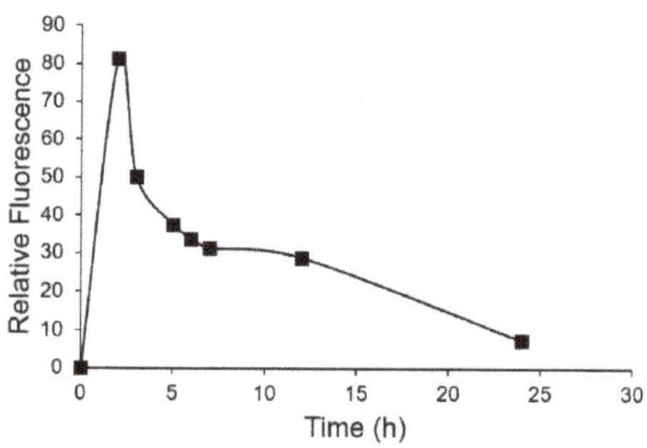

Figure 5. Quantitation of urine fluorescence after administration of UreterGlow-11. Mice were administered 10 nmol of various conjugates (n = 1 per time point) via tail vein injection. After varying times, mice were euthanized, and urine was collected from their bladders. The fluorescence of the isolated urine was quantitated.

Finally, because urine pH can vary from pH 4.5 to pH 8 [26], it was important to ensure that the UreterGlow-45 fluorescence did not vary with urine pH. As shown in Figure 6A, the emission spectrum of UreterGlow-45 was independent of pH between 2.5 and 10 and also showed no impact when dissolved in urine (Figure 6B). Taken together, these data collectively suggest that UreterGlow-11 should perform well as a ureter imaging agent during abdominal surgeries for cancer.

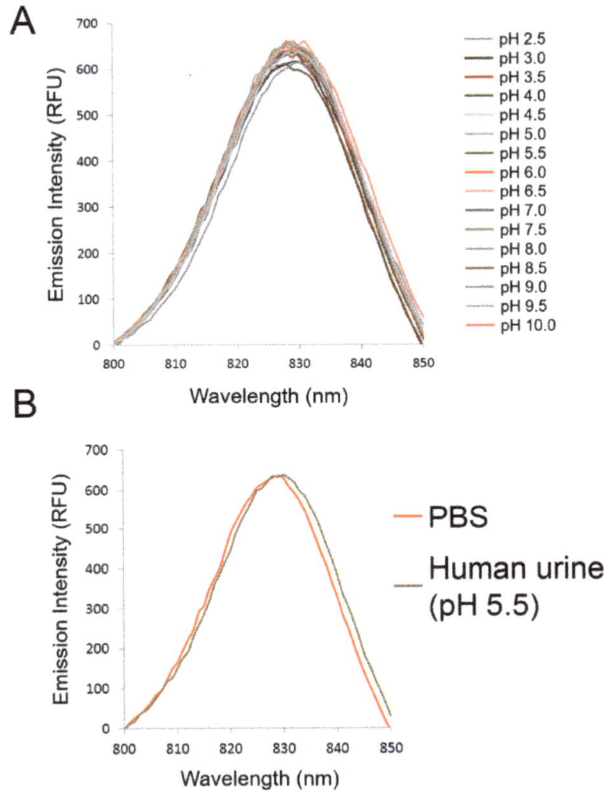

Figure 6. Sensitivity of UreterGlow-45 emission spectra in different pH buffers. (**A**) UreterGlow-11 (1 μM) was excited at 800 nm while dissolved in sodium acetate buffered saline at various pH levels, and the emission spectra were characterized. (**B**) UreterGlow-11 (1 μM) was excited at 800 nm while dissolved in PBS (pH 7.4) or human urine (pH 5.5), and the emission spectra were characterized.

4. Conclusions

Although this brief report described only the impact of two compositional variables on the properties of a NIR dye for intra-operative ureter imaging, many other modifications could also have been explored for further optimization. Thus, NIR dyes with other excitation and emission wavelengths could have been generated by the insertion of other heteroatoms at other locations in the UreterGlow conjugate. PEGs of intermediate lengths between 11 and 45 oxyethylene units could also have been examined for improved biodistribution and pharmacokinetic properties. And finally, a targeting ligand could have been designed that would enable sustained binding of the fluorescent conjugate to the epithelial cells lining the ureters. Thus, while the above improvements in ureter specificity, emission wavelength, and transit time through the ureters now renders UreterGlow-11 a good candidate for translation into the clinic, opportunities may remain for further optimization with an eventual goal of totally eliminating accidental damage to ureters during abdominal surgeries.

Supplementary Materials: The following are available online, Figure S1: Chemical Structures of Selected Dyes Used in Fluorescence-Guided Surgeries; Figure S2: Structure and LC-MS Characterization of UreterGlow-0; Figure S3: Structure and LC-MS Characterization of UreterGlow-3; Figure S4: Structure and LC-MS Characterization of UreterGlow-11; Figure S5: Structure and LC-MS Characterization of UreterGlow-45.

Author Contributions: Conceptualization, M.S. and P.S.L.; methodology, S.M.M., M.S. and P.S.L.; formal analysis S.M.M. and K.S.P.; data curation, K.S.P., writing-original draft preparation K.S.P. and P.S.L.; writing review and editing; K.S.P., P.S.L., M.S., S.M.M.; supervision P.S.L.; project administration M.S. and P.S.L.; funding acquisition P.S.L. All authors have read and agreed to the published version of the manuscript.

Funding: The authors gratefully acknowledge support from the Purdue University Center for Cancer Research, P30CA023168, the Purdue Institute for Drug Discovery, and a grant from On Target Laboratories LLC.

Institutional Review Board Statement: The study was conducted according to the guidelines of the Declaration of Helsinki, and approved by the Purdue Animal Care and Use Committee of Purdue University (protocol #1111000316 approved 2 February 2021).

Data Availability Statement: The data presented in this study are available in Putt KS. 2021. Supplemental Information Design of a near infrared fluorescent ureter imaging agent for prevention of ureter damage during abdominal surgeries; Zenodo http://doi.org/10.5281/zenodo.4987325.

Conflicts of Interest: The authors declare the following competing financial interest(s): This work was supported in part by a grant from On Target Laboratories. Philip Low is a cofounder and a member of the Board of Directors of on Target Laboratories.

Sample Availability: Samples of the compounds are not available from the authors.

References

1. Silva, G.; Boutros, M.; Wexner, S. Role of prophylactic ureteric stents in colorectal surgery. *Asian J. Endosc. Surg.* **2012**, *5*, 105–110. [CrossRef] [PubMed]
2. Packiam, V.T.; Andrew, C.; Joseph, P.; Charles, N.; Sarah, F.; Gregory, B. The Impact of Minimally Invasive Surgery on Major Iatrogenic Ureteral Injury and Subsequent Ureteral Repair during Hysterectomy: A National Analysis of Risk Factors and Outcomes. *Urology* **2016**, *98*, 183–188. [CrossRef]
3. Barber, M.D.; Visco, A.G.; Weidner, A.C.; Amundsen, C.L.; Bump, R.C. Bilateral uterosacral ligament vaginal vault suspension with site-specific endopelvic fascia defect repair for treatment of pelvic organ prolapse. *Am. J. Obstet. Gynecol.* **2000**, *183*, 1402–1411. [CrossRef]
4. Sandberg, E.M.; Cohen, S.L.; Hurwitz, S.; Einarsson, J.I. Utility of cystoscopy during hysterectomy. *Obstet. Gynecol.* **2012**, *120*, 1363–1370. [CrossRef]
5. Sharp, H.T.; Adelman, M.R. Prevention, Recognition, and Management of Urologic Injuries during Gynecologic Surgery. *Obstet. Gynecol.* **2016**, *127*, 1085–1096. [CrossRef] [PubMed]
6. Chou, M.T.; Wang, C.J.; Lien, R.C. Prophylactic ureteral catheterization in gynecologic surgery: A 12-year randomized trial in a community hospital. *Int. Urogynecol. J. Pelvic Floor Dysfunct.* **2009**, *20*, 689–693. [CrossRef]
7. Speicher, P.J.; Goldsmith, Z.G.; Nussbaum, D.P.; Turley, R.S.; Peterson, A.C.; Mantyh, C.R. Ureteral stenting in laparoscopic colorectal surgery. *J. Surg. Res.* **2014**, *190*, 98–103. [CrossRef] [PubMed]
8. Dadkhah, F.; Yari, H.; Ali Asgari, M.; Fallahnezhad, M.H.; Tavoosian, A.; Ghadian, A. Benefits and Complications of Removing Ureteral Stent Based on the Elapsed Time after Renal Transplantation Surgery. *Nephrourol. Mon.* **2016**, *8*, e31108. [CrossRef]
9. Fanning, J.; Fenton, B.; Jean, G.M.; Chae, C. Cost analysis of prophylactic intraoperative cystoscopic ureteral stents in gynecologic surgery. *J. Am. Osteopath. Assoc.* **2011**, *111*, 667–669.
10. Cherrick, G.R.; Stein, S.W.; Leevy, C.M.; Davidson, C.S. Indocyanine green: Observations on its physical properties, plasma decay, and hepatic extraction. *J. Clin. Investig.* **1960**, *39*, 592–600. [CrossRef] [PubMed]
11. Mahalingam, S.M.; Dip, F.; Castillo, M.; Roy, M.; Wexner, S.D.; Rosenthal, R.J.; Low, P.S. Intraoperative Ureter Visualization Using a Novel Near-Infrared Fluorescent Dye. *Mol. Pharm.* **2018**, *15*, 3442–3447. [CrossRef] [PubMed]
12. Siddighi, S.; Yune, J.J.; Hardesty, J. Indocyanine green for intraoperative localization of ureter. *Am. J. Obstet. Gynecol.* **2014**, *211*, 436.e1–436.e2. [CrossRef] [PubMed]
13. Rajanbabu, A.; Patel, V.J. Ureteric mapping with Indocyanine green: A new tool to prevent ureteral injury in complex gynecological surgery. *J. Endometr. Pelvic Pain Disord.* **2020**, *12*, 190–192. [CrossRef]
14. Lee, Z.; Moore, B.; Giusto, L.; Eun, D.D. Use of indocyanine green during robot-assisted ureteral reconstructions. *Eur. Urol.* **2015**, *67*, 291–298. [CrossRef]
15. Cabanes, M.; Boria, F.; Gutiérrez, A.H.; Zapardiel, I. Intra-operative identification of ureters using indocyanine green for gynecological oncology procedures. *Int. J. Gynecol. Cancer* **2019**, *30*, 278. [CrossRef] [PubMed]
16. Slooter, M.; Janssen, A.; Bemelman, W.; Tanis, P.; Hompes, R. Currently available and experimental dyes for intraoperative near-infrared fluorescence imaging of the ureters: A systematic review. *Tech. Coloproctol.* **2019**, *23*, 305–313. [CrossRef]
17. Korb, M.L.; Hartman, Y.E.; Kovar, J.; Zinn, K.R.; Bland, K.I.; Rosenthal, E.L. Use of monoclonal antibody-IRDye800CW bioconjugates in the resection of breast cancer. *J. Surg. Res.* **2014**, *188*, 119–128. [CrossRef]

18. Shrivastava, A.; Ding, H.; Kothandaraman, S.; Wang, S.-H.; Gong, L.; Williams, M.; Milum, K.; Zhang, S.; Tweedle, M.F. A High-Affinity Near-Infrared Fluorescent Probe to Target Bombesin Receptors. *Mol. Imaging Biol.* **2014**, *16*, 661–669. [CrossRef]
19. Kanduluru, A.K.; Srinivasarao, M.; Low, P.S. Design, Synthesis, and Evaluation of a Neurokinin-1 Receptor-Targeted Near-IR Dye for Fluorescence-Guided Surgery of Neuroendocrine Cancers. *Bioconjug. Chem.* **2016**, *27*, 2157–2165. [CrossRef] [PubMed]
20. Wayua, C.; Low, P.S. Evaluation of a cholecystokinin 2 receptor-targeted near-infrared dye for fluorescence-guided surgery of cancer. *Mol. Pharm.* **2014**, *11*, 468–476. [CrossRef] [PubMed]
21. Hamidi, M.; Rafiei, P.; Azadi, A. Designing PEGylated therapeutic molecules: Advantages in ADMET properties. *Expert Opin. Drug Discov.* **2008**, *3*, 1293–1307. [CrossRef] [PubMed]
22. Milla, P.; Dosio, F.; Cattel, L. PEGylation of proteins and liposomes: A powerful and flexible strategy to improve the drug delivery. *Curr. Drug Metab.* **2012**, *13*, 105–119. [CrossRef] [PubMed]
23. Mozar, F.S.; Chowdhury, E.H. Impact of PEGylated nanoparticles on tumor targeted drug delivery. *Curr. Pharm. Des.* **2018**, *24*, 3283–3296. [CrossRef] [PubMed]
24. De Boer, E.; Harlaar, N.J.; Taruttis, A.; Nagengast, W.B.; Rosenthal, E.L.; Ntziachristos, V.; van Dam, G.M. Optical innovations in surgery. *Br. J. Surg.* **2015**, *102*, e56–e72. [CrossRef] [PubMed]
25. Van den Bos, J.; Al-Taher, M.A.; Bouvy, N.D.; Stassen, L.P.S. Near-infrared fluorescence laparoscopy of the ureter with three preclinical dyes in a pig model. *Surg Endosc.* **2019**, *33*, 986–991. [CrossRef]
26. Bono, M.J.; Reygaert, W.C. Urinary Tract Infection. In *StatPearls [Internet]*; StatPearls Publishing: Treasure Island, FL, USA, 2021.

molecules

MDPI

Article

Crosslinking of CD38 Receptors Triggers Apoptosis of Malignant B Cells

M. Tommy Gambles [1,2], Jiahui Li [1,2], Jiawei Wang [1,2], Douglas Sborov [3], Jiyuan Yang [1,2,*] and Jindřich Kopeček [1,2,4,*]

1 Center for Controlled Chemical Delivery, University of Utah, Salt Lake City, UT 84112, USA; u0480778@utah.edu (M.T.G.); Jiahui.Li@pharm.utah.edu (J.L.); jiawei.wang@utah.edu (J.W.)
2 Department of Pharmaceutics and Pharmaceutical Chemistry, University of Utah, Salt Lake City, UT 84112, USA
3 Huntsman Cancer Institute, University of Utah, Salt Lake City, UT 84112, USA; Douglas.Sborov@hci.utah.edu
4 Department of Biomedical Engineering, University of Utah, Salt Lake City, UT 84112, USA
* Correspondence: jiyuan.yang@utah.edu (J.Y.); jindrich.kopecek@utah.edu (J.K.)

Abstract: Recently, we designed an inventive paradigm in nanomedicine—drug-free macromolecular therapeutics (DFMT). The ability of DFMT to induce apoptosis is based on biorecognition at cell surface, and crosslinking of receptors without the participation of low molecular weight drugs. The system is composed of two nanoconjugates: a bispecific engager, antibody or Fab′ fragment—morpholino oligonucleotide (MORF1) conjugate; the second nanoconjugate is a multivalent effector, human serum albumin (HSA) decorated with multiple copies of complementary MORF2. Here, we intend to demonstrate that DFMT is a platform that will be effective on other receptors than previously validated CD20. We appraised the impact of daratumumab (DARA)- and isatuximab (ISA)-based DFMT to crosslink CD38 receptors on CD38+ lymphoma (Raji, Daudi) and multiple myeloma cells (RPMI 8226, ANBL-6). The biological properties of DFMTs were determined by flow cytometry, confocal fluorescence microscopy, reactive oxygen species determination, lysosomal enlargement, homotypic cell adhesion, and the hybridization of nanoconjugates. The data revealed that the level of apoptosis induction correlated with CD38 expression, the nanoconjugates meet at the cell surface, mitochondrial signaling pathway is strongly involved, insertion of a flexible spacer in the structure of the macromolecular effector enhances apoptosis, and simultaneous crosslinking of CD38 and CD20 receptors increases apoptosis.

Keywords: CD38; drug-free macromolecular therapeutics; human serum albumin conjugates; morpholino oligonucleotides; daratumumab; isatuximab; multiple myeloma; lymphoma

Citation: Gambles, M.T.; Li, J.; Wang, J.; Sborov, D.; Yang, J.; Kopeček, J. Crosslinking of CD38 Receptors Triggers Apoptosis of Malignant B Cells. *Molecules* **2021**, *26*, 4658. https://doi.org/10.3390/molecules26154658

Academic Editors: Marina A. Dobrovolskaia and Kirill A. Afonin

Received: 4 June 2021
Accepted: 28 July 2021
Published: 31 July 2021

1. Introduction

The use of monoclonal antibodies (mAb) in the treatment of hematological malignancies has become an essential part of immunotherapy regimens [1]. Often mAb's are used in combination with small molecule chemotherapeutics to improve patient prognoses. Immunotherapy offers highly specific targeting to overexpressed cancer cell surface antigens. Once engaged with their target cell surface receptor, various mechanisms of action may occur to initiate cancer cell death. Immune effector cells can interact with the Fc domains leading to a variety of cell death events including complement-dependent cytotoxicity (CDC), antibody-dependent cellular cytotoxicity (ADCC), and antibody-dependent cellular phagocytosis (ADCP) [2]. Additionally, crosslinking of some receptor-bound antibodies leads to a clustering effect of surface receptors that triggers apoptotic mechanisms within the cell.

Crosslinking of cell-surface receptors has important biological consequences, including enhancing internalization of receptor-ligand complexes [3,4], changing the subcellular fate of the receptor-ligand complex from recycling to the lysosomal route [5], and phosphatidylserine translocation and apoptosis initiation [6].

The CD38 receptor is substantially expressed on multiple myeloma cells and at low levels on normal lymphoid and myeloid cells. It is also expressed on lymphoma cells. In addition to its receptor function, CD38 is an ectoenzyme that cleaves nicotinamide dinucleotide phosphate (NADP) and nicotinamide adenine dinucleotide (NAD^+) [7]. DARA (fully human IgG1-κ) [8] and ISA (chimeric IgG1-κ) [9] are two FDA approved antibodies for multiple myeloma (MM) treatment [7]. DARA's mechanisms of action include CDC, ADCC, and ADCP. In addition, FcγR-mediated crosslinking of tumor-bound DARA initiates cell death (Scheme 1) [6]. ISA's mechanism of action includes CDC, ADCC, and ADCP. Importantly, ISA has a strong apoptosis-inducing activity that is independent of crosslinking and inhibits the enzymatic activity of CD38 [10].

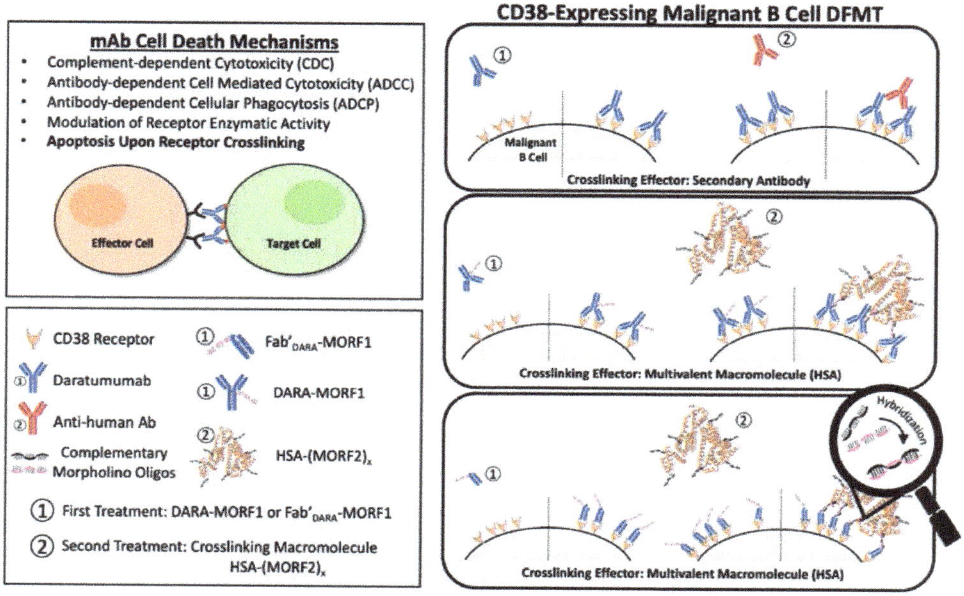

Scheme 1. CD38-targeting DFMT system depicted using DARA-MORF1 as the bispecific engager and HSA-$(MORF2)_y$ as the crosslinking effector molecule. DFMT is a two-step process involving step (1): specific antigen engagement by the bispecific engager followed by step (2): receptor crosslinking via the effector nanoconjugate. Crosslinking occurs via hybridization of complementary morpholino oligonucleotides conjugated to the bispecific engager and effector molecules.

Drug-free macromolecular therapeutics (DFMT) is a new paradigm for the treatment of malignancies [11–13]. Induction of apoptosis is mediated by receptor crosslinking facilitated by biorecognition of complementary peptide or oligonucleotide motifs; no low molecular weight drug is needed [11,12]. DFMT is comprised of two complementary nanoconjugates: (a) the bispecific engager: an antibody or Fab′ fragment conjugate with morpholino oligonucleotide MORF1; and (b) the crosslinking effector: *N*-(2-hydroxypropyl)methacrylamide (HPMA) copolymer or human serum albumin (HSA) modified with multiple copies of complementary oligonucleotide MORF2. Hybridization of MORF1/MORF2—mediated receptor crosslinking initiates apoptosis (Scheme 1). We have demonstrated the efficacy of DMFT on CD20 positive (CD20+) Raji B cells in vitro [14,15], in vivo on a disseminated non-Hodgkin lymphoma (NHL) model in SCID mice [16,17], and on patient cells diagnosed with various blood borne malignancies [18]. Apoptosis induction by DMFT is triggered by relocation of crosslinked CD20 complexes to lipid rafts resulting in calcium influx, mitochondrial depolarization, and caspase 3 activation [19].

The linkers used in the DFMT nanoconjugations are based on a bifunctional PEG. Intact nanoconjugates, therefore, have inert linkers consisting of short PEG dimers with amide and thioether bonds on either terminus. All antibodies used in this work are FDA approved products which have overcome much scrutiny in terms of toxicity, pharmacokinetics and biocompatibility [20–22]. Human serum albumin is a ubiquitous protein in human plasma and is gaining more and more interest in drug delivery systems because of its long circulation half-life and non-immunogenicity [23]. The morpholino oligomer strands are DNA analogues that have had their backbone chemistry altered to allow for protease resistance in vivo. Numerous oligonucleotides have been evaluated in clinical trials and proven their biocompatibility [24]. A detailed study of the immune properties of the crosslinking effector HSA-(MORF2)x is planned using protocols we developed when evaluating the peptide containing crosslinking effector P-(CCK)x [25].

Rituximab (RTX) and other Type I antibodies dramatically improved treatment of CD20+ B-cell hematological malignancies [26]. However, a number of patients develop resistance and due to polymorphism of Fcγ receptors on immunocompetent cells the hyper-crosslinking of RTX bound to CD20 is not efficient. This catalyzed the advance of Type II antibodies, such as Obinutuzumab (OBN), that do not need crosslinking. They induce apoptosis by actin rearrangement, lysosomal disruption, and homotypic cell adhesion [21].

Interestingly, when DFMT was applied to OBN, classified as a Type II antibody, a system that combines Type I and Type II mechanisms was developed [27]. The first nanoconjugate was OBN-MORF1 (OBN conjugated to one morpholino oligonucleotide MORF1); and the second nanoconjugate was HSA-$(MORF2)_y$ (HSA grafted with multiple copies of complementary morpholino oligonucleotide 2). Modification of OBN with one MORF1 does not impact the binding of OBN-MORF1 to CD20 and following binding to CD20 Type II effects occur. Further exposure to multivalent effector HSA-$(MORF2)_y$ results in crosslinking of CD20-OBN-MORF1 complexes, their clustering into lipid rafts and initiation of Type I effects. This new approach, called "clustered OBN (cOBN)" combines effects of both cell death-inducing mechanisms resulting in very high apoptotic levels [27].

Aiming to improve the efficacy of treatment of B cell malignancies in general and of multiple myeloma in particular, we evaluated the impact of crosslinking CD38 receptors on the mechanism and extent of apoptotic induction in four CD38 positive malignant B cells (Daudi, Raji, RPMI 8226, and ANBL-6). DFMT based on Fab'_{DARA}-MORF1, Fab'_{ISA}-MORF1, DARA-MORF1, and ISA-MORF1 as bispecific engagers and HSA-PEG_x-$(MORF2)_y$ as multivalent crosslinking effector were evaluated. The ultimate goal of our studies is to establish the cell and nanoconjugate structure-dependent participation of Type I and Type II apoptotic mechanisms in CD38 crosslinking-mediated apoptosis induction.

2. Results and Discussion

2.1. Nanoconjugates Synthesis and Cell Lines

DARA and ISA nanoconjugates were synthesized using procedures previously reported by our group (Figure 1A) [27–29]. First, whole antibody was selectively reduced targeting the interchain disulfide bonds located in the hinge region. In parallel to the reduction reaction, the 3′-amine-functionalized MORF1 was reacted with maleimide-PEG_2-NHS by aminolysis, resulting in maleimide-MORF1 intermediate. The latter was then coupled with freshly reduced whole antibody via thiol–ene click chemistry yielding the desired antibody-MORF1 nanoconjugate. The Fab′ fragment MORF1 analogues were generated by first digesting whole antibody with pepsin resulting in dual chain cleavage below the disulfide hinge region generating the divalent $F(ab')_2$ intermediate. $F(ab')_2$ was then reduced to generate two equivalents of Fab′ fragments which were further coupled via a thiol–ene click reaction with the maleimide-MORF1 intermediate yielding the desired Fab′-MORF1 nanoconjugate. The HSA-$(PEG)_x$-$(MORF2)_y$ nanoconjugate was synthesized in a similar fashion utilizing the bifunctional reactivity of maleimide-$(PEG)_x$-NHS (SM-PEG). Free lysine amine groups on the periphery of the HSA molecule were coupled with maleimide-PEG_x-NHS (x = 2, 8, or 24) yielding the multivalent maleimide functionalized

HSA-PEG$_x$-maleimide intermediate. The complementary morpholino, MORF2, was customized with an easily reducible disulfide bond on its 3′ terminus. HSA-PEG$_x$-maleimide was decorated with freshly reduced MORF2 molecules via thiol–ene maleimide click reactions. A valency greater than about 5 in HSA-(MORF2)$_y$ did not show significant increase in efficacy of CD20 receptor crosslinking [27]; so minor variations in valency (Figure 1C) should not have an impact on efficacy. Reaction intermediates and final nanoconjugates were characterized by size exclusion chromatography (SEC) (Figure 1B). Each nanoconjugate morpholino valency was characterized by UV-Vis spectrophotometry (Figure 1C) and BCA assay. The valence of MORF per macromolecule was calculated by (i) attaining the concentration of MORF in solution by UV-Vis spectrophotometry and (ii) determining concentration of protein in solution by BCA assay, then (iii) dividing the two concentrations yielding ratio of MORF per molecule. Hybridization between complementary nanoconjugates was confirmed by UV-Vis spectrophotometry by observing absorbance changes at λ = 260 nm of varying molar ratio solutions of MORF1:MORF2 (Figures S2 and S5); and size exclusion chromatography (Figures S1 and S4).

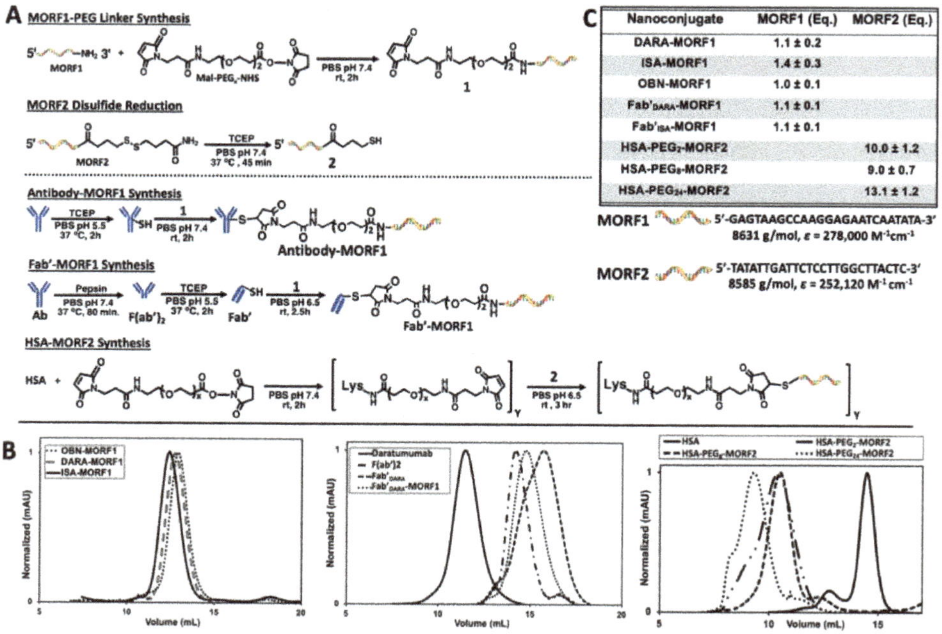

Nanoconjugate	MORF1 (Eq.)	MORF2 (Eq.)
DARA-MORF1	1.1 ± 0.2	
ISA-MORF1	1.4 ± 0.3	
OBN-MORF1	1.0 ± 0.1	
Fab'$_{DARA}$-MORF1	1.1 ± 0.1	
Fab'$_{ISA}$-MORF1	1.1 ± 0.1	
HSA-PEG$_2$-MORF2		10.0 ± 1.2
HSA-PEG$_8$-MORF2		9.0 ± 0.7
HSA-PEG$_{24}$-MORF2		13.1 ± 1.2

Figure 1. Nanoconjugate synthesis and characterization. (**A**) Synthetic route to produce antibody-MORF1, Fab′-MORF1 and HSA-(MORF2)$_y$ conjugates. Note: If not otherwise stated all conjugates contain a diethyleneglycol unit in the spacer inserted by SM-PEG2. Conjugates containing longer spacers are denoted as, e.g., Fab′-PEG$_8$-MORF1. (**B**) Nanoconjugate size exclusion chromatography profiles detected on a Superdex 200 10/300 GL column, PBS (pH 7.4) as eluant at 0.4 mL/min flow rate. (**C**) Characterization (MORF valence) of nanoconjugates determined using UV-Vis absorbance at λ = 260 nm and BCA assay.

Four CD38 positive cell lines were used: Raji (Burkitt's lymphoma), Daudi (Burkitt's lymphoma), RPMI 8226 (multiple myeloma), and ANBL-6 (multiple myeloma). CD38 negative cell line U266 (multiple myeloma) was used as control. Level of CD38 expression was estimated by DARA binding to each cell line at 4 °C followed by exposure to a fluorescently labeled anti-human secondary antibody. The level of CD38 expression was Daudi > RPMI 8226 > Raji > ANBL-6 >> U266 (Figure 2A).

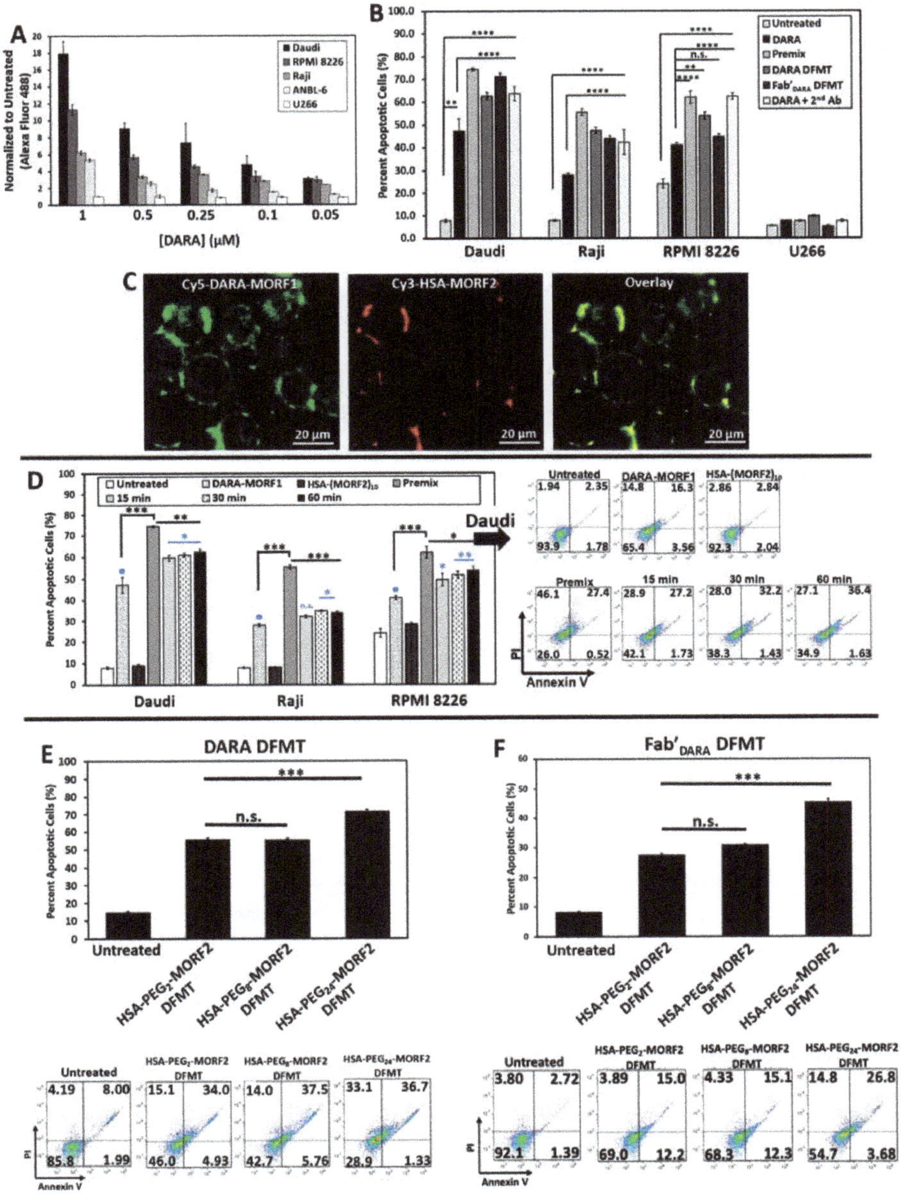

Figure 2. (**A**) DARA binding on various B cell lines as determined by secondary fluorescence of a Fluor 488-labeled anti-human goat antibody. (**B**) Apoptosis induced by DARA-MORF1 and Fab'$_{Dara}$-MORF1 DFMT treatments on various CD38+ cell lines and one CD38- cell line: U266. Cell viability was measured by Annexin V and Propidium Iodide staining and analyzed by flow cytometry. (**C**) Confocal microscopy of fluorescently labeled nanoconjugates: Cy5-DARA-MORF1 (left, green); Cy3-HSA-(MORF2)$_{10}$ (middle, red); overlay (right, yellow) of Cy5 and Cy3 channels on Daudi cells. (**D**) Effect of time lag (15, 30, and 60 min) between the administration of the bifunctional engager, DARA-MORF1 and the multivalent effector, HSA-(MORF2)$_{10}$ on apoptosis initiation. (**E**,**F**) Effect of HSA PEG linker length on DFMT apoptosis efficacy on Daudi cells. (**E**) DARA-MORF1 + HSA-PEG$_{2,8,24}$-(MORF2)$_y$. (**F**) Fab'$_{DARA}$-MORF1 + HSA-PEG$_{2,8,24}$-(MORF2)$_y$. Flow cytometry cell population distribution data for the HSA PEG linker length studies are also shown. All experiments were performed in triplicate. **** $p < 0.0001$, *** $p < 0.001$, ** $p < 0.01$, * $p < 0.05$, n.s., not significant by One-Way ANOVA and Tukey test.

2.2. DFMT Triggers Apoptosis in CD38+ Lymphoma and Myeloma Cell Lines by Consecutive Binding of Nanoconjugates

To validate the hypothesis that crosslinking of CD38 directly initiates apoptosis, we evaluated the levels of apoptosis initiation in Daudi, Raji, RPMI 8226, ANBL-6, and U266 cell lines by exposing them to DARA-MORF1 or Fab'$_{DARA}$-MORF1 (0.5 μM MORF1) for 1 h, followed (after washing and resuspending) to HSA-(MORF2)$_{10}$ (0.5 μM MORF2) for 24 h. High levels of apoptosis were achieved in the three CD38+ cell lines (Daudi cells exhibited the highest levels) as well as in controls, premix and Daratumumab + sec. antibody. As expected, CD38- U266 cells exhibited negligible levels of apoptosis. Interestingly, percentage of apoptotic cells for the various cell types correlated with the level of CD38 expression observed in the binding studies (Figure 2A,B).

We next investigated the biorecognition of nanoconjugates at the cell surface employing confocal fluorescence microscopy. Consecutive exposure of Raji cells to Cy5-DARA-MORF1 resulted in cell surface green signal; exposure of decorated cells to HSA-(MORF2)$_{10}$ showed red surface signal. Both signals were colocalized (yellow color) indicating successful biorecognition (hybridization) of MORF1/MORF2 at cell surface (Figure 2C).

DFMT is a two-step process: The first nanoconjugate a bispecific engager, DARA-MORF1 or Fab'$_{DARA}$-MORF1, binds to CD38 and decorates the cell surface with MORF1 moieties. After a time lag, the second nanoconjugate, a multivalent macromolecular effector, HSA-PEG$_x$-(MORF2)$_y$, hybridizes and crosslinks multiple CD38 receptors resulting in apoptotic response. One important factor related to the efficacy of the process is the potential internalization of CD38 following binding with the bispecific engager. It is known that surface CD38 is internalized after receptor binding [30,31]. The internalization is gradual with time and crosslinking enhances the rate of internalization on the Jurkat cell line [30]. To validate the two-step pretargeting approach, we compared apoptosis induction for different time lags between cells' (Raji, Daudi, and RPMI 8226) exposure to the two nanoconjugates; the second nanoconjugate HSA-(MORF2)$_{10}$ was administered after 15 min, 30 min, and 1 h after the administration of the bispecific engager (Figure 2D and Figure S7). Additionally, we exposed cells to a multivalent premix of both conjugates (control). In all three cell lines the length of the time lag had no impact on the level of apoptosis. Premixing nanoconjugates before cell exposure enhanced apoptotic levels when compared to two-step administration. The difference was largest in Raji cells and minor in Daudi and RPMI 8226 cells. This may be the effect of crosslinking enhanced internalization of the loaded CD38 receptor. The difference in apoptosis induction between premixed nanoconjugates and consecutive administration was minimal for the CD20 receptor [28], reflecting different internalization kinetics of CD20 vs. CD38 following receptor binding and crosslinking.

We described the advantages of the two-step administration previously, e.g., [32]. Importantly, a two-step approach permits pretargeting in vivo, a strategy commonly used in cancer radioimmunotherapy [33,34]. The experiments in this work were performed in vitro which makes the nanoconjugate premixture a meaningful control treatment group because hybridization is allowed to occur in an idealized setting and no washing step between treatments is needed. This provides a theoretical "maximum efficacy" for the in vitro experiments. For in vivo applications, one must consider important factors such as immune response, effector cell interactions and clearance and how each of these factors influence both the targeting of the system and the hybridization.

Pretargeting strategy (two-step treatment) permits the amplification of therapeutic efficacies and reduces adverse side reactions [35]. For example, in our previous work with DFMT targeted to CD20 we determined the time lag when the pretargeting agent (Fab'-MORF1) was mostly cleared from the blood and reached a steady plasma concentration, and, second, by determining the tumor targeting efficiency when using this time interval [14]. Results indicated a suitable timing for P-(MORF2)$_x$ administration at 5 h (in female SCID mice); at this time, Fab'-MORF1 was efficiently distributed to the tumors. Based on this result, we further performed therapy experiments in a disseminated B-NHL mouse model. When the optimized pretargeting time lag (5 h) was used, the therapeutic efficacy was significantly better than that of identical experimental conditions but with a 1 h interval. A low dose (58 μg × 3) of Fab′-MORF1 with a 5× excess P-$(MORF2)_x$ resulted in significantly delayed tumor growth and substantially improved animal survival [14]. The optimized therapeutic system surpassed rituximab in anticancer efficacy and completely eradicated lymphoma B-cells in 83% of the animals. This pretargeting approach may constitute a novel personalized nanotherapy to enable more efficient treatment and limit potential side effects associated with off-target binding.

The decreased adverse (off-target) effects of two-step administration in vivo are based on these phenomena: CD20 is a very slowly internalizing receptor. When Fab′-MORF1 is administered, the part bound to CD20 remains at the surface, whereas the off-target bound part is internalized (and degraded in the lysosomes) during the time-lag before the administration of the crosslinking effector. Thus, the crosslinking effector administered following a time-lag, finds the bispecific engager bound just to the target (CD20).

2.3. Impact of Spacer Length on Apoptosis Initiation

Spacer length is an important factor in the efficiency of nanoconjugates. PEG spacers enhanced the efficacy of DMFT systems based on flexible HPMA copolymer molecules [28] as well as multivalent liposomes [36]. Since HSA has a relatively rigid structure, introduction of a flexible spacer between HSA and MORF2 should enhance biorecognition and apoptosis induction. We used succinimidyl-PEG_x-maleimides, hetero-bifunctional crosslinkers with different numbers of repeating ethyleneglycol (EG) units to synthesize HSA-PEG_x-MORF2 conjugates with variable spacer length. In particular, NHS-PEG_2-maleimide (succinimidyl-[(*N*-maleimidopropionamido)-diethyleneglycol]ester), NHS-PEG_8-maleimide (succinimidyl-[(*N*-maleimidopropionamido)-octaethyleneglycol] ester), and NHS-PEG_{24}-maleimide (succinimidyl-[(*N*-maleimidopropionamido)-tetracosaethylene glycol] ester) (Thermo Scientific) were used for the synthesis of HSA-PEG_2-$(MORF2)_{10}$, HSA-PEG_8-$(MORF2)_9$, and HSA-PEG_{24}-$(MORF2)_{13}$. The characterization of conjugates is shown in Figure 1b (*right panel*). Apoptosis was determined on Raji cells by Annexin V/PI assay. Two bispecific engagers, DARA-MORF1 and Fab'_{DARA}-MORF1 were employed. Data in Figure 2E,F show no statistically significant difference between spacers containing PEG_2 (17.6 Å) and PEG_8 (39.2 Å). Increasing the length of the spacer to PEG_{24} (95.2 Å) resulted in statistically significant enhancement of apoptosis induction. This is valid for both bispecific engagers used. It appears that a relatively rigid carrier, such as HSA, needs a longer spacer to enhance efficacy when compared to a flexible HPMA copolymer carrier. In the latter DFMT system (Fab′-MORF1 + HPMA copolymer-MORF2) a statistically significant enhancement of apoptosis was observed when increasing the spacer length form PEG_2 to PEG_8 [28].

2.4. Prevention of Calcium Influx and Cholesterol Depletion from Lipid Rafts Lessen Apoptosis

Two important features were observed when initiating CD20 mediated apoptosis by DFMT. Crosslinking of CD20 receptors in Raji cells resulted in rapid rise in the Ca++ intracellular concentration [19]. Additionally, extracting cholesterol from cell membranes by β-cyclodextrin (β-CD) impacted receptor clustering as detected by STORM (stochastic optical reconstruction microscopy) [15]. Cholesterol is an important part of lipid rafts and contributes to mechanisms of anti-CD20 antibodies action [37,38]. Both phenomena seem to be correlated; transfer of loaded CD20 into lipid rafts promotes calcium influx [39].

We hypothesized that crosslinking of CD38 receptors by DFMT will have similar impact on the apoptosis initiation as observed with CD20 receptors. To this end, we preincubated Raji cells either with 0.02% β-CD (to extract cholesterol) or with 1 mM EGTA (to chelate extracellular calcium) before exposing them to DFMT. Following pretreatment, decrease in apoptotic levels was observed for both, DARA-based DFMT (DARA-MORF1 followed 1 h later by HSA-$(MORF2)_{10}$) and Fab'_{DARA}-based DFMT (Fab'_{DARA}-MORF1 followed 1 h later by HSA-$(MORF2)_{10}$). Data seem to suggest a higher impact of pretreatment on Fab'_{DARA}-based DFMT. When normalized to untreated, the percent apoptotic cells pretreated with β-CD decreased by 0.7-folds for DARA-based DFMT and by about 1.0-folds for Fab'_{DARA}-based DFMT. The percent apoptotic cells pretreated with EGTA decreased by 0.5-folds for DARA-based DFMT and by 1.2-folds for Fab'_{DARA}-based DFMT (Figure 3A).

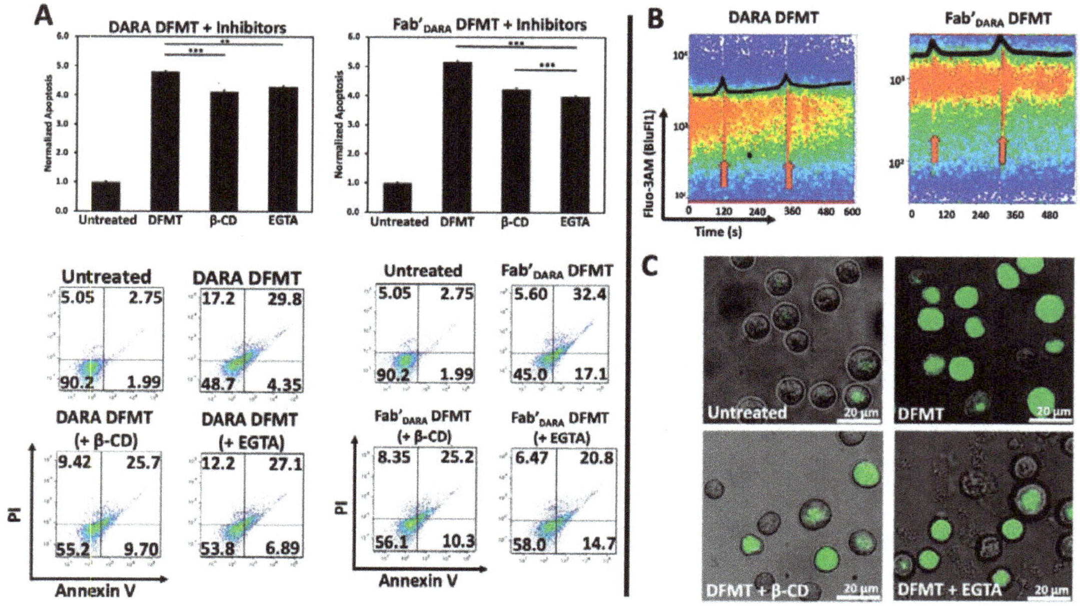

Figure 3. (**A**) Investigation of Raji cell apoptosis induced by DFMT with or without pretreatment with 0.02 wt% β-CD or 1 mM EGTA. (**B**) Flow cytometry time lapse fluorescence of calcium chelator Fluo-3AM after addition of nanoconjugates in DARA-based DFMT (DARA-MORF1 + HSA-$(MORF2)_{10}$ (left panel) and Fab'_{DARA}-based DFMT (Fab'_{DARA}-MORF1 + HSA-$(MORF2)_{10}$) (right panel). Red arrows indicate time of nanoconjugate addition to cell sample. (**C**) Confocal microscopy images of β-CD or EGTA treated Daudi cells compared to normal cells undergoing Fab'_{DARA}-based DFMT. *** $p < 0.001$, ** $p < 0.01$, n.s. nonsignificant, by One-Way ANOVA and Tukey test.

Calcium influx was also measured directly by using a fluorescent calcium chelator, Fluo-3AM, to observe calcium influx into Raji cells following DARA-based DFMT and Fab'_{DARA}-based DFMT (Figure 3B). Raji cells were used because the cell line is both CD20 and CD38 positive. Therefore, we could compare CD38-induced calcium influx with previously reported CD20-induced calcium influx data [19,27]. Red arrows indicate the time at which the nanoconjugates were added to the cell samples (Figure 3B). The DARA-MORF1 treatment had a distinct calcium signal spike immediately upon addition to the sample; however, the Fab'_{DARA}-MORF1 had a much less prominent Ca^{2+} influx event upon addition of the Fab′ nanoconjugate. Conversely, upon addition of HSA-$(MORF2)_{10}$ to both samples, the calcium spike was more pronounced for the Fab'_{DARA}-based DFMT than whole antibody-based DFMT corresponding with the calcium inhibition data (Figure 3B).

As mentioned above, partial inhibition of calcium influx impacted the efficacy of Fab'$_{DARA}$-based DFMT more than it impacted DARA-based DFMT. The calcium influx observed in Fab'$_{DARA}$-based DFMT treated cells corresponds with the results of the calcium inhibition experiments where calcium inhibition hampered the Fab'$_{DARA}$-based DFMT efficacy over DARA-based DFMT. The inhibition of calcium influx was also confirmed by confocal fluorescence microscopy (Figure 3C). Fluorescence was markedly lower in the EGTA and β-CD pretreated cells. Raji cells are CD38+/CD20+ so comparison of effects resulting from crosslinking of both receptors seem to indicate a stronger response following crosslinking of CD20 (comparing data of this manuscript with [19,27]).

2.5. DARA- and Fab'$_{DARA}$-based DFMT Induce Apoptosis via Mitochondrial Signaling Pathway

We next investigated the possible activation of the mitochondrial signaling pathway following crosslinking of decorated CD38 receptors on Daudi cells by multivalent macromolecular effector. The major features of the mitochondrial pathway include mitochondrial depolarization, cytochrome C release, caspase 3 activation, and bcl-2 downregulation [19,40].

Mitochondrial depolarization was assayed using the JC-1 mitochondrial membrane polarization sensor. In healthy mitochondria, membrane polarization remains intact and JC-1 aggregation occurs resulting in red fluorescence emission. As membrane potential diminishes, JC-1 can diffuse out of the mitochondria, thereby losing its red fluorescent signature as aggregates disperse into monomers. This solubilization event is observed by a change in fluorescent signature from red to green fluorescence. Therefore, red fluorescence indicates healthy mitochondria while green fluorescence indicates depolarized mitochondrial membranes. Mitochondrial membrane potential for DARA-based DFMT and Fab'$_{DARA}$-based DFMT was investigated using flow cytometry (Figure 4A) and confocal microscopy (Figure 4D) with and without the presence of EGTA and β-CD. The amount of observed mitochondrial depolarization was larger for DARA-based DFMT than Fab'$_{DARA}$-based DFMT, but both had higher mitochondrial membrane depolarization than DARA alone. This higher mitochondrial membrane depolarization observed in DFMT-treated cells compared to naked mAb is consistent with the higher apoptosis observed in the cell viability experiments.

B-cell lymphoma 2 (Bcl-2) and Bcl-2-associated X protein (Bax) expression levels were assayed by fluorescent immunostaining (Figure 4B). DFMT treated or untreated Daudi cells were incubated with fluorescently labeled antibodies specific to these two proteins: Bcl-2 mAb (100) Alexa Fluor® 488 and Bax mAb (2D2) Alexa Fluor® 647. Bcl-2 is located in the outer mitochondrial membrane and inhibits actions of pro-apoptotic proteins such as Bax. The expression level ratio of Bax to Bcl-2 is often used to indicate apoptotic states of cells [41,42]. DARA-based DFMT and Fab'$_{DARA}$-based DFMT treated cells were tested against one another and against untreated cells for Bcl-2/Bax expression. The enhanced Bax/Bcl-2 ratio, especially in DARA-based DMFT (Figure 4B) is the indication of mitochondrial signaling pathway involvement, as supported by data on cytochrome C release (Figure 4D) and caspase 3 activity (Figure 4C). Both, DARA-based DFMT and Fab'$_{DARA}$-based DFMT treated cells demonstrated 150–200% of caspase 3 activity when compared to untreated cells. The enhancement of activity was similar for both DFMT approaches. Apparently, the release of cytochrome C initiates the formation of the apoptosome with procaspase 9 and Apaf-1, followed by activation of procaspase 3 and the activation of the caspase cascade results in cell death [40,43].

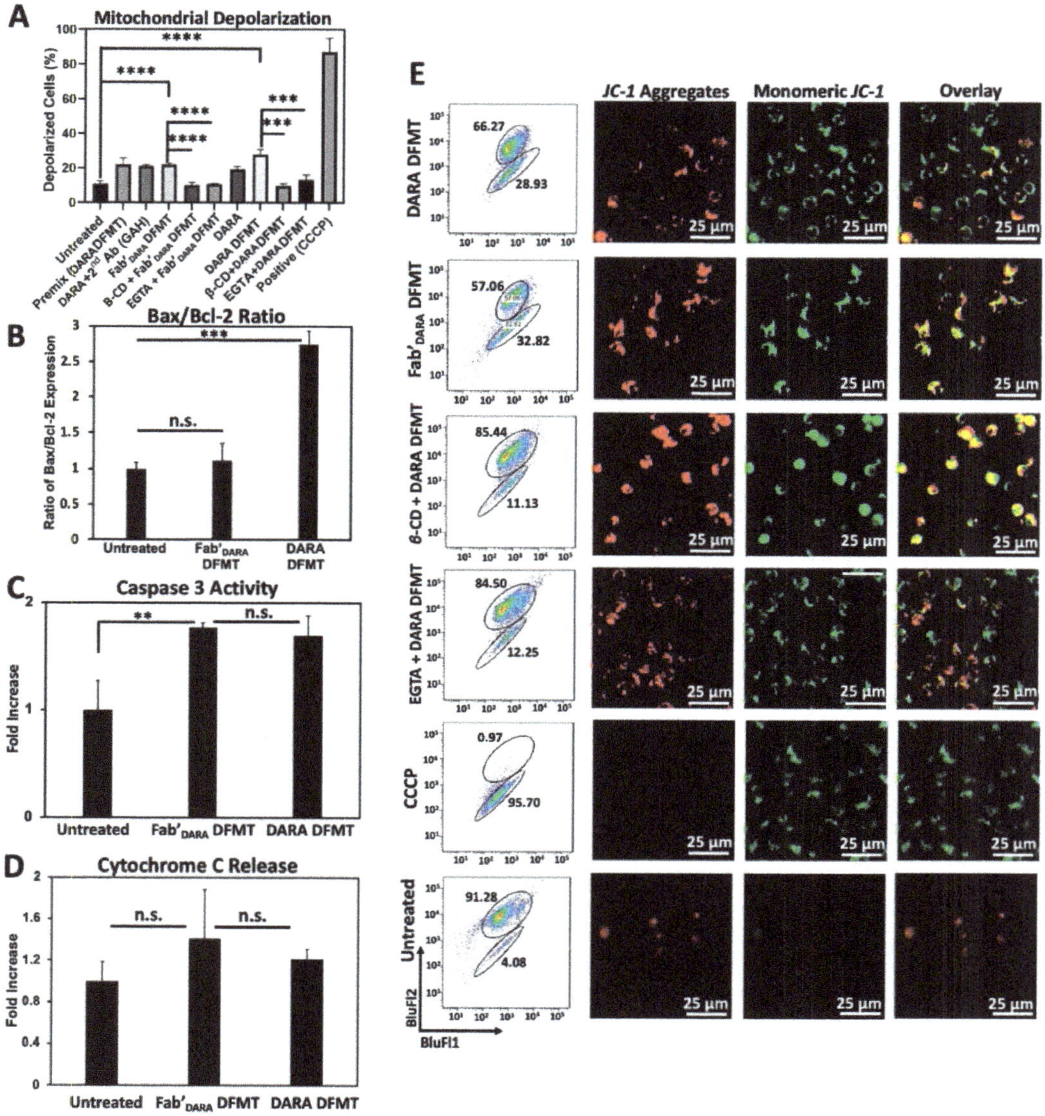

Figure 4. Investigation of mitochondrial pathway involvement in apoptosis induction in Daudi cells. (**A**) Compiled mitochondrial membrane depolarization of DARA DFMT and Fab′$_{DARA}$ DFMT treated Daudi cells with or without the presence of inhibitors (EGTA or β-CD). Controls include untreated cells, a DARA-based DFMT "Premix" group, DARA crosslinked with a secondary antibody (goat antihuman: GAH), and a positive control treatment with CCCP. Depolarization was measured by JC-1 red/green fluorescence by flow cytometry. (**B**) Bax/Bcl-2 expression ratio of DARA-based DFMT and Fab′$_{DARA}$-based DFMT treated Daudi cells. Expression was measured using immunostaining and flow cytometry. (**C**) Caspase 3 activity of DARA DFMT and Fab′$_{DARA}$ DFMT treated Daudi cells was measured using PhiPhiLux® assay kit and flow cytometry. (**D**) Cytochrome C release of DARA and Fab′$_{DARA}$ DFMT on Daudi cells. Increase in Cytochrome C release shown as fold increase over untreated cells. Determined by ELISA. (**E**) Confocal fluorescence microscopy visualization of mitochondrial depolarization of DARA DFMT with or without the presence of EGTA or β-CD. More red fluorescence indicates healthier mitochondria while more green fluorescence indicates the occurrence of depolarization. **** $p < 0.0001$, *** $p < 0.001$, ** $p < 0.01$, n.s. nonsignificant, by One-Way ANOVA and Tukey test.

2.6. Reactive Oxygen Species (ROS) Generation, Lysosomal Enlargement, Translocation to Lipid Rafts

ROS are generated by both Type I and Type II antibodies. In the mechanism of CD20 apoptosis initiation there is a clear distinction of mechanisms between Type I Abs, such as RTX [19] and Type II Abs, such as OBN [44]. Following receptor crosslinking, Type I antibodies induce apoptosis by receptor crosslinking followed by calcium influx, mitochondrial depolarization, and caspase activation. In contrast, Type II antibodies do not need crosslinking, they initiate apoptosis by actin remodeling, homotypic cell adhesion and lysosome disruption. Both types produce ROS.

ISA binds to an epitope independent to that of DARA and provides more enzymatic inhibition of the CD38 function [22]. ISA can induce apoptosis without crosslinking, however, lysosomal breakage and cathepsin B leakage was observed [45]. Initiation of direct apoptosis by DARA upon crosslinking and by ISA without crosslinking was reported in refs. [6,22,45]. In contrast, Moreno et al. did not observe it [46]. The discrepancy was explained by the level of CD38 expression—use of cells transduced with CD38 vs. cells with lower CD38 expression levels close to those in MM patients.

To compare mechanisms of DFMT based on CD38-targeting antibodies, DARA and ISA, with CD20-targeting OBN, Raji cells positive for both receptors were used. Three bispecific engagers, DARA-MORF1, ISA-MORF1, and OBN-MORF1, were employed. Following binding to the corresponding receptor, HSA-$(MORF2)_{10}$ was used for crosslinking.

The production of ROS by DARA, ISA, DARA-based and ISA-based DFMT was at the same level without statistically significant differences (Figure 5A). However, OBN and OBN-based DFMT produced considerably higher amounts of ROS. The highest ROS production was observed in OBN-based DFMT, an indication that both mechanisms (Type I and II) of apoptosis induction are operative.

OBN DFMT showed 1.5-fold increase in lysosome size compared to untreated cells (Figure 5B). Lysosomal enlargement and increased reactive oxygen species are indicative of Type II and coincides with data our group published previously [27].

Additionally, confocal microscopy was employed to investigate redistribution of receptor-bound nanoconjugates on the cell surface into lipid rafts (Figure 5C). Fluorescently labeled antibody-MORF1 conjugates and immunostaining of lipid compartments on the cell surface with cholera toxin subunit B (CTxB) were used to visualize clustering or aggregation of antibody-bound receptors on the cell membrane. Only OBN-based DFMT exhibited typical Type II behavior of intercell homotypic adhesion and lipid raft distribution at cell–cell adhesion sites (Figure 5C *top row*).

DARA-based and ISA-based DFMT treated cells showed no significant increase in lysosome size compared to untreated cells and about only half as much ROS production as OBN-based DFMT treated cells. The confocal imaging showed minimal homotypic cell–cell adhesion and even when some cell adhesion was observed, no pronounced lipid raft accumulation at cell adhesion sites was distinguishable. All of these results are consistent with Type I antibody characteristics for DARA- and ISA-based DFMT. Only OBN-based DFMT demonstrated Type II mechanism characteristics.

The purpose of the ROS assay, the lysosomal enlargement study and the homotypic cell adhesion experiments was to elucidate any Type II antibody mechanisms of action of ISA and ISA-based DFMT. DARA (Type I antibody) and DARA-based DFMT, OBN (Type II antibody) and OBN-based DFMT were used to compare and contrast ISA's apoptosis induction mechanisms. Type II antibodies are characterized by lysosomal disruption, ROS production and homotypic cell adhesion. Type II antibodies, like OBN, induce apoptosis upon binding to their receptor epitope on the cell surface without the need for crosslinking. It is currently unknown if ISA behaves as a Type II antibody; however, ISA has been shown to induce apoptosis without the need for crosslinking [10]. Therefore, we hypothesized ISA could induce similar lysosomal disruption, ROS production and homotypic cell adhesion as OBN. Our investigation proved otherwise. ISA and ISA-based DFMT did not increase

ROS, did not enlarge the lysosomes, nor did we observe any homotypic cell adhesion under confocal microscopy.

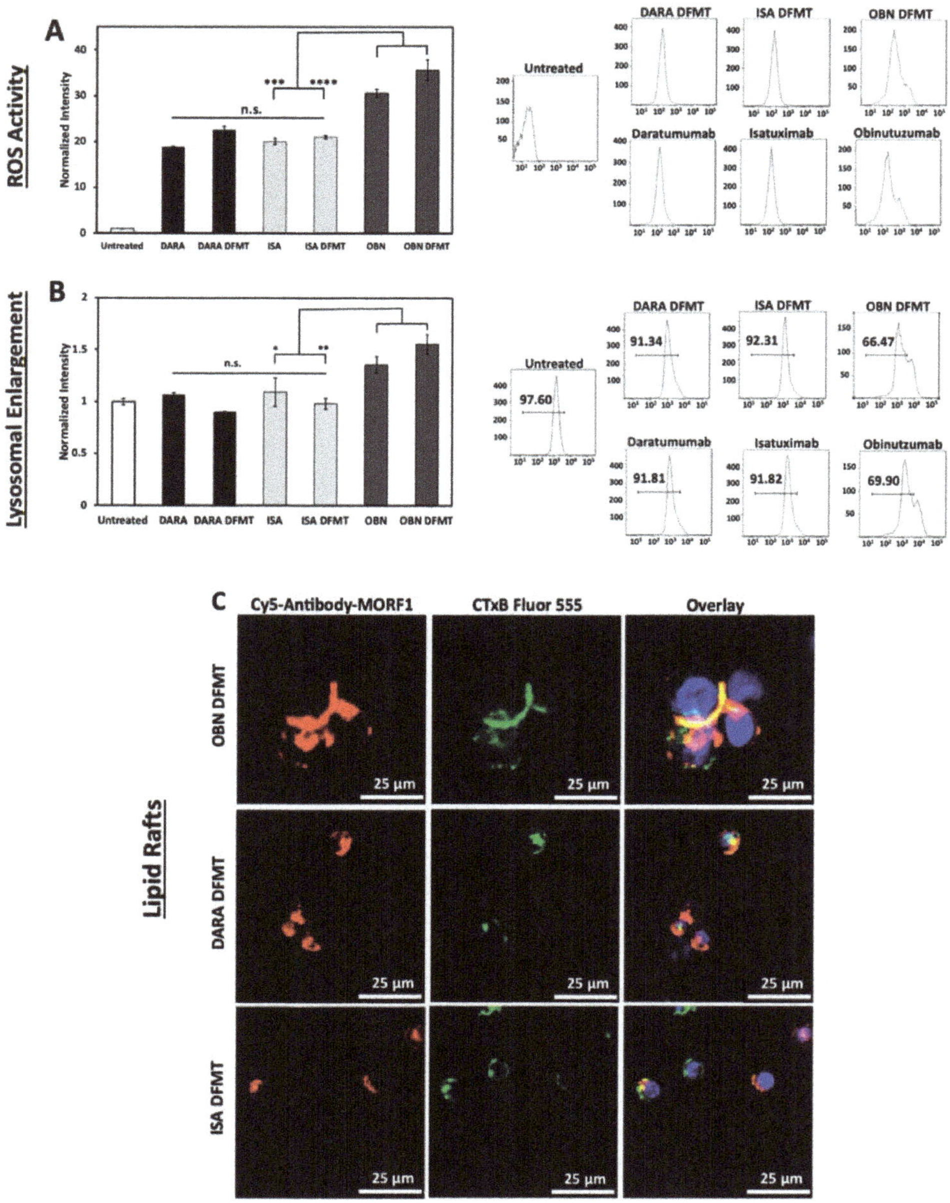

Figure 5. (**A**) Generation of ROS in DARA-, ISA-, and OBN-based DFMT treated Raji cells. The ROS production was measured by oxidation of 2′,7′-dichlorodihydrofluorescein diacetate (H_2DCFDA) and quantified with flow cytometry. (**B**) Lysosomal enlargement of DARA-, ISA-, and OBN-based DFMT was quantified by LysoTracker Green DND-26 and flow cytometry. (**C**) Confocal microscopy was employed to observe any redistribution of CD38 receptors into cholesterol-rich lipid rafts. Ab-MORF1 (red); lipid raft marker: cholera toxin subunit B (CTxB) Alexa Fluor 555 (green). Nuclei were stained with Hoechst (blue). **** $p < 0.0001$, *** $p < 0.001$, ** $p < 0.01$, * $p < 0.05$, n.s. nonsignificant by One-Way ANOVA and Tukey test.

To investigate ISA-based DFMT further, cellular apoptosis experiments were conducted on Daudi cells (Figure 6) and Raji cells (Figure S18) to examine CD38 receptor crosslinking effects when using ISA nanoconjugates. Under the same DFMT cell apoptosis treatment conditions as used for DARA DFMT (Figure 1B), ISA DFMT showed several key differences. First, the same concentration of ISA antibody induced roughly 20 percent more apoptosis than DARA antibody. This could indicate the ISA-CD38 binding epitope to inhibit CD38 enzymatic activity to a larger extent over DARA-CD38 binding, and/or binding of ISA stimulates apoptotic pathways within the cell to a larger extent. Second, Fab'$_{ISA}$-MORF1 control samples showed levels of apoptosis comparable with whole antibody. This is drastically different from Fab'$_{DARA}$-MORF1 controls that showed little to no apoptotic efficacy (Figure S6). Binding of either whole ISA antibody or Fab' fragment at the ISA binding site induces the therapeutic effect. Lastly, ISA DFMT did not enhance the overall apoptotic efficacy on either Daudi or Raji cells over antibody alone. However, Fab'$_{ISA}$ DFMT did produce significantly increased apoptosis over Fab'$_{ISA}$-MORF1 control and whole antibody.

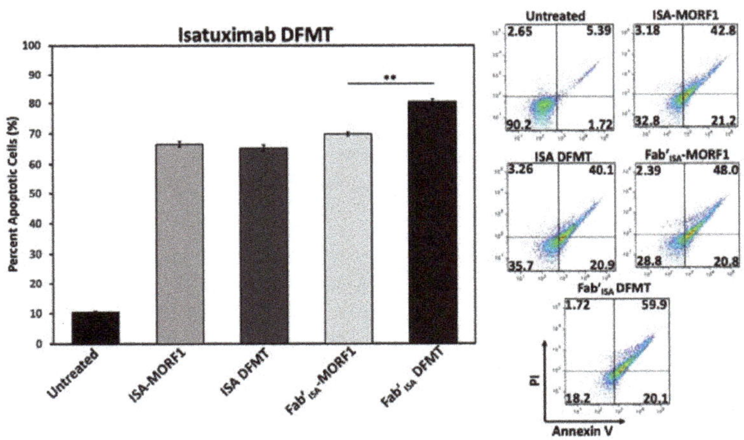

Figure 6. ISA-based DFMT performed on Daudi cells. Apoptosis was measured after ISA DFMT or Fab'$_{ISA}$ DFMT using Annexin V and Propidium Iodide staining. Flow cytometry was performed to assess percentage of apoptotic cells compared to an untreated sample group. Experiment was performed in triplicate and statistically analyzed by One-Way ANOVA and Tukey test. ** $p < 0.01$.

The findings with ISA suggest antibodies capable of triggering apoptosis directly upon binding to their target antigen will do so regardless of secondary crosslinking. However, the Fab' fragment crosslinking of this variety of antibody could lead to enhanced efficacy over antibody alone.

2.7. *Simultaneous Crosslinking of CD38 and CD20 Receptors Enhances Apoptosis*

Raji cells are CD38+/CD20+ and crosslinking of both receptors initiates apoptosis. We investigated if dual crosslinking with combination of DFMT based on Fab' fragments from DARA and RTX would enhance the efficacy of apoptosis induction. To this end we exposed Raji cells to three DFMT treatments (concentrations relate to MORF): (a) 0.5 μM Fab'$_{DARA}$-MORF1 + (1 h later) HSA-(MORF2)$_{10}$; (b) 0.5 μM Fab'$_{RTX}$-MORF1 + (1 h later) HSA-(MORF2)$_{10}$; (c) combination of 0.25 μM Fab'$_{DARA}$-MORF1 + (1 h later) HSA-(MORF2)$_{10}$ and 0.25 μM Fab'$_{RTX}$-MORF1 + (1 h later) HSA-(MORF2)$_{10}$.

Combination treatment resulted in substantially enhanced apoptotic level (Figure 7). In addition to more effective crosslinking of receptors, probably, more signaling pathways will be involved in apoptosis initiation via two receptors when compared with one. This will be evaluated in our future research.

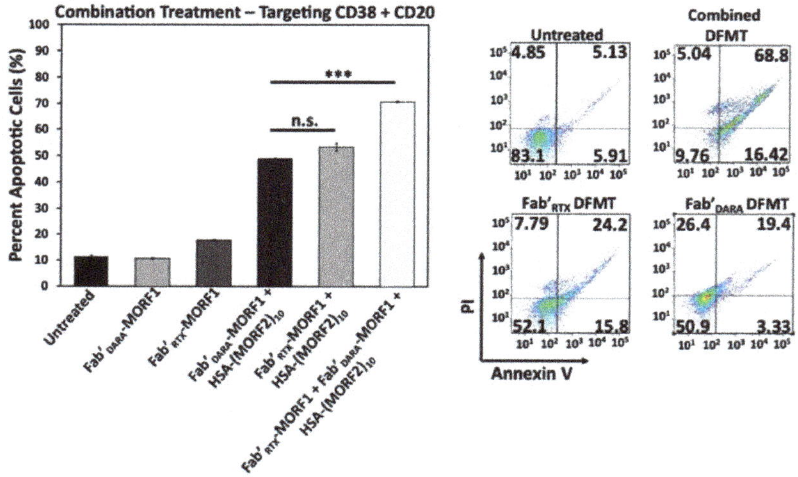

Figure 7. Combinational DFMT treatment of Raji cells with CD38-targeting Fab′$_{DARA}$ DFMT coupled with CD20-targeting Fab′$_{RTX}$ DFMT. The combined treatment was compared to individual therapies at the same MORF1 equivalence. Percent apoptotic cells were determined after DFMT treatment and analyzed with Annexin V/Propidium Iodide staining via flow cytometry. All experiments were performed in triplicate and statistically analyzed by One-Way ANOVA and Tukey test. *** $p < 0.001$, n.s. nonsignificant.

Notably, in relapsed non-Hodgkin lymphoma [47] and diffuse large B-cell lymphoma [48] successful treatment of patient derived xenografts was achieved by replacing anti-CD20 RTX with anti-CD38 DARA or DARA-drug conjugates. Our data combined with these results suggests that treatment of lymphoma patients with therapies involving both antibodies might be beneficial.

3. Methods and Materials

3.1. Materials

A pair of 25-mer phosphorodiamidate morpholino oligonucleotides, MORF1 and MORF2, were customized from Gene Tools (Philomath, OR, USA). In particular, MORF1 (5′-GAGTAAGCCAAGGAGAATCAATATA-3′) with 3′ primary amine modification, and its complementary MORF2 (5′-TATATTGATTCTCCTTGGCTTACTC-3′) with 3′-disulfide amide modification were used. Human serum albumin (HSA, chromatographically and fractionation purified with purity > 95%) was purchased from Innovative Research (Peary CourtNovi, MI, USA). DARA (Darzalex® 20 mg/mL, Janssen Biotech (Harsham, PA, USA), ISA, OBN, and RTX were obtained from Huntsman Cancer Hospital, University of Utah. Cy3-/Cy5-NHS (*N*-hydroxysuccinimide) were purchased from Lumiprobe (Hallandale Beach, FL, USA). Tris(2-carboxyethyl) phosphine (TCEP) and heterobifunctional crosslinkers NHS-PEG$_x$-maleimide (SM(PEG)$_x$, x = 2, 8, and 24) were purchased from Thermo Fisher Scientific (Rockford, IL, USA). Pepsin (from porcine gastric mucosa) was from Sigma-Aldrich (St. Louis, MO, USA). Anti-Bcl-2 (100) Alexa Fluor® 488 mAb and anti-Bax (2D2) Alexa Fluor® 647 mAb were purchased from Santa Cruz Biotechnology (Dallas, TX, USA). Fluo-3 AM, JC-1 (5,5′,6,6′-tetrachloro-1,1′,3,3′-tetraethylbenzimidazoylcarbocyanine iodide), and carbonyl cyanide 3-chlorophenylhydrazone (CCCP) were purchased from Invitrogen (Carlsbad, CA, USA). PhiPhiLux® kit was purchased from OncoImmunin (Gaithersburg, MD, USA). Lysosome tracker Green DND-26, H_2DCFDA (2′,7′-dichlorofluorescein), Cytochrome C ELISA kit (human) and cholera toxin subunit B Alexa Fluor 555 were purchased from Thermo Fisher. All solvents were purchased from Fisher Scientific as the highest purity available.

3.2. Synthesis and Characterization of Nanoconjugates

3.2.1. Antibody-MORF1 Nanoconjugates

Antibody-MORF1 nanoconjugates are synthesized in two steps as previously described [20]: first, a monoclonal antibody is reduced with TCEP to generate a sulfhydryl group, which is then conjugated with maleimide-modified MORF1. Here, is an example of synthesizing DARA-MORF1. DARA (225 μL, 4.5 mg) was buffer exchanged into 100 mM citrate buffer (pH 5.5) using an Amicon® 4 mL ultra-centrifugal filter unit (MWCO 30,000 Da). The DARA solution was then added to 20 mM TCEP (7 mg/mL, pH rebalanced to pH 5.5), and the reaction was kept for 2 h at 37 °C water bath with gentle shaking. The reduced DARA was obtained after removal of TCEP by washing with 10 mM PBS (pH 6.5) over 6 times. In parallel, maleimide-modified MORF1 (MORF1-MAL) was prepared by reaction of MORF1-NH_2 with 50 molar excess SM(PEG)2 (succinimidyl-[(*N*-maleimidopropionamido)-diethyleneglycol]ester). In brief, SM(PEG)2 (4.25 mg, 10 μmol) was dissolved in 50 μL DMSO, then added into 100 μL MORF1-NH_2 solution (1.8 mg, 200 nmol) in 10 mM PBS pH 7.4. The reaction was performed for 2 h at room temperature. Then, MORF1-MAL was isolated by removal of excess SM(PEG)2 with 6 times washing using 20 mM PBS pH 6.5 buffer via Amicon® 0.5 mL Ultra Centrifugal filter unit (MWCO 3000 Da). Finally, freshly reduced DARA was conjugated with MORF1-MAL in 500 μL 20 mM PBS pH 6.5 for 2.5 h at room temperature, with 1:1.5 molar ratio of [Ab]:[MORF1-MAL]. The resultant DARA-MORF1 was purified by ultracentrifugation using the Amicon® 4 mL Ultra Centrifugal filter unit (MWCO 30,000 Da). The purity of conjugate DARA-MORF1 was confirmed with SEC on AKTApure with a Superdex™ 200 10/300 GL column using PBS pH 7.4 as eluent. There was no detectable peak of either free MORF1 or free DARA. Furthermore, MORF1 content in the conjugate was quantified using NanoDrop (ND-1000 Spectrophotometer) at 260 nm (ε = 278,000 $M^{-1}cm^{-1}$ in 0.1 N HCl aq.), whereas the concentration of DARA was determined using a bicinchoninic acid (BCA) assay.

To prepare Cy5-labeled DARA-MORF1 (Cy5-DARA-MORF1), DARA was first labeled with Cy5-NHS. The reaction was performed with molar ratio of [Ab]:[Cy5] = 1:2 by adding Cy5-NHS solution in DMSO to DARA in 1 mL PBS pH 7.4. After 2 h reaction at room temperature, free dye was removed using a PD10 column (GE Healthcare). The collected product was then buffer exchanged using ultracentrifugation 4 times with 100 mM citric acid buffer pH 5.5. Then, procedures described above were followed.

3.2.2. Fab′-MORF1 Nanoconjugates

Fab′-MORF1 nanoconjugates are synthesized in multiple steps as previously described [28]. First, a monoclonal antibody is digested into $F(ab')_2$ in the presence of pepsin, followed by reduction with TCEP to generate Fab′ with a sulfhydryl group (Fab′-SH), which is then conjugated with maleimide-modified MORF1. Here, is an example of synthesizing Fab'_{DARA}-MORF1. Briefly, DARA was buffer exchanged into citrate buffer pH 4.0. Pepsin (10 *w*/*w*%) was added to DARA solution, and the reaction was kept at 37 °C. The digestion process was monitored on AKTApure till complete disappearance of DARA peak, while a new peak with lower molecular weight (at right side) showed up, which was related to $F(ab')_2$. Complete digestion occurred in 80 min. The $F(ab')_2$ was purified using ultracentrifugation with a 30,000 MWCO tube and stored at 4 °C in PBS 7.4. $F(ab')_2$ (4 mg, 4 mg/mL) was reduced with TCEP (4.6 mg, 20 mM) in 100 mM citric acid buffer pH 5.5 for 2 h at 37 °C and purified using ultracentrifugation with a 10,000 MWCO tube. The MORF1-MAL was synthesized in parallel as described above. MORF1-MAL (1.2 equiv.) was reacted with freshly reduced Fab′ (1 equiv.) for 3 h at room temperature in PBS pH 6.5 buffer. The final product was purified by ultracentrifugation using a 30,000 MWCO tube by washing 6–8 times with PBS pH 7.4 buffer. MORF1 content in Fab'_{DARA} conjugates was determined using UV-Vis absorbance at 260 nm; the concentration of antibody fragment was determined using BCA assay.

ISA-MORF1 and Fab'_{ISA}-MORF1 were synthesized as described above for DARA conjugates.

3.2.3. Multiple MORF2 Modified Human Serum Albumin (HSA-$(MORF2)_y$)

Two steps were conducted to conjugate complementary MORF2 to HSA as previously described [29]. First, amino groups from accessible lysine residues in HSA were converted to maleimide groups by using heterobifunctional crosslinker NHS-PEG_x-maleimide, then freshly reduced MORF2-SH was attached to HSA in multiple copies via thiol-ene reaction. In this study, SM(PEG)x (x = 2, 8, and 24) that differ in length were used in order to investigate spacer effect on MORF1-MORF2 biorecognition and induction of apoptosis. Here, is an example in which $SM(PEG)_2$ was used to synthesize nanoconjugate HSA-PEG_2-$(MORF2)_{10}$. Briefly, HSA (5 mg, 3.9 μmol NH_2 equiv.) was dissolved in 400 μL PBS pH 7.4 buffer. SM(PEG)2 (18.3 mg, 10 eq) in 150 μL DMSO was added into HSA solution. The reaction was kept stirring for 2 h at room temperature. The maleimide-modified HSA was then purified by ultracentrifugation using an Amicon® 4 mL ultra centrifugal filter unit (MWCO 30,000 Da). The number of maleimide groups per HSA molecule was determined by a modified Ellman's assay (for maleimide group) and BCA assay (for quantification of HSA).

In a parallel reaction, 3′-disulfide MORF2 (2.89 mg) was reduced with 3.5 mg/mL TCEP (10 mM) in 250 μL PBS pH 7.4 at 37 °C for 45 min, followed by purification via ultrafiltration using a Amicon® 0.5 mL ultracentrifugal filter unit (MWCO 3000 Da). Freshly reduced MORF2-SH was then reacted with HSA-PEG_2-MAL_x with molar ratio of [SH]:[MAL] = 2:1 in 500 μL PBS (pH 6.5) for 3 h at room temperature. HSA-PEG_2-MORF2 was purified by ultracentrifugation using a 30,000 MWCO ultra centrifugal unit washing 6–8 times with PBS pH 7.4 buffer. Purity was confirmed by AKTApure. The ratio of MORF2 per HSA molecule was determined by UV-Vis spectrophotometry (ε = 252,120 M^{-1} cm^{-1} in 0.1 N HCl) and BCA assay.

For synthesis of Cy3-labeled HSA-PEG_2-$(MORF2)_y$, HSA was first reacted with Cy3-NHS with molar ratio of [HSA]:[Cy3] = 2:1. For example, a solution of Cy3-NHS (70 μg, 100 nmol) in DMSO was added to a solution of HSA (3.35 mg, 50 nmol) in 1 mL PBS pH 7.4 buffer and reacted for 2 h at room temperature. Cy3-labeled HSA was purified using a PD-10 column using PBS pH 7.4 as eluent. Then, the synthesis proceeded as described above.

3.2.4. MORF1-MORF2 Hybridization

MORF1-MORF2 hybridization upon mixing Ab-MORF1 (or Fab′-MORF1) with HSA-$(MORF2)_y$ was determined by the changes of optical density at 260 nm (ND-1000 spectrophotometer) that was a reflection of the hypochromic effect. For example, DARA-MORF1 (or ISA-MORF1) and HSA-$(MORF2)_{10}$ solutions in PBS, pH 7.4 were mixed in different ratios with a constant total MORF (MORF1 + MORF2) concentration of 2.5 μM at room temperature. Then, 10 min post mixture, the optical density at 260 nm was recorded. All measurements were performed in triplicate.

In addition, MORF1-MORF2 hybridization among Ab-MORF1 (or Fab′-MORF1) and HSA-$(MORF2)_{10}$ was also confirmed by SEC by comparison with individual conjugates, Ab-MORF1 and HSA-$(MORF2)_{10}$.

3.3. *Cell Culture*

Human lymphoma cell lines (Daudi and Raji) and human MM cell lines (RPMI 8226 and U266) were purchased from the American Type Culture Collection (ATCC). All cells were cultured in RPMI-1640 medium supplemented with 10% fetal bovine serum (FBS), penicillin (100 units mL^{-1}) and streptomycin (0.1 mg/mL^{-1}) at 37 °C in a 5% CO_2 humidified atmosphere. Human MM cell line ANBL-6 was obtained from Dr. Diane Jelinek of Mayo Clinic (Rochester, MN, USA). The cells were cultured in IMDM with 10% FBS and interleukin-6 (1 ng/mL).

3.4. Cell Surface CD38 Expression and Binding Assay

DARA binding experiments were conducted by incubating Daudi, RPMI 8226, Raji, ANBL-6, and U266 cells with a range of DARA (primary antibody) concentrations, and subsequently exposing them to a fluorescently labeled, anti-human secondary antibody. Specifically, 2×10^5 cells were treated with DARA (1 µM, 0.5 µM, 0.25 µM, 0.1 µM or 50 nM; in PBS pH 7.4) in a 24-well plate in 400 µL media for 1 h at 4 °C. Then, the cells were washed with PBS and resuspended in PBS containing 1 *w*/*v*% BSA. An Alexa Fluor 488-labeled goat anti-human antibody (3 µL, 2 mg/mL) was added to each well and further incubated for 1 h at 4 °C. The cells were washed and resuspended in PBS and analyzed by flow cytometry. The fluorescence was normalized to an untreated control for each cell type and reported as a "fold-increase over untreated" value. All experiments were performed in triplicate.

3.5. MORF1-MORF2 Hybridization on Cell Surface—CD38 Crosslinking

Daudi cells (2×10^5) were treated in a 24-well plate (400 µL RPMI-1640 medium per well) with Cy5-labeled DARA-MORF1 (0.5 µM MORF1) for 30 min, followed by a Cy3-labeled HSA-$(MORF2)_{10}$ (0.5 µM [MORF2]) treatment for 1 h. Then, the cells were washed with PBS and cell nuclei were stained with Hoechst 33342 (5 µg/mL) for 5 min. The cells were washed and resuspended in PBS for imaging. Cy3 fluorescence was measured by 488 nm excitation with 530/30 nm band-pass filter. Cy5 fluorescence was measured by 633 nm excitation with 695/40 nm band-pass filter. An overlay of the respective fluorescent signals was used to visualize MORF1-MORF2 hybridization and co-localization of the nanoconjugates on the cell surfaces.

3.6. Apoptosis Assay

Apoptosis of DFMT-treated cells was quantified using Annexin V/Propidium Iodide staining and measured by flow cytometry. Each cell type's (Daudi, Raji, RPMI 8226, ANBL-6, and U266) apoptotic response to DARA, DARA-based DFMT, Fab'_{DARA}-based DFMT and DARA plus a secondary antibody were evaluated. Briefly, 2×10^5 cells were treated in a 24-well plate in 400 µL appropriate media. Cells were first incubated with DARA-MORF1 or Fab'_{DARA}-MORF1 (0.5 µM MORF1) for 1 h, followed by a PBS wash and subsequent treatment with HSA-MORF2$)_{10}$ (0.5 µM MORF2) in fresh media for 24 h. Then, the cells were washed with Binding (HEPES saline) Buffer and stained with Annexin V/Propidium Iodide for 20 min at 4 °C. Cells were washed and resuspended in Binding Buffer for flow cytometry analysis. Untreated cells were used to gate as Annexin V/PI -/- population. Each treated sample was then compared to untreated. Values reported are the percentage of Annexin V positive cells. All treatments were performed in triplicate with appropriate controls. One-way ANOVA ($\alpha = 0.05$) followed by Tukey Test analysis was used to determine significant differences in the reported data.

Details and Nomenclature of DFMT Apoptosis Assays

DARA DFMT: DARA-MORF1 (0.5 µM MORF1) 1 h followed by HSA-$(MORF2)_{10}$ (0.5 µM MORF2) in fresh media for 24 h; 2×10^5 cells/well; 24-well plate.

Fab'_{DARA} DFMT: Fab'_{DARA}-MORF1 (0.5 µM MORF1) 1 h followed by HSA-$(MORF2)_{10}$ (0.5 µM MORF2) in fresh media for 24 h; 2×10^5 cells/well; 24-well plate.

ISA DFMT: ISA-MORF1 (0.5 µM MORF1) 1 h followed by HSA-$(MORF2)_{10}$ (0.5 µM MORF2) in fresh media for 24 h; 2×10^5 cells/well; 24-well plate.

Fab'_{ISA} DFMT: Fab'_{ISA}-MORF1 (0.5 µM MORF1) 1 h followed by HSA-$(MORF2)_{10}$ (0.5 µM MORF2) in fresh media for 24 h; 2×10^5 cells/well; 24-well plate.

3.7. Apoptosis Inhibition

To validate the participation of calcium influx in apoptosis initiation, Daudi cells were pretreated with lipid raft inhibitor β-cyclodextrin (β-CD; inhibiting CD20 crosslinking) or Ca^{2+} chelating agent EGTA (ethyleneglycol-bis(β-aminoethyl ether)-*N*,*N*,*N*′,*N*′-tetraacetic acid) which had significantly reduced the calcium influx after DFMT treatment. Daudi cells were pretreated with either 1 mM Ca++ doped RPMI-1640 medium or 0.02% β-cyclodextrin containing RPMI-1640 medium for 1 h. The cells were washed with fresh medium and exposed to DARA-MORF1 or Fab'$_{DARA}$-MORF1 for 1 h at 37 °C, followed by exposure to HSA-(MORF2)$_{10}$. Cell viability was quantitated using Annexin V/PI labeling and flow cytometry.

3.8. Caspase 3

Caspase 3 activity was evaluated using a PhiPhiLux®-G_1D_2 kit (OncoImmunin, Gaithersburg, MD). The manufacturer's protocol was followed, including the final stage Propidium Iodide staining to assess cell membrane integrity. The reported values are "fold-increase over control" fluorescence measurements of the PhiPhiLux indicator on treated Daudi cells over untreated cells.

3.9. Cytochrome C

Levels of cytochrome C were evaluated in DARA and Fab'$_{DARA}$ DFMT-treated Daudi cells by ELISA. After treatment, Daudi cells (2×10^5 cells) were washed with cold PBS and cell pellets were subsequently lysed with 100 µL cell extraction buffer (1 mM phenylmethanesulfonyl fluoride (PMSF) and 1:100 protease inhibitor cocktail) for 30 min on ice. The extract was transferred to a microcentrifuge tube and centrifuged at 13,000 rpm for 10 min at 4 °C. The amount of cytochrome C in the lysate was measured using an ELISA kit (R&D Systems) according to the manufacturer's instructions. All samples were conducted in triplicate.

3.10. Mitochondrial Depolarization

JC-1 mitochondrial membrane potential sensor (Thermo Scientific) was used to evaluate the extent of mitochondrial depolarization of Daudi cells. After indicated treatments, the cells (2×10^5) were washed with PBS two times and resuspended in 100 µL PBS. JC-1 (4 µM) was added to each sample and incubated at 37 °C for 30 min. For the positive control group, CCCP (0.5 µM) was added and incubated simultaneously with JC-1 for 30 min. After washing by PBS, cells were resuspended in PBS and analyzed by flow cytometry using 488 nm excitation with 530/30 nm and 585/42 nm band-pass filters, or observed under confocal microscopy. All experiments were carried out in triplicate.

3.11. Calcium Influx by Confocal Microscopy

Daudi cells (2×10^5) were incubated with Fluo-3AM (5 µM) in 100 µL RPMI-1640 medium containing 2.5 mM Ca^{2+} for 30 min at 37 °C and compared with cells pretreated with 0.02 wt.% β-cyclodextrin or 1 mM EGTA. Following treatment, cells were washed with PBS and resuspended in RPMI-1640 medium containing 2.5 mM Ca^{2+} and observed under confocal microscopy.

3.12. Calcium Influx by Flow Cytometry

Raji cells (4×10^5 cells/well) were counted and stained with Fluo-3 AM (5 µM) for 30 min at 37 °C. After staining, the cells were washed with PBS and resuspended in 400 µL cell culture medium containing 2.5 mM Ca^{2+} and immediately taken for flow cytometry analysis (excitation at 488 nm and emission at 530 nm). Baseline fluorescence was measured for 100 s. Then, 1 µM DARA-MORF1 or Fab'$_{DARA}$-MORF1 was added to the sample. The fluorescence was measured continuously for another 200 s, followed by addition of HSA-(MORF2)$_{10}$ (1 µM). Fluorescence was monitored for another 600 s or until all cells were counted. Rituximab-based DFMT calcium influx was employed as control (Figure S6).

3.13. Bcl-2/Bax Detection

Following treatment, levels of expression of Bcl-2 and Bax were quantified by fluorescent immunostaining. The cells were sequentially fixed by 4% paraformaldehyde for 15 min at room temperature, permeabilized by 90% methanol for 30 min on ice, and immunostained by Alexa Fluor 488 conjugated anti-Bcl-2 mAb (1:50, Santa Cruz Biotechnology) and AF647 conjugated anti-Bax mAb (1:50, Santa Cruz Biotechnology) in 1% BSA buffer for 1 h at room temperature. After washing by cold PBS twice, the fluorescence was quantified by flow analysis. All experiments were carried out in triplicate.

3.14. *Lysosomal Enlargement*

Type II antibody-induced cell death involves lysosomal enlargement/breakage. Lysosome activity was tested by LysoTracker Green staining followed by flow cytometry. Raji cells (2×10^5) were exposed to naked antibody or antibody-MORF1 conjugate for 1 h at 37 °C, followed by exposure to HSA-$(MORF2)_{10}$. After 24 h treatment, cells were stained with LysoTracker Green DND-26 (200 nM) for 20 min at 37 °C. Fluorescence was quantified using flow cytometry. Each sample was prepared in triplicate.

3.15. *Reactive Oxygen Species Production*

Quantification of reactive oxygen production was performed on Raji cells (2×10^5) by oxidation of 2′,7′-dichlorodihydrofluorescein diacetate (H_2DCFDA). Cells (2×10^5 cell/well, 24-well plate) were treated with antibody (DAR, ISA, or OBN) (0.5 µM) or antibody (DARA, ISA, or OBN)-MORF1 conjugate followed by HSA-$(MORF2)_{10}$ (MORF1 = MORF2 = 0.5 µM). After 24 h treatment, cells were incubated with H_2DCFDA (5 µM) for 30 min at 37 °C. Cells were washed with PBS and analyzed with flow cytometry. Each sample was prepared in triplicate.

3.16. *Translocation of CD38 into Lipid Rafts*

The motility and translocation of CD38 receptors under the influence of DFMT was evaluated by cholera toxin B staining and observed under confocal microscopy. Briefly, Raji cells (2×10^5 cells/well) were loaded into a 24-well plate. The cells were treated with either 0.5 µM antibody alone or 0.5 µM Cy5-labeled antibody-MORF1 for 1 h at 37 °C followed by PBS wash and 0.5 µM HSA-$(MORF2)_{10}$ exposure for 2 h at 37 °C. Then, the cells were washed to remove unbound antibodies, and stained with Alexa Fluor-555 conjugated cholera toxin B subunit (10 µg/mL) for 1 h at 4 °C. The samples were immediately imaged by confocal microscopy.

3.17. *Statistical Analysis*

All statistical analysis was performed on Microsoft Excel. No samples were left out of any analysis calculations. Sample groups were compared using one-way ANOVA followed by Tukey test. $p < 0.05$ was considered statistically significant. All experiments were performed with at least $n = 3$ samples per group. Percent apoptotic cells was calculated by summation of quadrants one, two and four in the Annexin V/PI gated flow cytometry runs. Magnitude of fluorescence in immunostaining procedures was quantified by flow cytometry's geometric mean average function. The geometric mean average was either normalized to untreated control cells or presented as fold increase over untreated, as specified above.

4. Conclusions

The experiments presented herein illustrate the versatility of the DFMT system and demonstrate how it can be applied towards the treatment of several B cell malignancies including multiple myeloma, lymphoma and leukemia. Four novel antibody nanoconjugates as bispecific engagers (Fab'_{DARA}-MORF1, Fab'_{ISA}-MORF1, DARA-MORF1, and ISA-MORF1) and adaptations to the HSA-based multivalent crosslinking effector molecule (HSA-PEG_x-$(MORF2)_y$) were synthesized. Biorecognition of complementary engagers and effectors at cell surface mediated by MORF1/MORF2 hybridization resulted in crosslinking of CD38 receptors and apoptosis initiation in CD38+ cells: Daudi, Raji RPMI 8226 and ANBL-6. The level of apoptosis induction, Daudi > RPMI 8226 > Raji > ANBL-6 >> U266, correlated with CD38 expression. Additionally, insertion of a flexible PEG_{24} (95.2 Å) spacer into the effector conjugate HSA-PEG_{24}-$(MORF)_{13}$ significantly increased apoptosis of Raji cells when compared to effectors containing PEG_8 (39.2 Å) or PEG_2 (17.6 Å) moieties.

Beyond the synthesis and apoptosis efficacy studies with the new conjugates, a thorough investigation into the mechanisms of action of DARA, DARA DFMT, and Fab'_{DARA} DFMT on CD38+ cell lines were conducted. Preincubation of cells with β-cyclodextrin (to extract cholesterol) or EGTA (to complex extracellular Ca++) decreased levels of apoptosis. DARA-based and Fab'_{DARA}-based DMFT induced, following crosslinking of CD38 receptors, apoptosis via the mitochondrial signaling pathway as indicated by enhanced Bax/Bcl-2 expression ratio, ROS generation, cytochrome C release, and caspase 3 activation.

A comparison of ISA and ISA-based DFMT to DARA and a known Type II antibody, OBN was conducted. ISA induced apoptosis in Daudi and Raji cells. ISA DFMT did not enhance apoptosis when compared to ISA; however, crosslinking of the CD38-Fab'_{ISA}-MORF1 complex with HSA-$(MORF2)_{10}$ resulted in enhanced apoptotic levels. Additionally, Fab'_{ISA}-MORF1 induced apoptosis in Daudi and Raji cells on its own. Comparison with DFMT based on anti-CD20 Type II antibody OBN revealed that both DARA and ISA did not exhibit features related to Type II apoptotic mechanisms (lysosomal enlargement, homotypic cell adhesion).

Finally, simultaneous crosslinking of CD38 and CD20 receptors on Raji cells increases the level of apoptosis when compared to crosslinking of individual receptors. This finding suggests a therapeutic potential of lymphoma treatment with a mixture of antibodies.

Supplementary Materials: The following are available online. Figure S1. Size exclusion chromatography (SEC) of DARA-MORF1 and HSA-$(MORF2)_{10}$ hybridization. Figure S2. UV-Vis spectroscopy to determine DARA-MORF1 and HSA-$(MORF2)_{10}$ hybridization. Figure S3. SEC of Fab'_{ISA}-MORF1 and its intermediates. Figure S4. SEC of ISA-MORF1 and HSA-$(MORF2)_{10}$ hybridization. Figure S5. UV-Vis spectroscopy to determine ISA-MORF1 and HSA-$(MORF2)_{10}$ hybridization. Figure S6. Flow cytometry cell population shifts for DARA DFMT experiments. Figure S7. Flow cytometry cell population shifts for Fab'_{DARA} DFMT experiments. Figure S8. DARA DFMT on ANBL-6 cells. Figure S9. Flow cytometry readout of Rituximab DFMT to monitor calcium influx. Figure S10. Confocal microscopy of Fab'_{DARA} DFMT apoptosis inhibition by β-CD and EGTA. Figure S11. Bax/Bcl-2 expression of DFMT-treated Daudi cells. Figure S12. Cytochrome C calibration curve from ELISA assay. Figure S13. Flow cytometry of Bax and Bcl-2 expression in Daudi cells. Figure S14. Flow cytometry of caspase 3 population gating. Figure S15. Flow cytometry of caspase 3 post-measurement propidium iodide staining. Figure S16. Confocal microscopy of assessment of lipid raft redistribution using Cy5-labeled antibodies. Figure S17. Additional confocal microscopy of lipid raft redistribution induced by DFMT systems. Figure S18. ISA DFMT & Fab'_{ISA} DFMT apoptosis assessment on Raji cells.

Author Contributions: Conceptualization, J.K., J.Y., M.T.G.; methodology, M.T.G., J.L., J.W., J.Y.; validation J.K., J.Y., M.T.G., D.S.; investigation, M.T.G., J.L., J.W.; data analysis, M.T.G., J.K., J.Y., J.L.; writing—original draft preparation, M.T.G., J.L.; writing—review and editing, J.K., J.Y., D.S., M.T.G.; funding acquisition, J.K., J.Y., D.S.; supervision, J.K., J.Y. All authors have read and agreed to the published version of the manuscript.

Funding: Research was supported by NIH grant RO1 CA246716 from the National Cancer Institute (to JK) and by Huntsman Cancer Institute ET grant 39024 (to DS/JY).

Institutional Review Board Statement: Not applicable.

Informed Consent Statement: Not applicable.

Data Availability Statement: All data generated or analyzed during this study are included in the article.

Acknowledgments: We acknowledge support of funds in conjunction with grant P30 CS042014 awarded to Huntsman Cancer Institute. We thank core facilities, flow cytometry, confocal fluorescence microscopy, and mass spectrometry for support. We are indebted to Diane Jelinek of Mayo Clinic for kind gift of ANBL-6 cells.

Conflicts of Interest: The authors declare the following competing financial interest(s): J.Y. and J.K. are co-inventors on a pending US patent application (PCT/US2014/023784; assigned to the University of Utah) related to this work. J.K. is Chief Scientific Advisor and J.Y. Scientific Advisor for Bastion Biologics. Otherwise, the authors declare no competing financial interests.

Sample Availability: Samples of the compounds are not available from the authors.

References

1. Im, A.; Pavletic, S.Z. Immunotherapy in hematologic malignancies: Past, present, and future. *J. Hematol. Oncol.* **2017**, *10*, 94. [CrossRef] [PubMed]
2. Meyer, S.; Evers, M.; Jansen, J.H.M.; Buijs, J.; Broek, B.; Reitsma, S.E.; Moerer, P.; Amini, M.; Kretschmer, A.; Ten Broeke, T.; et al. New insights in Type I and Type II CD20 antibody mechanisms of action with a panel of novel CD20 antibodies. *Br. J. Haematol.* **2018**, *180*, 808–820. [CrossRef] [PubMed]
3. Moody, P.R.; Sayers, E.J.; Magnusson, J.P.; Alexander, C.; Borri, P.; Watson, P.; Jones, A.T. Receptor crosslinking: A general method to trigger internalization and lysosomal targeting of therapeutic receptor: Ligand complexes. *Mol. Ther.* **2015**, *23*, 1888–1898. [CrossRef]
4. Radford, D.C.; Yang, J.; Doan, M.; Li, L.; Dixon, A.S.; Owen, S.C.; Kopeček, J. Multivalent HER2-binding polymer conjugates facilitate rapid endocytosis and enhance intracellular drug delivery. *J. Control. Release* **2020**, *319*, 285–299. [CrossRef]
5. Li, L.; Li, Y.; Yang, J.; Radford, D.C.; Wang, J.; Janát-Amsbury, M.; Kopeček, J.; Yang, J. Inhibition of immunosuppresive tumors by polymer-assisted inductions of immunogenic cell death and multivalent PD-L1 crosslinking. *Adv. Funct. Mater.* **2020**, *30*, 1908961. [CrossRef]
6. Overdijk, M.B.; Jansen, J.H.; Nederland, M.; Lammerts van Bueren, J.J.; Groen, R.W.; Parren, P.W.; Leusen, J.H.; Boross, P. The therapeutic CD38 monoclonal antibody daratumumab induces programmed cell death via Fcγ receptor-mediated crosslinking. *J. Immunol.* **2016**, *197*, 807–813. [CrossRef] [PubMed]
7. van de Donk, N.W.C.J.; Usmani, S.Z. CD38 antibodies in multiple myeloma: Mechanism of action and modes of resistance. *Front. Immunol.* **2018**, *9*, 2134. [CrossRef]
8. Sanchez, L.; Wang, Y.; Siegel, D.S.; Wang, M.L. Daratumumab: A first-in-class CD38 monoclonal antibody for the treatment of multiple myeloma. *J. Hematol. Oncol.* **2016**, *9*, 51. [CrossRef]
9. Richter, J.; Sanchez, L.; Thibaud, S. Therapeutical potential of isatuximab in the treatment of multiple myeloma. *Semin. Oncol.* **2020**, *47*, 155–164. [CrossRef]
10. van de Donk, N.W.C.J.; Richardson, P.G.; Malavasi, F. CD38 antibodies in multiple myeloma: Back to the future. *Blood* **2018**, *131*, 13–29. [CrossRef]
11. Wu, K.; Liu, J.; Johnson, R.N.; Yang, J.; Kopeček, J. Drug-free macromolecular therapeutics: Induction of apoptosis by coiled-coil mediated crosslinking of antigens on cell surface. *Angew. Chem. Int. Ed.* **2010**, *49*, 1451–1455. [CrossRef] [PubMed]
12. Yang, J.; Li, L.; Kopeček, J. Biorecognition: A key to drug-free macromolecular therapeutics. *Biomaterials* **2019**, *190–191*, 11–23. [CrossRef]
13. Rütter, N.M.; Milošević, A. David, Say no to drugs: Bioactive macromolecular therapeutics without conventional drugs. *J. Control. Release* **2021**, *330*, 1191–1207. [CrossRef]
14. Chu, T.-W.; Zhang, R.; Yang, J.; Chao, M.P.; Shami, P.J.; Kopeček, J. A Two-step pretargeted nanotherapy for CD20 crosslinking may achieve superior anti-lymphoma efficacy to Rituximab. *Theranostics* **2015**, *5*, 834–846. [CrossRef] [PubMed]
15. Hartley, J.M.; Chu, T.-W.; Peterson, E.M.; Zhang, R.; Yang, J.; Harris, J.; Kopeček, J. Super-resolution imaging and quantitative analysis of membrane protein/lipid raft clustering mediated by cell surface self-assembly of hybrid nanoconjugates. *ChemBioChem* **2015**, *16*, 1725–1729. [CrossRef] [PubMed]
16. Wu, K.; Yang, J.; Liu, J.; Kopeček, J. Coiled-coil based drug-free macromolecular therapeutics: In vivo efficacy. *J. Control. Release* **2012**, *157*, 126–131. [CrossRef] [PubMed]
17. Li, L.; Yang, J.; Wang, J.; Kopeček, J. Amplification of CD20 crosslinking in Rituximab resistant B-lymphoma cells enhances apoptosis induction by drug-free macromolecular therapeutics. *ACS Nano* **2018**, *12*, 3658–3670. [CrossRef]
18. Wang, J.; Li, L.; Yang, J.; Clair, P.M.; Glenn, M.; Stephens, D.M.; Radford, D.C.; Kosak, K.M.; Deininger, M.W.; Shami, P.J.; et al. Drug-free macromolecular therapeutics induce apoptosis in cells isolated from patients with B cell malignancies with enhanced apoptosis induction by pretreatment with gemcitabine. *Nanomedicine NBM* **2019**, *16*, 217–225. [CrossRef]
19. Li, L.; Yang, J.; Wang, J.; Kopeček, J. Drug-free macromolecular therapeutics induce apoptosis via calcium influx and mitochondrial signaling pathway. *Macromol. Biosci.* **2018**, *18*, 1700196. [CrossRef]
20. Bhatnagar, V.; Gormley, N.J.; Luo, L.; Shen, Y.L.; Sridhara, R.; Subramaniam, S.; Shen, G.; Ma, L.; Shord, S.; Glodberg, K.B.; et al. FDA approval summary: Daratumumab for treatment of multiple myeloma after one prior therapy. *Oncologist* **2017**, *22*, 1347–1353. [CrossRef]
21. Gabellier, L.; Carton, G. Obinutuzumab for relapsed or refractory indolent non-Hodgkin's lymphomas. *Ther. Adv. Hematol.* **2016**, *7*, 85–93. [CrossRef]
22. Jiang, H.; Acharya, C.; An, G.; Zhong, M.; Feng, X.; Wang, L.; Dasilva, N.; Song, Z.; Yang, G.; Adrian, F.; et al. SAR650984 directly induces multiple myeloma cell death via lysosomal-associated and apoptotic pathways, which is further enhanced by pomalidomide. *Leukemia* **2016**, *30*, 399–408. [CrossRef] [PubMed]

23. Wang, S.; Liu, S.; Zhang, Y.; He, J.; Coy, D. Human serum albumin (HSA) and its applications as a drug delivery vehicle. *Health Sci. J.* **2020**, *14*, 1–8.
24. Benizri, S.; Gissot, A.; Martin, A.; Vialet, B.; Grinstaff, M.W.; Barthélémy, P. Bioconjugated oligonucleotides: Recent developments and therapeutic applications. *Bioconjug. Chem.* **2019**, *30*, 366–383. [CrossRef]
25. Kverka, M.; Hartley, J.M.; Chu, T.-W.; Yang, J.; Heidchen, R.; Kopeček, J. Immunogenicity of coiled-coil based drug-free macromolecular therapeutics. *Biomaterials* **2014**, *35*, 5886–5896. [CrossRef] [PubMed]
26. Seyfizadeh, N.; Seyfizadeh, N.; Hasenkamp, J.; Huerta-Yepez, S. A molecular perspective on rituximab: A monoclonal antibody for B cell non-Hodgkin lymphoma and other affections. *Crit. Rev. Oncol. Hemat.* **2016**, *97*, 275–290. [CrossRef]
27. Li, L.; Wang, J.; Li, Y.; Radford, D.C.; Yang, J.; Kopeček, J. Broadening and enhancing functions of antibodies by self-assembling multimerization at cell surface. *ACS Nano* **2019**, *13*, 11422–11432. [CrossRef]
28. Zhang, L.; Fang, Y.; Yang, J.; Kopeček, J. Drug-free macromolecular therapeutics: Impact of structure on induction of apoptosis in Raji B cells. *J. Control. Release* **2017**, *263*, 139–150. [CrossRef] [PubMed]
29. Li, L.; Yang, J.; Soodvilai, S.; Wang, J.; Opanasopit, P.; Kopeček, J. Drug-free albumin-triggered sensitization of cancer cells to anticancer drugs. *J. Control. Release* **2019**, *293*, 84–93. [CrossRef]
30. Funaro, A.; Reiniš, M.; Trubiani, O.; Santi, S.; Di Primio, R.; Malavasi, F. CD38 functions are regulated through internalization step. *J. Immunol.* **1998**, *160*, 2238–2247.
31. Ghose, J.; Viola, D.; Terrazas, C.; Caserta, E.; Troadec, E.; Khalife, J.; Gulsen Gunes, E.; Sanchez, J.; McDonald, T.; Marcucci, G.; et al. Daratumumab induces CD38 internalization and impairs myeloma cell adhesion. *Oncoimmunology* **2018**, *7*, e1486948. [CrossRef]
32. Chu, T.-W.; Kopeček, J. Drug-free macromolecular therapeutics—A new paradigm in polymeric nanomedicines. *Biomater. Sci.* **2015**, *3*, 908–922. [CrossRef]
33. Goodwin, D.A.; Meares, C.F. Advances in pretargeting biotechnology. *Biotechnol. Adv.* **2001**, *19*, 435–450. [CrossRef]
34. Sharkey, R.M.; Karacay, H.; Cardillo, T.M.; Chang, C.H.; McBride, W.J.; Rossi, E.A.; Horak, I.D.; Goldenberg, D.M. Improving the delivery of radionuclides for imaging and therapy of cancer using pretargeting methods. *Clin. Cancer Res.* **2005**, *11*, 7109s–7121s. [CrossRef] [PubMed]
35. Liu, G.; Dou, S.; Ruscowski, M.; Hnatowich, D.J. An experimental and theoretical evaluation of the influence of pretargeting antibody on the tumor accumulation of effector. *Mol. Cancer Res.* **2008**, *7*, 1025–1032. [CrossRef] [PubMed]
36. Niwa, T.; Kasuya, Y.; Suzuki, Y.; Ichikawa, K.; Yoshida, H.; Kurimoto, A.; Tanaka, K.; Morita, K. Novel immunoliposome technology for enhancing the activity of the agonistic antibody against the tumor necrosis factor receptor superfamily. *Mol. Pharm.* **2018**, *15*, 3729–3740. [CrossRef]
37. Semac, I.; Palomba, C.; Kulangara, K.; Klages, N.; van Echten-Deckert, G.; Borisch, B.; Hoessli, D.C. Anti-CD20 therapeutic antibody rituximab modifies the functional organization of rafts/microdomains of B lymphoma cells. *Cancer Res.* **2003**, *63*, 534–540.
38. Li, Y.C.; Park, M.J.; Ye, S.-K.; Kim, C.-W.; Kim, Y.-N. Elevated levels of cholesterol-rich lipid rafts in cancer cells are correlated with apoptosis sensitivity induced by cholesterol-depleting agents. *Am. J. Pathol.* **2006**, *168*, 1107–1118. [CrossRef]
39. Janas, E.; Priest, R.; Wilde, J.I.; White, J.H.; Malhorta, R. Rituxan (anti-CD20 antibody)-induced translocation of CD20 into lipid rafts is crucial for calcium influx and apoptosis. *Clin. Exp. Immunol.* **2005**, *139*, 439–446. [CrossRef]
40. Kaufmann, S.H.; Hengartner, M.O. Programmed cell death: Alive and well in the next millenium. *Trends Cell Biol.* **2001**, *11*, 526–534. [CrossRef]
41. Saxena, A.; Viswanathan, S.; Moshynska, O.; Tandon, P.; Sankaran, K.; Sheridan, D.P. Mcl-1 and Bcl-2/Bax ratio are associated with treatment response but not with Rai stage in B-cell chronic lymphocytic leukemia. *Am. J. Hematol.* **2004**, *75*, 22–33. [CrossRef]
42. Fletcher, A.I.; Meusburger, S.; Hawkins, C.J.; Riglar, D.T.; Lee, E.F.; Fairlie, W.D.; Huang, D.C.S.; Adams, J.M. Apoptosis is triggered when prosurvival Bcl-2 proteins cannot restrain Bax. *Proc. Natl. Acad. Sci. USA* **2008**, *105*, 18081–18087. [CrossRef]
43. Kim, R.; Emi, M.; Tanabe, K. Role of mitochindria as the gradens of cell death. *Cancer Chemother. Pharmacol.* **2006**, *57*, 545–553. [CrossRef]
44. Alduaij, W.; Ivanov, A.; Honeychurch, J.; Cheadle, E.; Potluri, S.; Lim, S.H.; Shimada, K.; Chan, C.H.; Tutt, A.; Beers, S.A.; et al. Novel type II anti-CD20 monoclonal antibody (GA101) evokes homotypic adhesion and actin-dependent, lysosome-mediated cell death in B-cell malignancies. *Blood* **2010**, *117*, 4519–4529. [CrossRef]
45. Deckert, J.; Wetzel, M.-C.; Bartle, L.M.; Skaletskaya, A.; Goldmacher, V.S.; Valée, F.; Zhou-Liu, Q.; Ferrari, P.; Pouznieux, S.; Lahoute, C.; et al. SAR650984, A novel humanized CD38-targeting antibody, demonstrates potent antitumor activity in models of multiple myeloma and other CD38+ hematologic malignancies. *Clin. Cancer Res.* **2014**, *20*, 4574–4583. [CrossRef] [PubMed]
46. Moreno, L.; Perez, C.; Zabaleta, A.; Manrique, I.; Alignani, D.; Ajona, D.; Blanco, D.; Lasa, M.; Maiso, P.; Rodriguez, I.; et al. The mechanism of action of the anti-CD38 monoclonal antibody isatuximab in multiple myeloma. *Clin. Cancer Res.* **2019**, *25*, 3176–3187. [CrossRef] [PubMed]
47. Vočková, P.; Svatoň, M.; Karolová, J.; Pokorná, E.; Vokurka, M.; Klener, P. Anti-CD38 therapy with daratumumab for relapsed/refractory CD20-negative diffuse large B-cell lymphoma. *Folia Biol. (Praha)* **2020**, *66*, 17–23.
48. Lidický, O.; Klener, P.; Machová, D.; Vočková, P.; Pokorná, E.; Helman, K.; Mavis, C.; Janoušková, O.; Etrych, T. Overcoming resistence to rituximab in relapsed non-Hodgkin lymphomas by antibody-polymer drug conjugates actively targeted by anti-CD38 daratumumab. *J. Control. Release* **2020**, *328*, 160–170. [CrossRef] [PubMed]

molecules

MDPI

Article

An In Vitro Assessment of Immunostimulatory Responses to Ten Model Innate Immune Response Modulating Impurities (IIRMIs) and Peptide Drug Product, Teriparatide

Claire K. Holley [1,†], Edward Cedrone [1,†], Duncan Donohue [2,†], Barry W. Neun [1], Daniela Verthelyi [3], Eric S. Pang [4,*] and Marina A. Dobrovolskaia [1,*]

1 Nanotechnology Characterization Laboratory, Cancer Research Technology Program, Frederick National Laboratory for Cancer Research Sponsored by the National Cancer Institute, Frederick, MD 21702, USA; claire.holley@nih.gov (C.K.H.); edward.cedrone@nih.gov (E.C.); neunb@mail.nih.gov (B.W.N.)

2 Statistics Department, Data Management Services Inc., Frederick National Laboratory for Cancer Research Sponsored by the National Cancer Institute, Frederick, MD 21702, USA; duncan.donohue@nih.gov

3 Laboratory of Innate Immunity, Division of Biotechnology Review and Research-III, Office of Biotechnology Products, Center for Drug Evaluation and Research, U.S. Food and Drug Administration, Silver Spring, MD 20993, USA; daniela.verthelyi@fda.hhs.gov

4 Division of Therapeutic Performance, Office of Research and Standards, Office of Generic Drugs, Center for Drug Evaluation and Research, U.S. Food and Drug Administration, Silver Spring, MD 20993, USA

* Correspondence: eric.pang@fda.hhs.gov (E.S.P.); marina@mail.nih.gov (M.A.D.)

† These authors contributed equally.

Citation: Holley, C.K.; Cedrone, E.; Donohue, D.; Neun, B.W.; Verthelyi, D.; Pang, E.S.; Dobrovolskaia, M.A. An In Vitro Assessment of Immunostimulatory Responses to Ten Model Innate Immune Response Modulating Impurities (IIRMIs) and Peptide Drug Product, Teriparatide. *Molecules* **2021**, *26*, 7461. https://doi.org/10.3390/molecules26247461

Academic Editor: Aleksandra Misicka-Kesik

Received: 2 November 2021
Accepted: 6 December 2021
Published: 9 December 2021

Abstract: Understanding, predicting, and minimizing the immunogenicity of peptide-based therapeutics are of paramount importance for ensuring the safety and efficacy of these products. The so-called anti-drug antibodies (ADA) may have various clinical consequences, including but not limited to the alteration in the product's distribution, biological activity, and clearance profiles. The immunogenicity of biotherapeutics can be influenced by immunostimulation triggered by the presence of innate immune response modulating impurities (IIRMIs) inadvertently introduced during the manufacturing process. Herein, we evaluate the applicability of several in vitro assays (i.e., complement activation, leukocyte proliferation, and cytokine secretion) for the screening of innate immune responses induced by ten common IIRMIs (*Bacillus subtilis* flagellin, FSL-1, zymosan, ODN2006, poly(I:C) HMW, poly(I:C) LMW, CLO75, MDP, ODN2216, and *Escherichia coli* O111:B4 LPS), and a model biotherapeutic Forteo™ (teriparatide). Our study identifies cytokine secretion from healthy human donor peripheral blood mononuclear cells (PBMC) as a sensitive method for the in vitro monitoring of innate immune responses to individual IIRMIs and teriparatide (TP). We identify signature cytokines, evaluate both broad and narrow multiplex cytokine panels, and discuss how the assay logistics influence the performance of this in vitro assay.

Keywords: cytokines; innate immunity; immunogenicity; peptides; teriparatide

1. Introduction

Repeated administration of therapeutic drug products was shown to trigger unwanted immune responses and the production of antibodies capable of neutralizing both the therapeutic protein and its endogenous counterparts [1–3]. Antibodies to recombinant biotechnology therapeutics come in a variety of isotypes (e.g., IgM vs. IgG vs. IgE), allotypes (e.g., reflecting genetic differences between IgG of biologically unrelated individuals), idiotypes (e.g., reflecting binding to specific epitopes within antibody variable sites), and may ultimately lead to different functional consequences for the host (e.g., binding, PK-altering, neutralizing, hypersensitivity- or anaphylaxis-triggering, and cross-reactive neutralizing). Such anti-drug antibodies (ADA) may lead to severe and, when not timely and properly treated, potentially lethal clinical consequences, loss of treatment efficacy, and the formation of autoimmunity [4–8]. The frequency of different ADA types and their clinical impact have a reverse relationship, in that binding antibodies occur most frequently and have low clinical impact whereas cross-reacting neutralizing antibodies are rare but have the highest clinical significance.

The immunogenic risk of biotherapeutics and ADA response can be influenced by a multitude of factors. One such factor is the presence of innate immune response modulating impurities (IIRMIs) that might be inadvertently introduced during product manufacturing [4,5]. IIRMIs may have little or no impact on the function of the resulting drug product but may influence the host immune response [4,9–13]. While it is nearly impossible to predict the immunogenicity of a specific biotherapeutic without directly assessing the related immune responses in vivo [3], the presence of IIRMIs contributing to the immunogenicity via priming the immune cells could be identified using in vitro methods detecting innate immunostimulatory responses, including the production of inflammatory cytokines (e.g., IL-1, IFNs, IL-8, TNFα, etc.) and activation of the complement system. Therefore, there is an urgent need in understanding the applicability to, and performance of, in vitro assays in detecting IIRMIs presence in drug products.

Herein, we report the results of an in vitro study analyzing the applicability of several in vitro assays (i.e., complement activation, leukocyte proliferation, and cytokine secretion) to the screening of innate immune responses induced by ten common IIRMIs, including *Bacillus subtilis* flagellin, FSL-1, zymosan, ODN2006, poly(I:C) HMW, poly(I:C) LMW, CLO75, MDP, ODN2216, and *Escherichia coli* O111:B4 LPS, as well as model therapeutic Forteo™ (teriparatide or TP). The selected assays were chosen due to the known roles of the complement system, cytokines, and activated leukocytes in the process of immunogenicity [14]. While these immunostimulatory biomarkers do not directly predict immunogenicity, they serve as important prerequisites to it, which, when monitored in vitro, may allow for the detection of biologically active contaminants contributing to the process of immunogenicity by priming the immune cells [14].

2. Results

2.1. Initial In Vitro Characterization and Assay Selection

Forteo™ is a peptide-based therapeutic formulation where the active peptide, teriparatide (TP), is produced using recombinant DNA technology. To characterize the whole product, we first established that TP and its corresponding formulation buffer (FB) had no detectable endotoxin and β-glucans that could activate innate immune responses [15–17] using a commercial turbidity *Limulus* Amebocyte Lysate (LAL) assay and Factor-C depleted LAL (Glucatell) assay respectively (Table 1 and Table S1).

Detection of impurities in cell-based assays requires cells that are sensitive to the presence of IIRMI and can elicit a quantifiable response. Previous studies have shown that very low levels of impurities that trigger pattern recognition receptors (PRRs) can stimulate a local innate immune response at the site of inoculation and suggested that cell-based assays could be used to detect these types of impurities in the products. Since retaining cell viability throughout the assay is critical, we first determined whether TP would alter cell viability and determined the highest concentration of TP that could be used in a PBMC-based study where the cells were in culture for 24 h. As shown in Figure S1, when PBMC were cultured in the presence of TP at concentrations ranging from 0.025 to 25 μg/mL, the viability of the cells was retained, but higher concentrations of the product reduced cell viability to 60% (Figure S1). Based on this data, the highest non-toxic concentration (25 μg/mL) was chosen as the top concentration to be used for subsequent in vitro experiments, including the assessment of TP and/or IIRMI activation of C3a complement, leukocyte proliferation, and cytokine secretion (Table 1).

Table 1. Initial Characterization of Teriparatide. Teriparatide (TP) purity and capability of triggering innate immunity activation in vitro, either due to the presence of innate immune response modulating impurities (IIRMIs) in the drug formulation or due to the presence of the drug itself, was assessed through the following assays. Results were below the level of detection, so these assays were not used for future TP immunity experiments. LAL = *Limulus* Amoebocyte Lysate Assay; LLOQ = lower limit of quantification; STE = Sterility Endotoxin assay; ITA = Immuno-Toxicity Assay; CBA = Cell Based Assay; ELISA = Enzyme-Linked Immuno-Sorbent Assay; AO = Acridine Orange; PI = Propidium Iodine.

Purpose	Assay Type (NCL Protocol)	Main Findings
Endotoxin Detection	LAL (STE-1.2)	Endotoxin contamination is below the assay LLOQ
β-Glucan Detection	Glucatell (STE-4)	β-glucan contamination is below the assay LLOQ
Cell Viability/Teriparatide Cytotoxicity	AO/PI staining	>85% viability for TP <25 μg/mL ~60% viability for 50 μg/mL TP 25 μg/mL TP chosen for future experiments
Leukocyte Proliferation	CBA (ITA-6)	TP did not induce leukocyte proliferation IIRMIs induced low levels of leukocyte proliferation TP suppressed IIRMI-induced leukocyte proliferation The assay is not chosen for future studies
Complement Activation	ELISA (ITA-5.2)	TP resulted in complement activation Levels of IIRMIs contamination in drug product are insufficient for the complement activation The assay is not chosen for future studies

Next, we determined whether TP, in concentrations that do not interfere with cell viability, can reduce the response to potential impurities. Using an array of purified TLR agonists at concentrations that are close to those shown to elicit a local innate immune response in vivo, we examined whether the presence of TP in the culture would modulate the response to the PRR-agonists. As shown in Figure S2, while PRR-agonist exposure triggered low levels of leukocyte proliferation in a dose-dependent manner, the response was abrogated in TP-treated cultures (Table 1).

In addition to inducing cell proliferation, the activation of innate immune cells could also induce complement activation. Therefore, we next explored whether TP would activate complement. Treatment with TP resulted in an activation of the complement system as evidenced by an increase in detectable C3a split products; this activation was comparable to that detected in Cremophor-EL and Feraheme-treated plasma samples, used as positive controls (Figure S3). Concentrations of IIRMIs capable of inducing detectable complement activation are typically higher than what may potentially be present in drug products as undesirable contaminants. For example, concentrations of zymosan and lipopolysaccharide (LPS) required to produce detectable complement activation are 10 mg/mL or >500 μg/mL, respectively [18,19]. Therefore, this assay was not selected for subsequent experiments (Table 1).

2.2. In Vitro Cytokine Responses to Teriparatide

PBMCs treated with TP alone noticeably induced PGE-2 and IL-8 production (Figure S4). TP-induced PGE-2 production directly correlated with TP concentrations added to PBMC cultures (Figure 1A). Such correlation for IL-8 induction was only observed in 3 of the 10 tested PBMC cultures (Figure 1B). Cultures from the remaining donors showed increased IL-8 levels at the second-highest concentration (2.5 μg/mL) but not at the highest concentration (25 μg/mL). The reduced levels of IL-8 secreted after incubation with highest concentration of TP (25 μg/mL) suggest a level of PBMC exhaustion resulting from high stimulation over the course of 24 h.

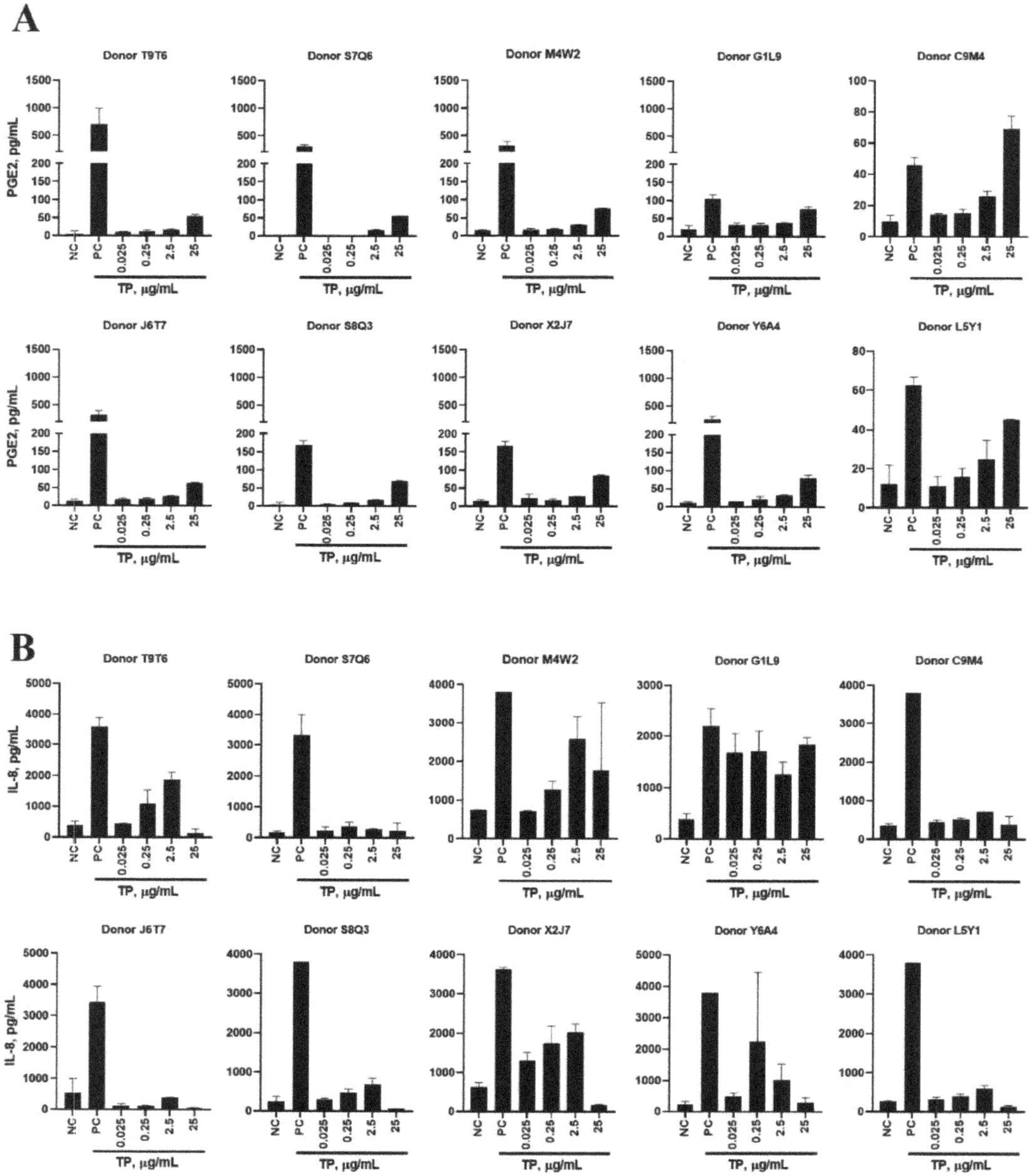

Figure 1. A 16-plex Induction of Prostaglandin-E_2 and Interleukin-8 by Teriparatide. PBMCs from 10 healthy human donors were treated with 0.025, 0.25, 2.5, and 25 µg/mL teriparatide (TP), compared to a PBS negative control (NC) and LPS/PHA-M/ODN positive control (PC) for 24 h. Supernatants were analyzed for the presence of (**A**) PGE-2 or (**B**) IL-8 by 16-plex multiplex ELISA. Each bar shows mean and standard deviation (N = 2).

2.3. *Teriparatide Effects on Cytokine Expression Are Due to the Formulation Buffer (FB)*

To understand whether the induction of PGE-2 and IL-8 observed in TP-treated cultures (Figure S4) was due to the active pharmaceutical ingredient (API) or FB, we conducted a follow-up experiment in which TP was tested side-by-side with FB at equivalent dilutions that resulted in equivalent concentrations of the FB; these dilutions were performed in

PBS. We also performed TP dilutions in the FB and tested them in the same cultures with PBS-diluted FB and TP. The results of this experiment demonstrated that PGE-2 and IL-8 responses to TP were due to the FB (Figure 2A,B and Figure S5).

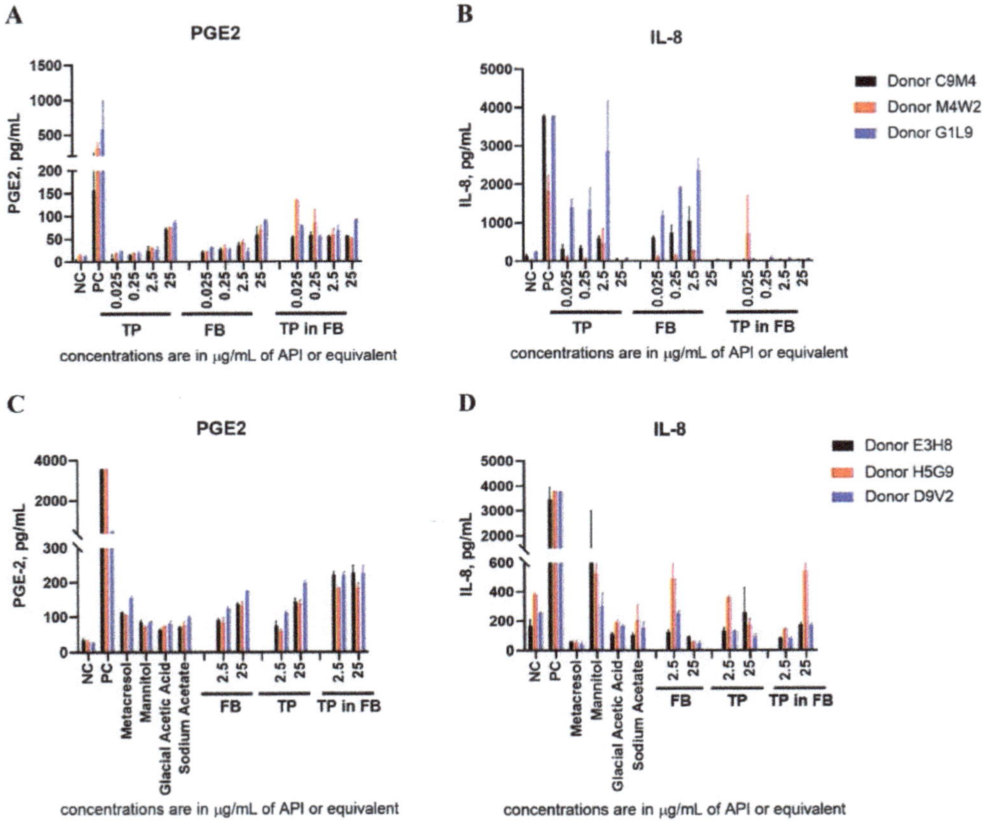

Figure 2. Formulation Buffer is Responsible for Prostaglandin-E_2 and Interleukin-8 Cytokine Response to Teriparatide. (**A**,**B**) PBMCs from three healthy human donors were used to test teriparatide (TP) at 0.025, 0.25, 2.5, and 25 μg/mL API, diluted in either PBS or Formulation Buffer (FB), compared to complete FB diluted in PBS to achieve the equivalent API concentrations, compared to a PBS negative control (NC) and LPS/PHA-M/ODN positive control (PC). Each bar shows a mean response and a standard deviation (N = 3); (**C**,**D**) PBMCs from another set of three healthy donors were used to test the components of FB (metacresol, mannitol, glacial acetic acid, and sodium acetate) at concentrations equivalent to 25 μg/mL of API in TP, in comparison to complete FB, TP diluted in PBS, and TP diluted in FB. Each bar shows a mean response and a standard deviation (N = 2).

Next, we hypothesize that metacresol, a preservative of FB, was the cause of the cytokine response to TP, because an earlier study in THP-1 cells reported that this excipient, at a concentration comparable to that present in our cultures (0.2 mg/mL), induced chemokine MCP-1 (but not TNFα, IL-1, or IL-6) [20]. To verify this hypothesis, metacresol and other components of FB (mannitol, glacial acetic acid, and sodium acetate) at concentrations equivalent to that of API in TP were added to PBMC cultures and the supernatants were analyzed for the presence of cytokines (Figure 2C,D and Figure S6). The result of this experiment demonstrated that, in addition to metacresol, all other individual components of the FB contribute to the cytokine response observed with TP. Contrary to our hypothesis about the potential inflammatory nature of metacresol, the highest cytokine response, specifically IL-8, was observed upon application of mannitol (Figure 2D), a response which has previously been reported on in vitro PBMCs and in vivo endothelial cells to deleterious effect [21,22].

2.4. In Vitro Cytokine Responses to Individual IIRMIs

Since IIRMIs activated a broad and often overlapping spectrum of cytokines (Figure 3, Figures S7 and S8), we next performed a global analysis using Euclidian distance and Ward's clustering for the dendrogram and constructed a heatmap of normalized values averaged across all donors and replicates (Figure 4 and Figure S9). This normalization included scaling each cytokine reading across all collected values by dividing each value by that cytokine's standard deviation obtained across all donors, which brought all cytokines onto roughly the same scale. The benefit of using this approach is that one can compare cytokines directly across all 10 donors, while keeping cytokines with very large values from swamping the comparative global analyses. These analyses revealed a pure red band representing the negative control and a bright yellow vertical band representing the positive control. These analyses also identified groupings of cytokines with similar response patterns across all IIRMI treatments (Figure 4). For example, chemokines IL-8 and MIP-1α showed very similar patterns; PGE-2 showed such a high response to the two higher concentrations of zymosan that it overshadowed the positive control; IL-2 and IL-17 were very similar in that they did not appear to be strongly induced by any IIRMI (Figure 4). Alternatively, these analyses also gave us insight on how various IIRMIs, and their concentrations, clustered with respect to the cytokine response patterns that they induced (Figure S9). These analyses demonstrated that the highest concentrations of each IIRMI often clustered together. For example, zymosan and CLO75 clustered together at the bottom of the heatmap. The analyses also highlighted a group of IIRMIs that seemed to have virtually no cytokine response, including the lowest concentrations of poly(I:C) LMW, ODN2006, poly(I:C) HMW, and ODN2216. Finally, these analyses also highlighted a group of IIRMIs, and their concentrations located in the center of the heat map, that predominantly activated IL-8 (Figure S9).

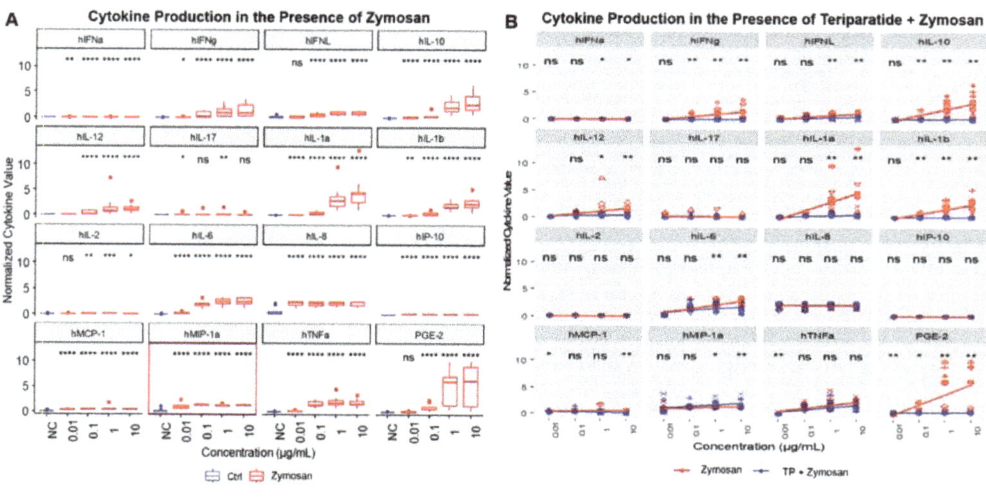

Figure 3. Normalized Cytokine Response to Zymosan and/or Teriparatide and Selection of One Signature Cytokine: PBMCs from 10 healthy human donors were treated with (**A**) zymosan alone or (**B**) zymosan in combination with 25 μg/mL TP for 24 h. Supernatants were analyzed for the presence of cytokines by multiplex ELISA. The signature cytokine (red box) is the one for which the IIRMI concentration, when compared to the PBS negative control (NC), results in a $p < 0.05$. The data for which statistical significance was not observed are marked with ns. Statistical significance is shown with an asterisk as follows: * $p < 0.05$; ** $p < 0.01$; *** $p < 0.001$; and **** $p < 0.0001$. Similar results for the other nine IIRMIs are available in Figures S7 and S8.

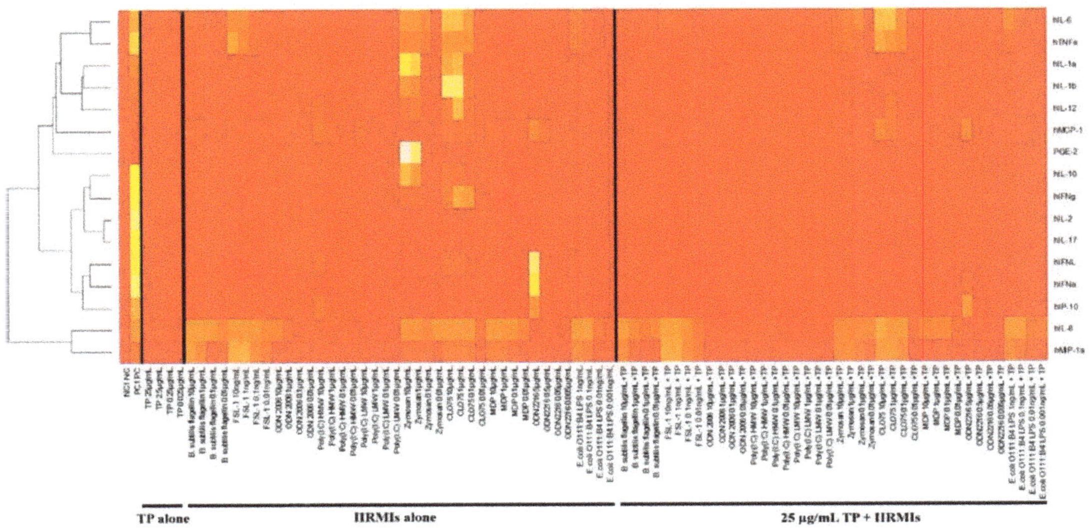

Figure 4. Innate Immune Response Modulating Impurity Treatment and Concentration Patterns via Euclidian Distance and Ward's Clustering. PBMCs from 10 healthy human donors were treated with various concentrations of IIRMIs, alone and in combination with 25 µg/mL Teriparatide (TP), compared to a PBS negative control (NC) and LPS/PHA-M/ODN positive control (PC), for 24 h. Supernatants were analyzed for the presence of cytokines by multiplex ELISA. Shown is the mean response of normalized values averaged across all donors, clustered based on cytokine response. Dendrograms were created using complete linkage clustering on the Euclidian distance matrices. Similar results for IIRMI clustering available in Figure S9.

Further Pearson's correlation analysis allowed for clustering the cytokine responses based on how well cytokine values correlated across all treatment groups (Figure S10). This analysis revealed that the strongest correlations were between IL-6 and TNFα, IL-8 and MIP-1α, as well as IL-1α, IL-1β, and IL-12 (Figure S10A). We also observed that different concentrations of the same IIRMI tended to correlate well, especially at the higher concentration ranges. For example, higher concentrations of ODN2216, poly(I:C) LMW, and poly(I:C) HMW showed a distinct cluster which was anti-correlated with the higher concentrations of zymosan) and to a lesser degree CLO75 (Figure S10B).

2.5. Identification of Signature Cytokines

A two-sided paired Wilcoxon test was used to compare cytokines induced by individual concentrations of IIRMIs with negative control samples pooled across all donors (Figure 3A). For each IIRMI, a signature cytokine was identified by determining the lowest IIRMI concentration which, when compared to the baseline, resulted in both an elevation of the cytokine and the lowest p-value (i.e., at least $p < 0.05$) (Figure 3A, red box). For each IIRMI concentration, if two cytokines achieved a level of significance, the lower (more significant) p-value won. Since many test samples, especially negative controls, resulted in cytokine levels below the assay lower limit of detection, we used a non-parametric test for statistical analysis. This approach ranks cytokine significance values rather than the magnitude of cytokine difference. Therefore, the "winning" cytokine was not always the one that appeared the best with regards to mean differences, but rather the one that both had the fewest overlaps between treated samples and controls and was consistent between individual donors (Table 2; Figure S7). Other cytokines with statistically significant elevation above the baseline, at the same IIRMI concentration as the signature cytokine, were also observed (Table 2; Figure S7).

Table 2. Cytokines Induced by Innate Immune Response Modulating Impurities. Individual IIRMIs, their cognate pattern recognition receptors (PRRs), and signature cytokines detected after treatment with IIRMI are summarized. Using a two-sided Wilcoxon test, a signature cytokine was identified for each IIRMI by determining the lowest IIRMI concentration, which, when compared to the baseline, resulted in an elevation of the cytokine, and had the lowest ranking *p*-value (i.e., at least $p < 0.05$). IIRMI = innate immune response modulating impurities; TLR = Toll-Like Receptor; IL = interleukin; IFN = interferon; MCP = monocyte chemoattractant protein; MIP = macrophage inflammatory protein; NOD = nucleotide-binding oligomerization domain; TNF = tumor necrosis factor; PGE = prostaglandin; LPS = lipopolysaccharide; CLO = thiazoloquinolone derivative; MDP = muramyldipeptide; ODN = oligo deoxyribonucleotide; LMW = low molecular weight; HMW = high molecular weight; FSL = Pam2CGDPKHPKSF, a synthetic lipopeptide derived from *Mycoplasma salivarium*.

IIRMI	PRR	Signature Cytokine	Lowest Conc. of IIRMI at Which Signature Cytokine Is Detected	Other Cytokines Statistically Higher than the Baseline at the Lowest IIRMI Conc. that Induced Signature Cytokine
B. subtilis flagellin	TLR5	IL-1β	0.01 μg/mL	IFNα, IL-10, IL-1α, IL-2, IL-6, IL-8, IL-1, MCP-1, MIP-1α, TNFα
FSL-1	TLR2/TLR6	IL-1α	10 pg/mL	IFNα, IFNγ, IFNλ, IL-10, IL-12, IL-1b, IL-2, IL-6, IL-8, IP-10, MCP-1, MIP-1α, TNFα, PGE-2
ODN2006 Class B	TLR9	IFNα	1 μg/mL	IFNγ, IL-1α, IL-10, IL-2, IL-6, IL-8, IP-10, MCP-1, MIP-1α, TNFα
Poly(I:C) HMW	TLR3	IP-10	0.1 μg/mL	IFNα, IFNγ, IFNλ, IL-10, IL-12, IL-1α, IL-2, IL-6, IL-8, IP-10, MCP-1, MIP-1α, TNFα
Poly(I:C) LMW	TLR3	MCP-1	1 μg/mL	IFNγ, IL-12, IL-6, IP-10, MIP-1α
Zymosan	TLR2/Dectin 1	MIP-1α	0.01 μg/mL	IFNα, IFNγ, IFNλ, IL-10, IL-12, IL-17, IL-1α, IL-1β, IL-6, IL-8, IP-10, MCP-1, TNFα
CLO75	TLR8	IL-10	0.01 μg/mL	IL-1α, IL-1β, IL-2, IL-6, IL-8, IP-10, MCP-1, MIP-1α, TNFα
MDP	NOD2	IL-8	0.01 μg/mL	IFNα, IL-10, IL-12, IL-6, IP-10, MCP-1, MIP-1α, TNFα
ODN2216	TLR9	IL-6	0.005 μg/mL	IL-6, IL-8
E. coli O111:B4 LPS	TLR4	IL-1α	1 pg/mL	IFNα, IFNγ, IL-10, IL-12, IL-1β, IL-2, IL-6, IL-8, IP-10, MCP-1, MIP-1α, TNFα, PGE-2

2.6. Selection of the Cytokine Panel Specific to Teriparatide and Individual IIRMIs

In order to understand whether the 16-cytokine panel could be narrowed down to three or four cytokines that would be representative of all 10 IIRMIs, we performed additional analysis using the same approach as described above but focused on the top three "winning" cytokines for each IIRMI (Table 3). For this analysis, IIRMIs were grouped based on the intracellular location of their cognate PRRs. Interestingly, all IIRMIs that activate membrane-tethered TLRs consistently induced two cytokines (IL-1α and MIP-1α (Table 3). This finding suggests that any of these two cytokines could be used as a biomarker for the detection of IIRMIs triggering membrane-tethered PRRs. In contrast, no such consistency was observed for IIRMIs that activate endosomal TLRs. Therefore, a combination of cytokines MCP-1 and IL-8 or MCP-1 and IL-6 would be required to suggest the presence of IIRMIs triggering endosomal TLRs (Table 3). One of the following cytokines—IL-6, IL-8, or IP-10—could be used to suggest the presence of IIRMIs triggering cytosolic PRRs (Table 3).

Table 3. Selection of three signature cytokines induced by individual Innate Immune Response Modulating Impurities. A two-sided unpaired Wilcoxon test was used to select the top three cytokines for each IIRMI, which had consistent responses between all donors and the lowest p-value. Starting with the lowest concentration for each IIRMI, if three cytokines did not achieve significance of $p \leq 0.05$, the next highest concentration was evaluated until three cytokines were chosen. If more than three cytokines achieved $p \leq 0.05$ at the selected concentration, the three with the lowest (most significant) p-values were selected. The top three cytokines selected for any IIRMI are shown as "TRUE" while the remaining less significant cytokines are shown as "FALSE". IIRMIs are grouped based on the intracellular localization of their cognate pattern-recognition receptors (PRRs) and color-coded as follows: BLUE-cellular membrane, RED-endosome, GREEN-cytosol. TRUE values in each group are highlighted in bold and the same color code as that used for corresponding IIRMIs.

IIRMI	IFNα	IFNγ	IFNλ	IL-10	IL-12	IL-17	IL-1α	IL-1β	IL-2	IL-6	IL-8	IP-10	MCP-1	MIP-1α	TNFα	PGE-2
*B. subtilis*flagellin	FALSE	FALSE	FALSE	FALSE	FALSE	FALSE	TRUE	TRUE	FALSE	FALSE	FALSE	FALSE	FALSE	TRUE	FALSE	FALSE
FSL-1	FALSE	FALSE	FALSE	FALSE	FALSE	FALSE	TRUE	TRUE	FALSE	FALSE	FALSE	FALSE	FALSE	TRUE	FALSE	FALSE
Zymosan	FALSE	FALSE	FALSE	FALSE	FALSE	FALSE	TRUE	FALSE	FALSE	FALSE	FALSE	FALSE	TRUE	TRUE	FALSE	FALSE
*E. coli*O111:B4 LPS	FALSE	FALSE	FALSE	FALSE	FALSE	FALSE	TRUE	TRUE	FALSE	FALSE	FALSE	FALSE	FALSE	TRUE	FALSE	FALSE
ODN2006	TRUE	FALSE	FALSE	FALSE	FALSE	FALSE	FALSE	FALSE	FALSE	FALSE	TRUE	FALSE	TRUE	FALSE	FALSE	FALSE
Poly(I:C) HMW	FALSE	FALSE	FALSE	FALSE	FALSE	FALSE	FALSE	FALSE	FALSE	TRUE	FALSE	TRUE	TRUE	FALSE	FALSE	FALSE
Poly(I:C) LMW	FALSE	TRUE	FALSE	FALSE	FALSE	FALSE	FALSE	FALSE	FALSE	FALSE	FALSE	TRUE	TRUE	FALSE	FALSE	FALSE
CLO75	FALSE	FALSE	FALSE	TRUE	FALSE	FALSE	FALSE	FALSE	FALSE	TRUE	TRUE	FALSE	FALSE	FALSE	FALSE	FALSE
ODN2216	TRUE	FALSE	FALSE	FALSE	FALSE	FALSE	FALSE	FALSE	FALSE	FALSE	FALSE	TRUE	TRUE	FALSE	FALSE	FALSE
MDP	FALSE	FALSE	FALSE	FALSE	FALSE	FALSE	FALSE	FALSE	FALSE	TRUE	TRUE	TRUE	FALSE	FALSE	FALSE	FALSE

For the subsequent experiments, we focused on a seven-cytokine panel which includes a combination of the following signature cytokines (IL-1α, MIP-1α, IP-10, MCP-1, IL-6, and IL-8) representing all tested IIRMIs, and one cytokine (PGE-2) representing the response to TP (Figure 5, Figures S11 and S12). Interestingly, the majority of IIRMIs that activate membrane-tethered TLRs, induced MCP-1 production, with the remaining IIRMI, zymosan, instead inducing MIP-1α expression rather than the expected IL-1α. For the cytosolic PRRs, the overwhelming response was IL-8 expression. For the endosomal PRRs, we again observed that there was no cytokine consistency, with the highest cytokine expression covering MIP-1α, IL-8, and IL-6, with two of the five IIRMIs inducing high levels of MCP-1. For the majority of the IIRMIs, there was little or no IL-1α, IP-10, or PGE-2 expression detected (Figure 5, Figures S11 and S12). However, PGE-2 production had a dose dependent response when PBMCs were treated with increasing concentrations of TP alone (Figure 6). This effect was previously observed (Figure 1A) indicating that PGE-2 is still a hallmark cytokine for tracking TP immunostimulatory activity.

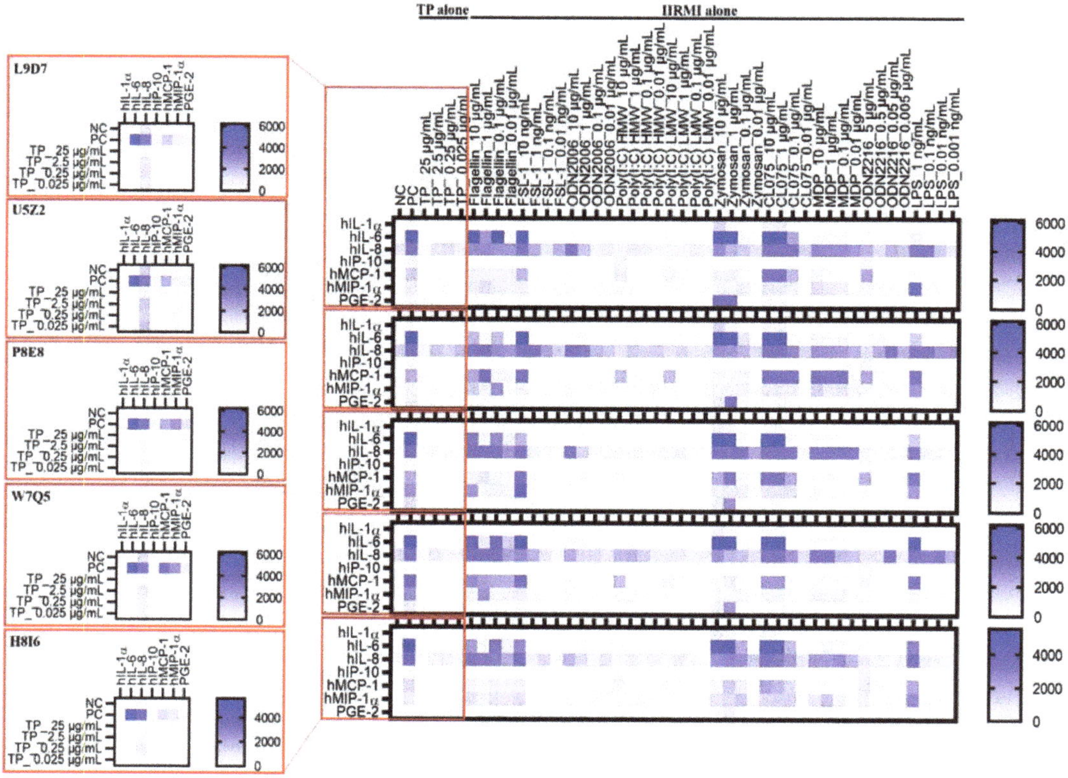

Figure 5. Seven-plex Induction of Cytokines in PBMCs. PBMCs collected from 10 healthy human donors were treated with 0.025, 0.25, 2.5, and 25 μg/mL TP (red box) or IIRMIs alone, compared to a PBS negative control (NC) and LPS/PHA-M/ODN positive control (PC), for 24 h. Supernatants were analyzed for the presence of cytokines by multiplex ELISA. Shown is the mean response (N = 2). Shown here are the data generated using PBMC cultures of five representative donors. The data generated using PBMCs of the remaining five donors are presented in Figure S11. Normalized data for each treatment set in all ten donors are also presented in Figure S12.

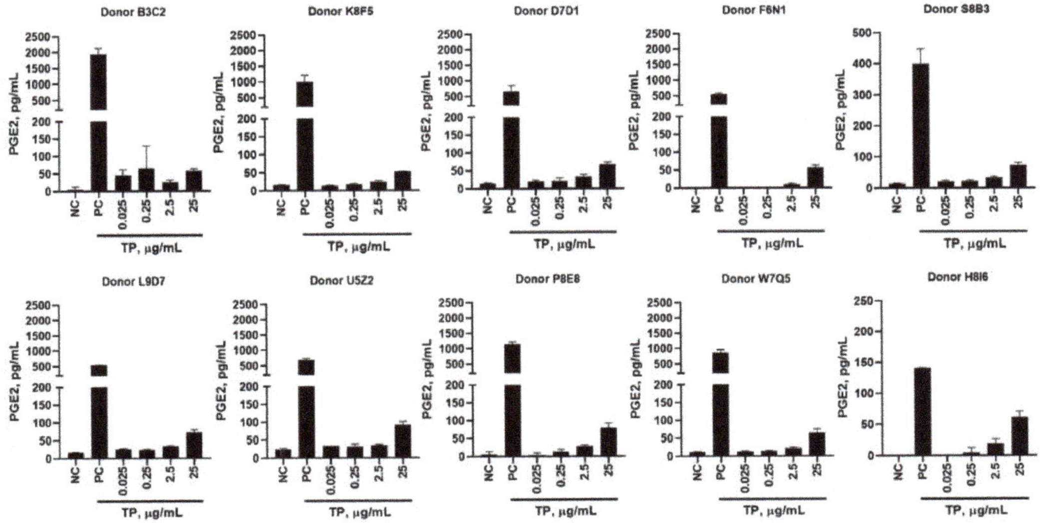

Figure 6. Seven-plex Induction of Prostaglandin-E_2 by Teriparatide. PBMCs from 10 healthy human donors were treated with 0.025, 0.25, 2.5, and 25 µg/mL teriparatide (TP), compared to a PBS negative control (NC) and LPS/PHA-M/ODN positive control (PC), for 24 h. Supernatants were analyzed for the presence of PGE-2 by 7-plex multiplex ELISA. Each bar shows mean and standard deviation (N = 2).

2.7. Teriparatide Affects Expression of IIRMI-Induced Cytokines

The presence of TP in cell cultures affected the induction of cytokines by individual IIRMIs (Table 4; Figure S8). Euclidian distance and Ward's clustering analysis demonstrated that the patterns for chemokines IL-8 and MIP-1α did not change with the addition of TP (right half of the plot) (Figure 4). In contrast, the group of IL-1β, IL-1α, and IL-12, which showed strong responses to the higher concentration of zymosan and CLO75, was strongly inhibited by the addition of TP. The loss of response with TP was also seen at the highest concentration of IIRMI for cytokines IFNλ and IFNα. PGE-2 induced by two higher concentrations of zymosan was also lost with the addition of TP (Figure 4).

Table 4. Teriparatide Affects Cytokines Induced by Innate Immune Response Modulating Impurities (IIRMIs). Individual IIRMIs and IIRMI-triggered cytokines in which expression is affected by the presence of 25 µg/mL of teriparatide (TP) are summarized in the table. In the presence of TP, all cytokines shown in the table are inhibited, except for the cytokines highlighted with an asterisk (*); levels of these cytokines are higher in the presence of TP. Statistical analysis included a two-sided Wilcoxon test.

IIRMI	IIRMI-Induced Cytokines Affected by TP
B. subtilis flagellin	IFNα, IL-1α, IL-1β, IL-6, MIP-1α, TNFα, PGE-2 *
FSL-1	IFNα, IFNγ, IFNλ, IP-10, IL-1α, IL-1β, IL-2, IL-6, TNFα
ODN2006 Class B	IFNα, IP-10, TNFα, PGE-2 *
Poly(I:C) HMW	IFNγ, IFNλ, IL-12, IP-10, MIP-1α, TNFα, PGE-2 *
Poly(I:C) LMW	IFNγ, IP-10, MCP-1, MIP-1α, TNFα, PGE-2 *
Zymosan	IFNα, IFNγ, IFNλ, IL-10, IL-12, IL-1α, IL-1β, IL-6, MCP-1, MIP-1α *, PGE-2
CLO75	IFNα, IFNγ, IFNλ, IL-10, IL-1α, IL-1β, IP-10, PGE-2 *
MDP	IL-1α, IL-1β, MIP-1α, TNFα, PGE-2 *
ODN2216	IFNα, IFNγ, IFNλ, IL-1α, MIP-1α, TNFα, PGE-2 *
E. coli O111:B4 LPS	IFNγ, IFNλ, IL-1β

2.8. Teriparatide Effects on IIRMI-Induced Cytokines Are Due to the Formulation Buffer (FB)

To understand whether the suppression of IIRMI-induced cytokines by TP was due to the API or FB, we conducted a follow-up experiment in which four concentrations of TP were tested side-by-side with the second highest concentration of IIRMIs alone, as well as IIRMIs in combination with either 25 μg/mL TP or equivalent 25 μg/mL FB. The results of this experiment demonstrated that changes in the expression of IIRMI-induced cytokines by TP were due to the FB (Figure S13).

2.9. Donor's Genetic Background Determines the Magnitude of Cytokine Response to IIRMIs

We observed that PBMCs from some donors demonstrated more robust (i.e., higher magnitude) responses to TP than cultures from other healthy donors (Figure S4). To understand whether such differences were due to the genetic background of the PBMC donor or variability in the day-to-day handling of donor's blood and PBMCs, we recalled one highest responder (donor G9L1) and two average responders (donors M4W2 and C9M4) for the second time, repeated the TP treatments, and compared the results between two experiments. The results were consistent between the two experiments despite some variability in the individual cytokine levels observed in all donors (Figure S14).

2.10. Influence of Assay Logistics on Cytokine Responses to TP and IIRMIs

We further examined the influence of blood handling and storage conditions on resultant cytokine responses to IIRMIs or TP. Blood from ten healthy donors was separated into six treatment groups: freshly isolated and freshly treated PBMCs; freshly isolated PBMCs cultured for 24 h prior to the treatment; freshly isolated and cryopreserved PBMCs; PBMCs isolated from blood refrigerated for 24 h or 48 h before PBMC isolation; and whole blood cultures. All groups were then dosed with IIRMIs or TP for 24 h. We then measured PBMC recovery and viability for these treatment groups (Figure 7) as well as the levels of our seven key cytokines, IL-1α, MIP-1α, IP-10, MCP-1, IL-6, and IL-8 (Figures 5, 8, S11, S12 and S15).

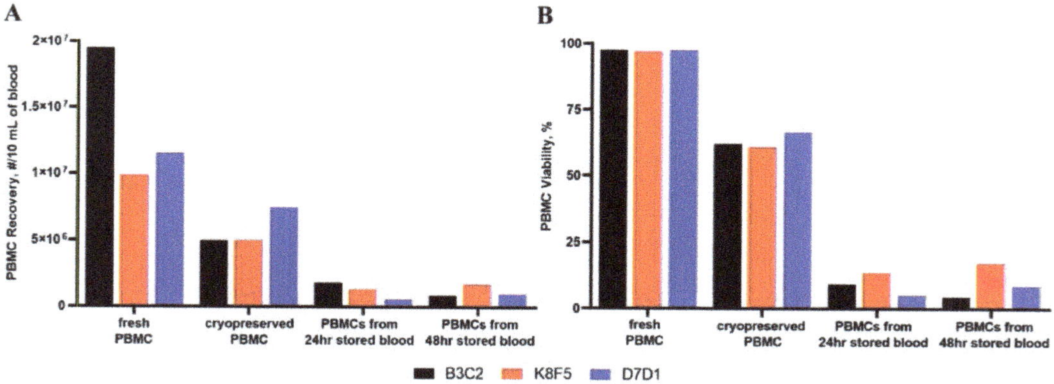

Figure 7. The Effect of Storage Conditions on PBMC Viability and Cell Recovery. To simulate various handling and storage conditions used in research, PBMCs from three healthy human donors were examined after fresh isolation, cryopreservation, and isolation from refrigerated blood (24 h or 48 h). Cell viability was then assessed using AO/PI. (**A**) Number of PBMCs recovered under the various storage/handling conditions. (**B**) Viability of stored PBMCs compared to their freshly isolated PBMC counterparts. Each bar shows the mean result and standard deviation (N = 3).

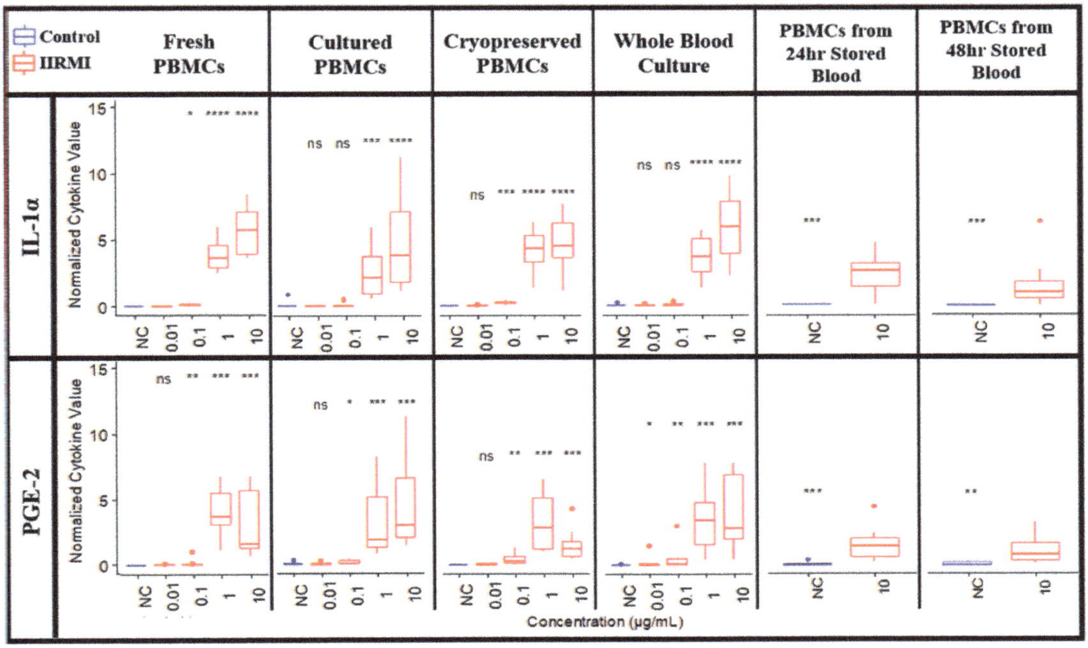

Figure 8. IL-1α and PGE-2 Responses to Zymosan are Affected by PBMC and Blood Handling Conditions. PBMCs from 10 healthy human donors were exposed to various common laboratory handling conditions (isolated from fresh blood, cultured for 24 h, cryopreserved, isolated from blood refrigerated for 24 h or 48 h, and whole blood cultures) before being treated with IIRMIs for 24 h. Supernatants were analyzed for the presence of cytokines. Shown are the mean cytokine responses to zymosan (red), compared to a PBS negative control (NC, blue). The data for which statistical significance was not observed are marked with ns. Statistical significance is shown with an asterisk as follows: * $p < 0.05$; ** $p < 0.01$; *** $p < 0.001$; and **** $p < 0.0001$. Additional results, including zymosan-induced levels of the remaining five cytokines and the cytokine responses for the other nine IIRMIs, are available in Figure 5 (fresh PBMCs) and Figure S15 (all other experimental conditions).

Compared to freshly isolated PBMCs, PBMC viability is reduced to approximately 63% after 48 h of cryopreservation, as compared to the very low viability (~10%) of PBMCs isolated from anti-coagulated blood after refrigeration storage for 24 h or 48 h (Figure 7B). Due to the loss of 90% of usable PBMCs from the stored blood samples, we were only able to treat the remaining cells with a limited selection of IIRMIs for comparison to the other treatment/storage conditions. In addition, this loss of available cells can potentially skew the resultant cytokine production (Figure 7A).

As previously discussed, there was very little general expression of IL-1α, IP-10, or PGE-2 detected even for freshly isolated PBMCs (Figure 5, Figure 8, Figures S11, S12 and S15). Interestingly, the highest levels of IL-1α and PGE-2 were observed after zymosan stimulation in whole blood cultures, indicating that other components of blood may be responsible for increasing the levels of these cytokines.

For the other four cytokines (MCP-1, MIP-1α, IL-8, and IL-6), cultured PBMCs and cryopreserved PBMCs had similar but reduced levels of cytokines compared to freshly isolated PBMCs. Cytokines from refrigerated blood further reduced cytokine levels, even at the highest IIRMI concentrations. This was especially true for IL-6, which were reduced to almost nothing even in the presence of strong LPS or zymosan stimulation (Figures 5, 8, S11, S12 and S15).

3. Discussion

Based on the results of these initial characterization studies (Table 1), complement activation and leukocyte-proliferation assays were not chosen for subsequent studies as they cannot adequately detect potential differences in IIRMI contamination between different batches of product. These assays, however, could be helpful in studies investigating different formulations of the same API. Examples may include when a product is reformulated, or when a generic or follow-on product elects to have differences in formulation compared to an innovator (reference) product. Therefore, we focused the rest of the study on the cytokine secretion by PBMC after in vitro exposure to TP and formulations containing IIRMIs.

Our study suggested that PGE-2 could be used as a signature cytokine for tracking TP induction of innate immune responses (Figures 1A and 6). We further found that this response is mediated by the FB rather than API. Further investigation found that, unlike our hypothesis about the influence of metacresol, all the FB ingredients contributed to the resultant cytokine response (Figure 2A,C; Figures S5 and S6).

IIRMIs activated a broad and often overlapping spectrum of cytokines (Figures 3, S4, S7 and S8). This finding is consistent with the current literature about PRRs and their cognate ligands [23–25]. Using Euclidian distance and Ward's clustering analyses, we obtained insight on the patterns of IIRMI stimulation and the resultant induced cytokine responses. From these results, we identified groupings of cytokines with similar response patterns across all IIRMI treatments (Figure 4), as well as several cytokines which did not appear to be strongly induced by any IIRMI. These analyses also demonstrated that the highest concentrations of each IIRMI often clustered together (Figure S9).

Further Pearson's correlation analysis allowed for clustering the cytokine responses based on how well cytokine values correlated across all treatment groups and donors (Figure S10A). Strong correlations patterns identified during these analyses were consistent with the currently available literature about the function of these cytokines and the cells that produce them. Specifically, IL-6 and TNFα are produced by monocytes and T-cells, and are responsible for pyrogenicity; IL-8 and MIP-1α are chemokines produced by monocytes and responsible for neutrophils and mixed leukocyte recruitment; IL-1α, IL-1β, and IL-12 are produced by monocytes and DCs and are responsible for the inflammation, fever, and activation of specific subsets of lymphocytes (i.e., IL-1β promotes T_H17 differentiation, whereas IL-12 supports T_H1 differentiation, NK and T-cell activation to increase IFNγ synthesis and increased cytotoxicity); in addition, IL-1α is a danger signal that indicates damaging effects of IIRMIs that induce its secretion [14]. According to our expectations from the global heatmaps (Figure 4 and Figure S9), Pearson's correlation between the different concentrations of the same IIRMI was consistent with the current knowledge about type and intracellular localization of PRRs stimulated by these IIRMIs (Figure S10B). Specifically, ODN2216, poly(I:C) LMW, and poly(I:C) HMW activate endosomal TLRs (TLR9 and TLR3), whereas zymosan triggers membrane-tethered PRRs (TLR2 and Dectin1) [25,26]. In contrast, CLO75, which showed a lower degree of anti-correlation, is also located in the endosome but is specific to a different PRR (TLR8) [27].

To understand whether our 16-cytokine panel could be narrowed down to three or four cytokines that would be representative of all 10 IIRMIs, we examined the top three "winning" cytokines identified for each IIRMI (Table 3). Due to the overlapping nature of the induced cytokines, we identified two possible panels of three cytokines which would provide at least one positive result for all 10 IIRMIs and potentially could be used by users who do not have access to more than a 3- or 4-plex cytokine detection panel. These panels include the following markers: 1) IL-1α (or MIP-1α), IP-10, and IL-8; or 2) IL-1α (or MIP-1α), MCP-1 and IL-8 (or IL-6).

The results from the subsequent 7-plex panel containing IL-1α, MIP-1α, IP-10, MCP-1, IL-6, IL-8, and PGE-2 (Figure 5, Figures S11 and S12), suggest that our initial 16-cytokine panel can be reduced to a four-cytokine panel, specifically containing MCP-1, MIP-1α, IL-8, and IL-6, which would be representative of all 10 IIRMIs, which can further be expanded to a five-cytokine TP-specific panel, which includes the TP-signature cytokine, PGE-2, in addition to the four IIRMI-specific cytokines.

TP did not significantly increase the levels of cytokines induced by IIRMIs. TP was, instead, found to decrease the levels of most IIRMI-induced cytokines (Figure 4). Reduced levels of IIRMI-induced cytokine responses in the presence of TP were also the result of the FB rather than the API (Figure S13). Collectively, this finding and the data demonstrating the induction of TP signature cytokine PGE-2 by the FB suggests that the assessment of potential IIRMI contamination of the API could be more informative for comparison of RLD and generic formulations. This data also suggests that a change in the formulation buffer may result in a change in the signature cytokine of the whole product.

The more robust cytokine responses to TP demonstrated by some donors suggests that day-to-day variability in phlebotomy and handling of whole blood and PBMCs may result in quantitative differences (i.e., influence the magnitude of the responses) but would not change the overall qualitative trends and resultant conclusions of the study. However, the genetic background of donors that donate their blood for in vitro experiments does appear to be an important factor in qualitative determination of the PBMC response to individual IIRMIs (Figure S14).

Overall, the PBMC handling and blood storage conditions have a significant effect on the detectable levels of cytokines, with freshly isolated PBMCs being the most preferred condition since it allows for more adequate detection of cytokines as a result of innate immunity activation (Figure 5, Figure 8, Figures S11, S12 and S15).

4. Materials and Methods

4.1. Materials

Feraheme (FH) (AMAG Pharmaceuticals, Waltham, MA) and Forteo™ (teriparatide, TP) (Eli Lilly, Indianapolis, IN, USA), were obtained from NIH Pharmacy. All *Limulus* amebocyte lysate (LAL) reagents, LAL grade (endotoxin free) water, Glucatell kits, Glucashield buffer, and *E. coli* lipopolysaccharide (LPS) were from Associates of Cape Cod (East Falmouth, MA, USA). Veronal Buffer was obtained from Boston BioProducts (Ashland, MA, USA). Phosphate Buffered Saline (PBS), RPMI-1640 media, fetal bovine serum (FBS), penicillin and streptomycin solution, L-glutamine, Ficoll-Paque Premium was from GE Life Sciences (Marlborough, MA, USA). Hank's balanced salt solution (HBSS) was from Gibco (Gaithersburg, MD). All IIRMIs—*B. subtilis* flagellin, FSL-1, ODN2006 Class B, poly(I:C) HMW, poly(I:C) LMW, zymosan, CLO75, MDP, ODN2216, and *E. coli* O111:B4 LPS—were from Invivogen (San Diego, CA, USA). Acridine orange (AO)/propidium iodide (PI) staining solution were purchased from Nexcelom Bioscience (Lawrence, MA, USA). The 16-plex and 7-plex cytokine multiplex kits were supplied by Quansys Biosciences (Logan, UT, USA). Cobra venom factor (CVF), Heat Aggregated Gamma Globulins (HAGG), and MicroVue EIA kits were purchased from Quidel Corporation (San Diego, CA, USA). Glacial acetic acid, sodium acetate, mannitol, sodium hydroxide (NaOH), hydrochloric acid (HCl), MTT (3-(4,5-dimethyl-2-thiazolyl)-2,5-diphenyl-2H-tetrazolium bromide), glycine, sodium chloride, dimethyl sulfoxide (DMSO), Phytohemagglutinin (PHA-M), and Cremophor (Cre) were purchased from Sigma-Aldrich (Burlington, MA, USA). Metacresol was from USP (Frederick, MD, USA).

4.2. Innate Immune Response Modulating Impurities

Ten model innate immune response modulating impurities (IIRMIs) were tested at four concentrations (Table 5) either alone or in combination with teriparatide (TP). Eight IIRMIs (*B. subtilis* flagellin, FSL-1, zymosan, ODN2006, poly(I:C) HMW, poly(I:C) LMW, CLO75, and MDP) were selected based on preliminary studies in HEK-TLR reporter cells [9,10]; two other IIRMIs (ODN2216 and *E. coli* O111:B4 LPS) were selected based on the Nanotechnology Characterization Laboratory (NCL) (https://ncl.cancer.gov/, accessed on 15 October 2020) prior experience using them as immunological assay cascade positive controls. Taken together, these ten IIRMIs bind Dectin 1, TLRs 2, 3, 4, 5, 6, 8, 9, and NOD2, as summarized in Table 5.

Table 5. Innate Immune Response Modulating Impurities used in the present study. IIRMIs and their final concentrations tested in vitro are summarized. LPS = lipopolysaccharide; CLO = thiazoloquinolone derivative; MDP = muramyldipeptide; ODN = oligo deoxyribonucleotide; LMW = low molecular weight; HMW = high molecular weight; FSL = Pam2CGDPKHPKSF, a synthetic lipopeptide derived from *Mycoplasma salivarium*.

Reagent	PRR	Final Concentrations per mL
B. subtilis flagellin	TLR5	10 μg, 1 μg, 100 ng, 10 ng
FSL-1	TLR2/TLR6	10 ng, 1 ng, 100 pg, 10 pg
ODN2006 Class B	TLR9	10 μg, 1 μg, 100 ng, 10 ng
Poly(I:C) HMW	TLR3	10 μg, 1 μg, 100 ng, 10 ng
Poly(I:C) LMW	TLR3	10 μg, 1 μg, 100 ng, 10 ng
Zymosan	TLR2/Dectin 1	10 μg, 1 μg, 100 ng, 10 ng
CLO75	TLR8	10 μg, 1 μg, 100 ng, 10 ng
MDP	NOD2	10 μg, 1 μg, 100 ng, 10 ng
ODN2216	TLR9	5 μg, 500 ng, 50 ng, 5 ng
E. coli O111:B4 LPS	TLR4	1 ng, 100 pg, 10 pg, 1 pg

4.3. Endotoxin Detection

Endotoxin levels were evaluated using the kinetic turbidity *Limulus* Amebocyte Lysate (LAL) Assay according to NCL protocol STE-1.2 [28,29]. Briefly, 100 μL of TP (at 250 μg/mL) and the equivalent amount of its formulation buffer (FB) were each mixed with 100 μL of LAL reagent in a glass tube, then measured via spectrophotometer at 660 nm for at least 7200 sec for appropriate development. Using a standard curve prepared with Control Standard Endotoxin of known potency, we calculated the concentration of endotoxin present in the TP and FB solutions.

4.4. β-Glucan Detection

Levels of β-glucans were evaluated using Glucatell® kit as detailed in NCL protocol STE-4 [15,30]. Briefly, 50 μL of TP (at 250 μg/mL) and the equivalent amount of its formulation buffer (FB) were each mixed with 50 μL of Glucatell reagent in a 96-well plate and incubated at 37 °C. The reaction was stopped through the addition of 50 μL of 1N HCl-sodium nitrite solution, 50 μL of ammonium sulfamate solution, and then 50 μL of NEDA solution to each well. Color development was immediately observed and measured at 540–550 nm using a spectrophotometer. Using a β-(1,3)-D-glucan standard curve, we calculated the concentration of β-(1,3)-D-glucan present in the TP and FB solutions.

4.5. Donor Blood

Blood from healthy human donors was collected in vacutainers containing either Li-heparin or K_2-EDTA (BD Biosciences, San Jose, CA, USA) under the NCI-Frederick protocol OH9-C-N046. At the time of blood collection, donors were not on any medications and have never been exposed to the model Forteo™ teriparatide formulation.

4.6. Peripheral Blood Mononuclear Cell (PBMC) Isolation and Culture

Fresh donor blood anti-coagulated with Li-heparin was mixed with an equal volume of room-temperature PBS. The blood/PBS mixture was then slowly layered on top of Ficoll-Paque solution in a 4:3 ratio. The sample was centrifuged for 30 min at 900× *g*, 18–20 °C, without brake. After centrifugation, the upper layer containing plasma and platelets was removed and discarded. The mononuclear cell layer was isolated and washed using an excess (approximately three times volume) of Hank's Balanced Salt Solution (HBSS) and centrifuged for 10–15 min at 400× *g*, 18–20 °C. After washing, the supernatant was discarded, and the wash step was repeated once more. The remaining mononuclear cells were then resuspended in complete RPMI-1640 medium, containing 10% FBS (heat inactivated), 2 mM L-glutamine, 100 U/mL penicillin, and 100 μg/mL streptomycin. Cell viability was then determined using the acridine orange (AO)/propidium iodide (PI) dual-fluorescence viability method, in which an equal volume of staining solution, containing AO (live cells, green) and PI (dead cells, red) was added to cells and analyzed in <60 s using a fluorescent Cellometer instrument. The details of the protocols are publicly available through NCL protocol ITA-10 and were previously described [31,32].

4.7. PBMC Cryopreservation

Isolated PBMCs were resuspended at a concentration of 5–7.5 $\times$ 10^6 cells/mL in freezing media (10% DMSO in FBS), placed into cryopreservation tubes, and stored in a freezing container containing isopropanol for controlled freezing at −80 °C.

4.8. Whole Blood Cell (WBC) Culture

Fresh donor blood anti-coagulated with Li-heparin was mixed 1:4 with room-temperature PBS (e.g., 10 mL blood added to 30 mL PBS). The blood/PBS mixture was then added directly to 96-well plate for treatment and culture at 37 °C. The details of the protocols are publicly available through NCL protocol ITA-10 and were previously described [31,32].

4.9. Teriparatide Cytotoxicity Analysis

PBMCs in complete 1640-RPMI were incubated with 0–50 μg/mL teriparatide (TP) for 24 h. Cell viability was then determined using the AO/PI staining method [33].

4.10. Leukocyte Proliferation

PBMCs were cultured at in the presence of controls, 0.025–25 μg/mL TP, four concentrations of IIRMIs (Table 5), or four concentrations of IIRMI + 25 μg/mL TP for 72 h. The proliferation of leukocytes was determined according to NCL protocol ITA-6 [34].

4.11. Complement Activation

These experiments were conducted according to NCL protocol ITA5.2 [35]. Briefly, K_2-EDTA plasma from individual donors was pooled and incubated with controls or 0.025–83.3 μg/mL TP, and veronal buffer for 30 min at 37 °C. Following incubation, the samples were analyzed for the presence of complement split product C3a using a commercial multiplex ELISA kit. In this experiment, Cobra venom factor (CVF) and Heat Aggregated Gamma Globulins (HAGG) were used as the assay positive controls (PC). Cremophor (Cre) and Feraheme (FH) were included as additional controls as they are known to cause complement-mediated toxicity in sensitive patients [36–39].

4.12. Cytokine Production

These experiments followed NCL protocol ITA-10 [31,32]. PBMCs were cultured at in the presence of PBC negative control, LPS/PHA-M/ODN positive control, 0.025–25 μg/mL TP, four concentrations of IIRMIs (Table 5), or four concentrations of IIRMI + 25 μg/mL TP for 24 h in a humidified 37 °C, 5% CO_2 incubator. After incubation, the plates were centrifuged for 5 min at 700$\times$ g to pellet the PBMCs. The supernatants were collected for cytokine analysis using custom 16-plex or 7-plex multiplex plates from Quansys Biosciences (Logan, UT, USA). The cytokines present in the multiplex panel included type I interferon (IFNα), type II interferon (IFNγ), type III interferon (IFNλ), interleukins (IL-1α, IL-1β, IL-2, IL-6, IL-8, IL-10, IL-12, IL-17), interferon-gamma inducible protein (IP-10), tumor necrosis factor alpha (TNFα), prostaglandin-E_2 (PGE-2), macrophage inflammatory protein (MIP-1α), and monocyte chemoattractant protein (MCP-1). Cytokine levels were each quantified against a standard curve of calibrator controls (provided in the Quansys kit).

4.13. Statistical Analysis

All experiments were performed with at least two independent samples, tested in duplicate (%CV < 25). Unless otherwise stated, results show the mean and standard deviation generated from these independent samples. For the cytokine multiplex assay, the analysis was performed using custom R scripts. Cytokine concentration values above the detection limit ("ADL") were set to the upper detection limit, and values below the detection limit ("BDL") were set to zero. Statistical analysis of cytokine data was performed using normalized values. The normalization included scaling each cytokine reading across all collected values by dividing each value by that cytokine's standard deviation obtained across all donors. The normalization brought all cytokines onto roughly the same scale. The benefit of using this approach is that one can compare cytokines directly on graphs across 10 donors, and it keeps cytokines with very large values from swamping global analyses. As an initial quality control (QC) step, we looked at the negative control (NC) vs. positive control (PC) values for each cytokine using the Wilcoxon non-parametric test on replicate-averaged cytokine-normalized values. All cytokines showed significantly higher PC than NC values. Additionally, we looked at the correlation between pairs of replicate runs (a vs. b for each treatment) and observed a good correlation for most pairs. Unless otherwise noted, comparisons between cytokine levels were made on normalized values using two-sided Wilcoxon tests.

5. Conclusions

Cytokine secretion by human PBMCs may be used to assess the innate immune responses to IIRMIs, formulation components, and whole products containing peptide and protein therapeutics. While the whole product needs to be analyzed, the results of our study emphasize that the components of FB are not immunologically inert and can contribute to both the cytokine stimulation by the whole product and inhibition of the IIRMI-mediated cytokines. Statistical analysis helps to identify signature cytokines and select cytokine panel appropriate for the given peptide drug product and any prospective generics and biosimilars. It is expected that signature cytokines maybe different between different products due to differences in formulation components, potential IIRMI contamination, immunological properties of API, and interactions among them, which collectively may lead to both quantitative and qualitative differences. Importantly, the logistics of blood storage and handling may influence the results, and, therefore, should be carefully investigated during assay validation phase.

Supplementary Materials: The following are available online, Table S1: Endotoxin and β-glucan Levels in Teriparatide Formulation, Figure S1: PBMC Viability in the Presence of Teriparatide, Figure S2: In vitro Leukocyte Proliferation in the Presence of Teriparatide and/or Innate Immune Response Modulating Impurities, Figure S3: In vitro Complement Activation Induced by Teriparatide, Figure S4: 16-plex Induction of Cytokines in PBMCs, Figure S5: Formulation Buffer is Responsible for the Cytokine Response to Teriparatide, Figure S6: Metacresol and Mannitol are Responsible for the Formulation Buffer Cytokine Response, Figure S7: Normalized 16-plex Cytokine Response to Innate Immune Response Modulating Impurities and Selection of One Signature Cytokine, Figure S8: Normalized 16-plex Cytokine Response in the Combined Presence of Innate Immune Response Modulating Impurities and Teriparatide, Figure S9: Innate Immune Response Modulating Impurity-Induced Cytokine Response Patterns via Euclidian Distance and Ward's Clustering, Figure S10: Cytokine Analysis via Pearson's Correlation, Figure S11: 7-plex Induction of Cytokines in PBMCs, Figure S12: Normalized 7-plex Cytokine Response to Innate Immune Response Modulating Impurities, Figure S13: Formulation Buffer Affects Cytokines Induced by Innate Immune Response Modulating Impurities, Figure S14: Reproducibility of Cytokine Response to Teriparatide in PBMC Cultures, Figure S15: Normalized Cytokine Responses to Innate Immune Response Modulating Impurities are Affected by PBMC and Blood Handling Conditions.

Author Contributions: C.K.H., E.C. and B.W.N. conducted experiments. D.D. performed statistical analysis. C.K.H., E.C. and D.D. contributed equally to this study. M.A.D., E.S.P. and D.V. conceived, anddesigned the study. M.A.D. managed the study. All authors analyzed data and wrote the manuscript. All authors have read and agreed to the published version of the manuscript.

Funding: This study was supported in whole by federal funds from the U.S. Food and Drug Administration, Center for Drug Evaluation and Research/Office of Generic Drugs, under an Inter-Agency Agreement with NCL/NCI/NIH (IAA #224-19-3008S).

Institutional Review Board Statement: Blood from healthy human donors was collected under the NCI-Frederick protocol OH9-C-N046.

Informed Consent Statement: Informed consent was obtained from all human subjects involved in donating blood for this study.

Data Availability Statement: Materials used in this study are available from commercial vendors; the data used to produce figures in this manuscript maybe requested by contacting principal investigators of the study.

Conflicts of Interest: The authors declare no conflict of interest.

References

1. Lee, S.; Raw, A.; Yu, L.; Lionberger, R.; Ya, N.; Verthelyi, D.; Rosenberg, A.; Kozlowski, S.; Webber, K.; Woodcock, J. Scientific considerations in the review and approval of generic enoxaparin in the United States. *Nat. Biotechnol.* **2013**, *31*, 220–226. [CrossRef] [PubMed]
2. Wang, J.; Lozier, J.; Johnson, G.; Kirshner, S.; Verthelyi, D.; Pariser, A.; Shores, E.; Rosenberg, A. Neutralizing antibodies to therapeutic enzymes: Considerations for testing, prevention and treatment. *Nat. Biotechnol.* **2008**, *26*, 901–908. [CrossRef] [PubMed]
3. Van Regenmortel, M.H. Antigenicity and immunogenicity of synthetic peptides. *Biologicals* **2001**, *29*, 209–213. [CrossRef]
4. Jawa, V.; Cousens, L.P.; Awwad, M.; Wakshull, E.; Kropshofer, H.; De Groot, A.S. T-cell dependent immunogenicity of protein therapeutics: Preclinical assessment and mitigation. *Clin. Immunol.* **2013**, *149*, 534–555. [CrossRef] [PubMed]
5. Kessler, M.; Goldsmith, D.; Schellekens, H. Immunogenicity of biopharmaceuticals. *Nephrol. Dial. Transpl.* **2006**, *21* (Suppl. 5), v9–v12. [CrossRef]
6. Büttel, I.C.; Chamberlain, P.; Chowers, Y.; Ehmann, F.; Greinacher, A.; Jefferis, R.; Kramer, D.; Kropshofer, H.; Lloyd, P.; Lubiniecki, A.; et al. Taking immunogenicity assessment of therapeutic proteins to the next level. *Biologicals* **2011**, *39*, 100–109. [CrossRef] [PubMed]
7. Chamberlain, P.; Mire-Sluis, A.R. An overview of scientific and regulatory issues for the immunogenicity of biological products. *Dev. Biol.* **2003**, *112*, 3–11.
8. Shankar, G.; Shores, E.; Wagner, C.; Mire-Sluis, A. Scientific and regulatory considerations on the immunogenicity of biologics. *Trends Biotechnol.* **2006**, *24*, 274–280. [CrossRef]
9. Haile, L.A.; Puig, M.; Kelley-Baker, L.; Verthelyi, D. Detection of innate immune response modulating impurities in therapeutic proteins. *PLoS ONE* **2015**, *10*, e0125078. [CrossRef]
10. Haile, L.A.; Puig, M.; Polumuri, S.K.; Ascher, J.; Verthelyi, D. In Vivo Effect of Innate Immune Response Modulating Impurities on the Skin Milieu Using a Macaque Model: Impact on Product Immunogenicity. *J. Pharm. Sci.* **2017**, *106*, 751–760. [CrossRef]
11. Polumuri, S.K.; Haile, L.A.; Ireland, D.D.C.; Verthelyi, D. Aggregates of IVIG or Avastin, but not HSA, modify the response to model innate immune response modulating impurities. *Sci. Rep.* **2018**, *8*, 11477. [CrossRef]
12. Verthelyi, D.; Wang, V. Trace levels of innate immune response modulating impurities (IIRMIs) synergize to break tolerance to therapeutic proteins. *PLoS ONE* **2010**, *5*, e15252. [CrossRef]
13. Carpenter, J.; Cherney, B.; Lubinecki, A.; Ma, S.; Marszal, E.; Mire-Sluis, A.; Nikolai, T.; Novak, J.; Ragheb, J.; Simak, J. Meeting report on protein particles and immunogenicity of therapeutic proteins: Filling in the gaps in risk evaluation and mitigation. *Biologicals* **2010**, *38*, 602–611. [CrossRef] [PubMed]
14. Abbas, A.K.; Lichtman, A.H.; Pillai, S. *Cellular and Molecular Immunology E-Book*; Elsevier Health Sciences: London, UK, 2014.
15. Neun, B.W.; Cedrone, E.; Potter, T.M.; Crist, R.M.; Dobrovolskaia, M.A. Detection of Beta-Glucan Contamination in Nanotechnology-Based Formulations. *Molecules* **2020**, *25*, 3367. [CrossRef] [PubMed]
16. Dobrovolskaia, M.A. Pre-clinical immunotoxicity studies of nanotechnology-formulated drugs: Challenges, considerations and strategy. *J. Control. Release* **2015**, *220*, 571–583. [CrossRef]
17. Dobrovolskaia, M.A.; Vogel, S.N. Toll receptors, CD14, and macrophage activation and deactivation by LPS. *Microbes Infect.* **2002**, *4*, 903–914. [CrossRef]
18. Castellheim, A.; Lindenskov, P.H.; Pharo, A.; Fung, M.; Saugstad, O.D.; Mollnes, T.E. Meconium is a potent activator of complement in human serum and in piglets. *Pediatr. Res.* **2004**, *55*, 310–318. [CrossRef] [PubMed]

19. Szebeni, J.; Wassef, N.M.; Rudolph, A.S.; Alving, C.R. Complement activation by liposome-encapsulated hemoglobin in vitro: The role of endotoxin contamination. *Artif. Cells Blood Substit. Immobil. Biotechnol.* **1995**, *23*, 355–363. [CrossRef] [PubMed]
20. Weber, C.; Kammerer, D.; Streit, B.; Licht, A.H. Phenolic excipients of insulin formulations induce cell death, pro-inflammatory signaling and MCP-1 release. *Toxicol. Rep.* **2015**, *2*, 194–202. [CrossRef]
21. Shapiro, L.; Dinarello, C.A. Hyperosmotic stress as a stimulant for proinflammatory cytokine production. *Exp. Cell Res.* **1997**, *231*, 354–362. [CrossRef]
22. Malek, A.M.; Goss, G.G.; Jiang, L.; Izumo, S.; Alper, S.L. Mannitol at clinical concentrations activates multiple signaling pathways and induces apoptosis in endothelial cells. *Stroke* **1998**, *29*, 2631–2640. [CrossRef] [PubMed]
23. Mogensen, T.H. Pathogen recognition and inflammatory signaling in innate immune defenses. *Clin. Microbiol. Rev.* **2009**, *22*, 240–273. [CrossRef]
24. Ozato, K.; Tsujimura, H.; Tamura, T. Toll-like receptor signaling and regulation of cytokine gene expression in the immune system. *Biotechniques* **2002**, *33*, S66–S75. [CrossRef]
25. Schaefer, T.M.; Desouza, K.; Fahey, J.V.; Beagley, K.W.; Wira, C.R. Toll-like receptor (TLR) expression and TLR-mediated cytokine/chemokine production by human uterine epithelial cells. *Immunology* **2004**, *112*, 428–436. [CrossRef] [PubMed]
26. Esteban, A.; Popp, M.W.; Vyas, V.K.; Strijbis, K.; Ploegh, H.L.; Fink, G.R. Fungal recognition is mediated by the association of dectin-1 and galectin-3 in macrophages. *Proc. Natl. Acad. Sci. USA* **2011**, *108*, 14270–14275. [CrossRef]
27. Du, J.; Wu, Z.; Ren, S.; Wei, Y.; Gao, M.; Randolph, G.J.; Qu, C. TLR8 agonists stimulate newly recruited monocyte-derived cells into potent APCs that enhance HBsAg immunogenicity. *Vaccine* **2010**, *28*, 6273–6281. [CrossRef] [PubMed]
28. Neun, B.W.; Dobrovolskaia, M.A. Detection of Endotoxin in Nano-formulations Using Limulus Amoebocyte Lysate (LAL) Assays. *J. Vis. Exp.* **2019**. [CrossRef]
29. Neun, B.W.; Dobrovolskaia, M.A. NCL Method STE-1.2: Detection and Quantification of Gram Negative Bacterial Endotoxin Contamination in Nanoparticle Formulations by Kinetic Turbidity LAL Assay. Available online: https://ncl.cancer.gov/sites/default/files/protocols/NCL_Method_STE-1.2.pdf (accessed on 15 October 2020).
30. Neun, B.W.; Dobrovolskaia, M.A. NCL Method STE-4: Detection and Quantification of β-(1,3)-D-Glucan Contamination in Nanoparticle Formulations by Factor C Depleted LAL (Glucatell®) Assay. Available online: https://ncl.cancer.gov/sites/default/files/NCL_Method_STE-4.pdf (accessed on 15 October 2020).
31. Potter, T.M.; Neun, B.W.; Rodriguez, J.C.; Ilinskaya, A.N.; Dobrovolskaia, M.A. Analysis of Pro-inflammatory Cytokine and Type II Interferon Induction by Nanoparticles. *Methods Mol. Biol.* **2018**, *1682*, 173–187. [CrossRef]
32. Potter, T.M.; Cedrone, E.; Neun, B.W.; Dobrovolskaia, M.A. NCL Method-10: Preparation of Human Whole Blood and Peripheral Blood Mononuclear Cell Cultures to Analyze Nanoparticle Potential to Induce Cytokines In Vitro. Available online: https://ncl.cancer.gov/sites/default/files/NCL_Method_ITA-10.pdf (accessed on 15 October 2020).
33. Nexcelom. Viability Using Acridine Orange/Propidium Lodide. Available online: https://www.nexcelom.com/applications/cellometer/viability/viability-using-ao-pi/ (accessed on 15 October 2020).
34. Potter, T.M.; Cedrone, E.; Skoczen, S.L.; Neun, B.W.; Dobrovolskaia, M.A. NCL Method ITA-6.1: Leukocyte Proliferation Assay. Available online: https://ncl.cancer.gov/sites/default/files/NCL_Method_ITA-6.1.pdf (accessed on 15 October 2020).
35. Neun, B.W.; Cedrone, E.; Dobrovolskaia, M.A. NCL Method STE 5.2: Analysis of Complement Activation by Single-Plex EIA or Multiplex ELISA. Available online: https://ncl.cancer.gov/sites/default/files/NCL_Method_ITA-5.2.pdf (accessed on 15 October 2020).
36. Hempel, J.C.; Poppelaars, F.; Gaya da Costa, M.; Franssen, C.F.; de Vlaam, T.P.; Daha, M.R.; Berger, S.P.; Seelen, M.A.; Gaillard, C.A. Distinct in vitro Complement Activation by Various Intravenous Iron Preparations. *Am. J. Nephrol.* **2017**, *45*, 49–59. [CrossRef]
37. Jackman, J.A.; Mészáros, T.; Fülöp, T.; Urbanics, R.; Szebeni, J.; Cho, N.J. Comparison of complement activation-related pseudoallergy in miniature and domestic pigs: Foundation of a validatable immune toxicity model. *Nanomedicine* **2016**, *12*, 933–943. [CrossRef]
38. Szebeni, J. Complement activation-related pseudoallergy: A new class of drug-induced acute immune toxicity. *Toxicology* **2005**, *216*, 106–121. [CrossRef] [PubMed]
39. Szebeni, J.; Alving, C.R.; Savay, S.; Barenholz, Y.; Priev, A.; Danino, D.; Talmon, Y. Formation of complement-activating particles in aqueous solutions of Taxol: Possible role in hypersensitivity reactions. *Int. Immunopharmacol.* **2001**, *1*, 721–735. [CrossRef]

molecules

Article

Change in Lipofectamine Carrier as a Tool to Fine-Tune Immunostimulation of Nucleic Acid Nanoparticles

Hannah S. Newton [1,†], Yasmine Radwan [2,†], Jie Xu [1], Jeffrey D. Clogston [1], Marina A. Dobrovolskaia [1,*] and Kirill A. Afonin [2,*]

1 Nanotechnology Characterization Laboratory, Cancer Research Technology Program, Frederick National Laboratory for Cancer Research Sponsored by the National Cancer Institute, Frederick, MD 21701, USA; hannah.newton@nih.gov (H.S.N.); jie.xu@nih.gov (J.X.); clogstonj@mail.nih.gov (J.D.C.)
2 Nanoscale Science Program, Department of Chemistry, University of North Carolina Charlotte, Charlotte, NC 28223, USA; yradwan@uncc.edu
* Correspondence: marina@mail.nih.gov (M.A.D.); kafonin@uncc.edu (K.A.A.)
† These authors contributed equally to this work.

Abstract: Nucleic acid nanoparticles (NANPs) require a carrier to allow for their intracellular delivery to immune cells. Cytokine production, specifically type I and III interferons, allows for reliable monitoring of the carrier effect on NANP immunostimulation. Recent studies have shown that changes in the delivery platform (e.g., lipid-based carriers vs. dendrimers) can alter NANPs' immunorecognition and downstream cytokine production in various immune cell populations. Herein, we used flow cytometry and measured cytokine induction to show how compositional variations in commercially available lipofectamine carriers impact the immunostimulatory properties of NANPs with different architectural characteristics.

Keywords: nucleic acid nanoparticles; lipofectamine; cytokine; interferons

Citation: Newton, H.S.; Radwan, Y.; Xu, J.; Clogston, J.D.; Dobrovolskaia, M.A.; Afonin, K.A. Change in Lipofectamine Carrier as a Tool to Fine-Tune Immunostimulation of Nucleic Acid Nanoparticles. *Molecules* **2023**, *28*, 4484. https://doi.org/10.3390/molecules28114484

Academic Editor: Aldo Galeone

Received: 27 April 2023
Revised: 27 May 2023
Accepted: 30 May 2023
Published: 1 June 2023

1. Introduction

Nucleic acid nanoparticles (NANPs) are therapeutic nucleic acids designed to assemble into various geometric shapes with distinct physicochemical properties and have a host of diagnostic and therapeutic benefits in a wide array of diseases [1–4]. Physicochemical characterization and immunological evaluation of various RNA and DNA NANPs have been performed to fully understand their structure–activity relationship and help bridge gaps that hinder the clinical translation of these novel nanomaterials [3,5–7].

It has been shown that RNA and DNA NANPs require a carrier for their intracellular delivery to immune cells [5–7]. Without a delivery agent, NANPs have repetitively been shown to remain invisible to the immune system and do not stimulate immune responses [5,7,8]. However, upon delivery with, for example, Lipofectamine 2000 (L2K), NANPs are recognized by peripheral blood mononuclear cells (PBMCs), more so by monocytes than lymphocytes [5,7]. Furthermore, NANPs induce interferon (IFN) response in PBMCs, particularly type I (IFNα; IFNβ; IFNω) and III (IFNλ) IFN responses, and NANP composition and structure define the degree of response—for example, RNA NANPs stimulate greater immune response as compared to their DNA counterparts [5–8]. Within the RNA NANP category, the potency of IFN responses is influenced by nanoparticle architectures, shape, and size. For example, 3D RNA cubes are more immunostimulatory than 2D RNA rings, and 1D RNA fibers are the least immunostimulatory NANPs of all [5,8]; likewise, RNA hexagons are more potent than RNA triangles [5].

Moreover, various delivery platforms can tailor NANPs' immunorecognition and subsequent function, including cytokine induction [7–9]. For example, NANPs' delivery with dendrimers influences their uptake and PBMC cytokine induction when compared to L2K-assisted deliveries. NANPs delivered using cationic dendrimers induce pro-inflammatory cytokines and danger signals but not type I and III IFNs. In contrast, the same NANPs delivered with L2K induce the IFN response with no/low cytokines and danger signals [7]. To further examine the role of the delivery carrier in the qualitative and quantitative outcomes of NANPs' interactions with the primary human immune cells, we investigate two different commercial lipofectamine carriers.

Lipofectamine is a 3:1 (*w*/*w*) formulation of 2,3-di-oleyloxy-N- [2(spermine-carboxamido)-ethyl]-N,N-dimethyl-l-propan-aminium (DOSPA) and dioleoylphosphatidylethanolamine (DOPE) [10]. While L2K and Lipofectamine MessengerMAX (LMM) are both lipofectamines, their composition and chemical structures were optimized to improve the transfection of different types of nucleic acids. L2K is marketed as a more versatile transfection reagent with superior co-transfection performance and the ability to deliver a variety of nucleic acids [11]. LMM, on the other hand, is optimized and recommended for delivery of mRNA without genomic integration [12]. We hypothesized that fine structural variations in lipofectamine might further contribute to controlling the magnitude of NANP-mediated immunostimulation.

Herein, we present results indicating that the type of lipofectamine, L2K vs. LMM, alters NANPs' immunostimulation and cytokine production, thereby providing additional tools to researchers for controlling the magnitude of the IFN response.

2. Results

2.1. Assembly of NANPs and Formation of Lipoplexes

A representative panel of NANPs—RNA fibers, RNA rings, RNA cubes, and DNA cubes—were selected to address the effect of NANPs' composition and architectural parameters on their delivery with lipofectamines and immunorecognition. The assembly of NANPs took place in endotoxin-free conditions. The successful assembly of NANPs was confirmed using non-denaturing polyacrylamide gel electrophoresis (native-PAGE) and visualized via atomic force microscopy (AFM), as shown in Figure 1A.

In addition, eight lipoplexes formed between L2K or LMM, and each of the tested NANPs were visualized using transmission electron microscopy (TEM) and compared to free lipofectamines. The change in morphology of the carrier alone compared to the carriers complexed with NANPs suggests that NANPs were successfully complexed in L2K and LMM, as demonstrated in Figure 1B.

2.2. Monocytes Have Greater NANP Uptake Than Lymphocytes Regardless of Lipofectamine Carrier

To compare the ability of LMM vs. L2K to serve as carriers for NANPs, representative Alexa Fluor 488 (AF488) fluorescent DNA or RNA NANPs (AF488-DNA cubes; AF488-RNA cubes; and AF488-RNA rings) were incubated overnight with PBMCs at a final concentration of 10 nM. The uptake (and/or association with the cellular plasma membrane) of the fluorescent NANPs in both lymphocyte and monocyte populations was determined using flow cytometry. Lymphocyte and monocyte populations were defined via forward and side scatter. The NANP-association with the cells was measured in two ways: (i) the percentage of AF488+ lymphocytes or monocytes, i.e., the proportion of cells that have NANP-associated fluorescence, and (ii) the degree of geometric mean fluorescence intensity (gMFI) in each AF488+ population, i.e., the magnitude of NANP uptake/association by individual cells. Representative gating of the lymphocyte and monocyte populations, along with the AF488+ gating, is shown in Figure 2A.

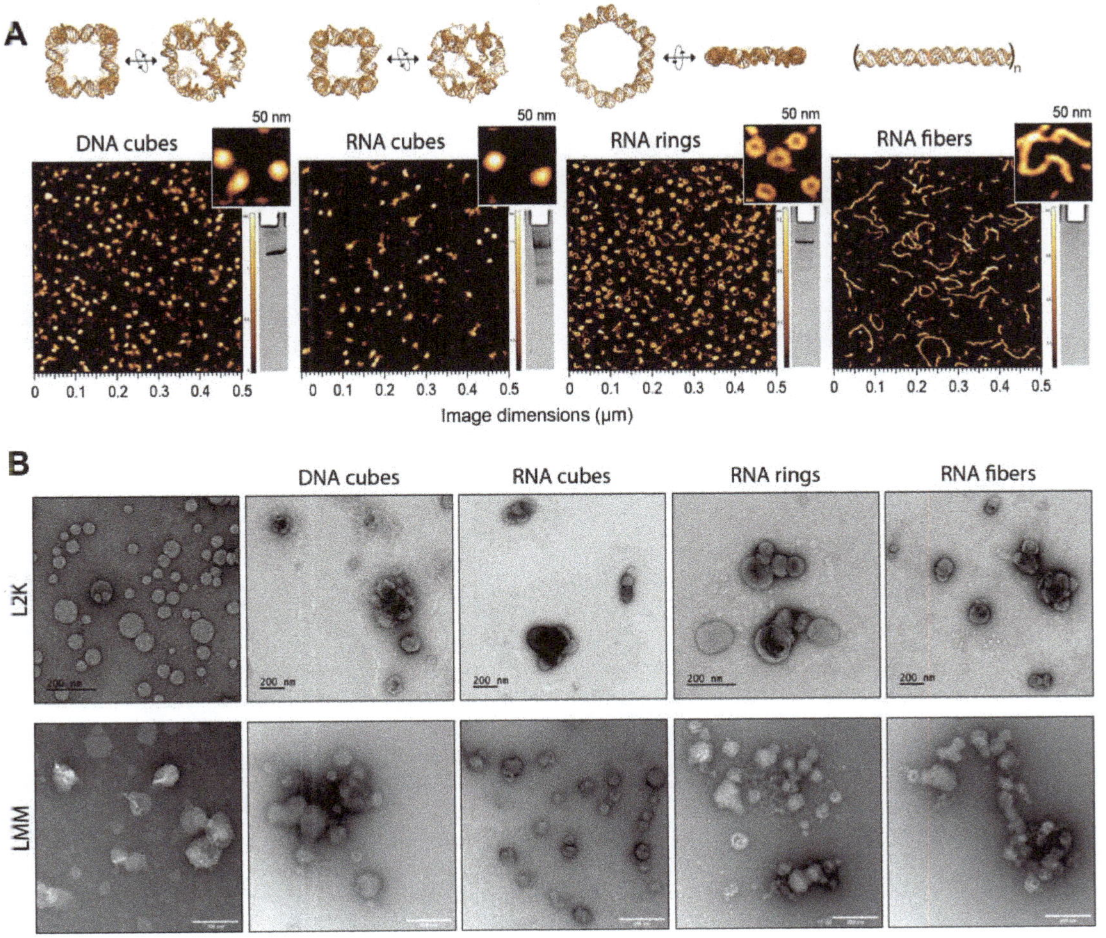

Figure 1. Characterization of NANPs and their lipoplexes. (**A**) 3D models and AFM images of representative NANPs. (**B**) TEM images of NANPs complexed with either L2K (upper panel) or LMM (lower panel).

As previously established by our group, AF488-labeled NANPs have different levels of fluorescence due to the differences in labeling efficiencies of individual oligos. Therefore, the experimental results should not be compared across different NANP types and should be considered qualitatively [5]. Nonetheless, our results were in agreement with previous data from our group, which showed lipofectamine leads to NANP uptake predominately by the monocyte population (Figure 2B,C) [5,7]. Both the percentage of AF488+ monocytes (~60–90%) and the gMFI of AF488+ monocytes (~10 K–40 K arbitrary units (a.u.)) were more significant than the results in the lymphocyte population (~10–50% and 800–1600 a.u., respectively), regardless of lipofectamine type (Figure 2B,C). However, there were a few significant differences when we compared L2K- versus LMM-mediated NANP uptake within a particular NANP type. The only difference between the percentage of AF488+ populations was in lymphocytes, where LMM led to a higher uptake percentage of AF488-RNA rings than L2K (Figure 2B). Furthermore, for the magnitude of NANP uptake, LMM led to lower gMFI for DNA and RNA cubes in the lymphocyte population, while LMM led to lower gMFI for RNA cubes and RNA rings in the monocyte population (Figure 2B,C).

However, while these differences may be statistically significant, biological significance may not follow.

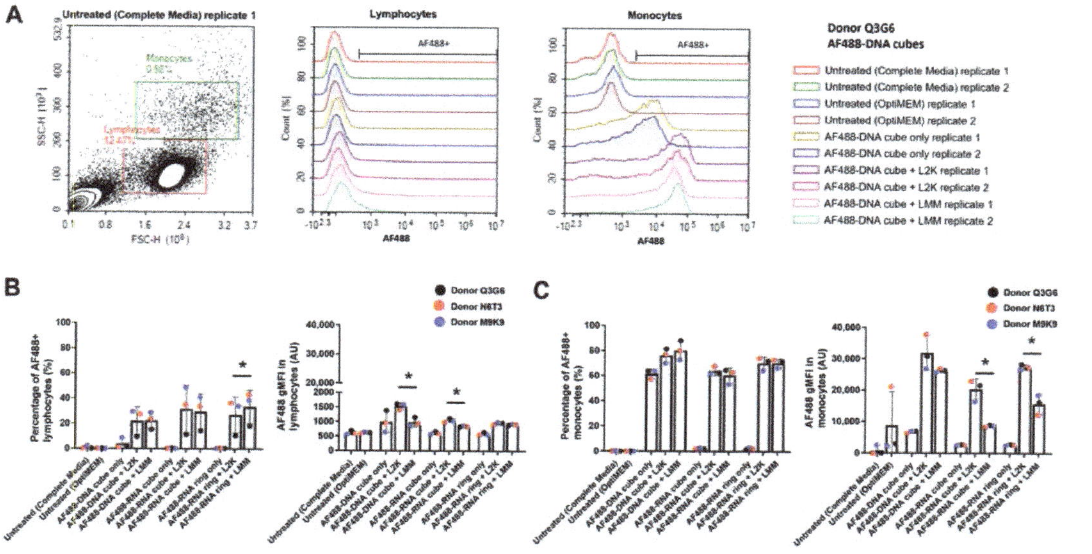

Figure 2. Monocytes have greater NANP uptake than lymphocytes regardless of lipofectamine carrier. PBMCs were treated with 10 nM AF488-NANPs for 20 h, fixed, and acquired on the flow cytometer. Cells were gated for lymphocyte and monocyte populations based on side and forward scatter and then gated on the AF488 signal. (**A**) Representative gating strategy for one healthy donor (Q3G6) showing the raw data for the negative controls and AF488-DNA cubes. (**B**) Lymphocytes and (**C**) monocytes were assessed for uptake of AF488-labeled NANPs. The percentage of cells positive for the AF488+ signal (left plots) and the geometric mean fluorescence intensity (gMFI) of AF488+ cells (right plots) were assessed for both populations. Each bar graph represents the mean data ± standard deviation from three healthy donors. Each dot represents the mean for each individual donor (run in duplicate). An asterisk (*) indicates $p \leq 0.05$ for paired *t*-test between lipofectamine carriers for a particular NANP type. L2K—Lipofectamine™ 2000 reagent; LMM—Lipofectamine™ MessengerMAX™ reagent.

2.3. RNA Fibers Delivered with LMM Carrier Decrease IFN Production in PBMCs

To determine if the lipofectamine-carrier-type-induced changes in NANP uptake affected PBMC biologically, IFN response was determined. Multiplex analysis was used to assess type I (IFNα; IFNβ; IFNω) and type III (IFNλ) interferon production in PBMCs after overnight treatment with 10 nM NANPs delivered using either L2K or LMM. The IFN panel was used in our earlier studies, which identified IFNs as biomarkers of immunostimulation of NANPs delivered using lipofectamine carriers [5,7]. The cytokine levels are presented as a heat map (Figure 3A) and as a bar graph (Figure 3B). It was determined that treatment with RNA cubes led to IFN levels similar to the positive control (ODN2216), and this finding agrees with previous studies (Figure 3) [5]. Furthermore, PBMC treatment with DNA rings, RNA fibers, and RNA rings generally led to lower IFN responses than the positive control.

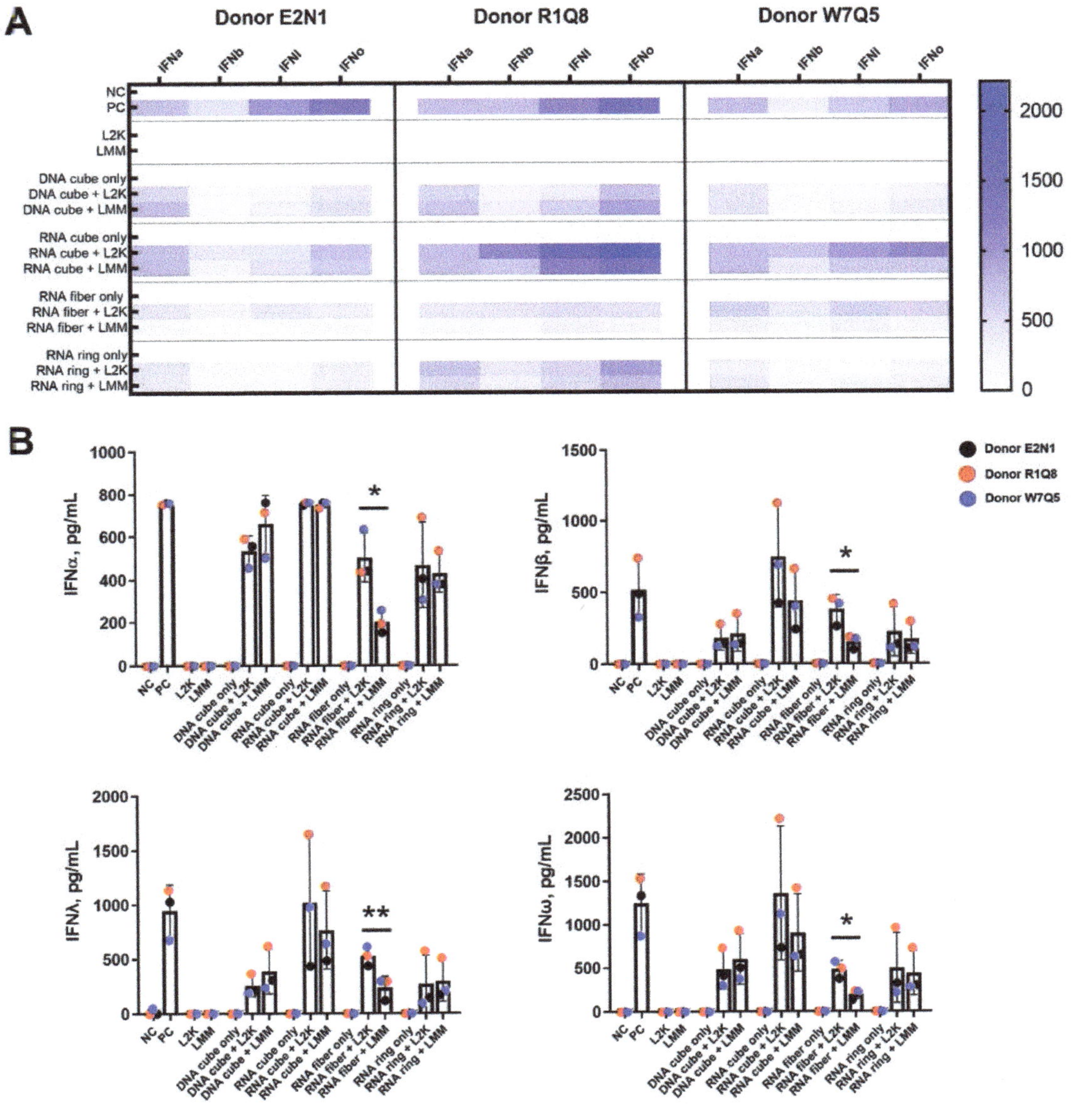

Figure 3. Lipofectamine carrier type alters PBMC IFN production in response to incubation with RNA fibers. PBMCs were treated with 10 nM NANPs for 20 h, and supernatants were collected and analyzed via multiplex for IFN production (IFNα; IFNβ; IFNλ; IFNω). (**A**) Heat map of the different IFN production levels of three healthy donors. Data points for each donor were run in duplicate. (**B**) Bar graphs representing the IFN production levels of three healthy donors. Each bar graph represents the mean data ± standard deviation from three healthy donors. Each dot represents the mean for each individual donor (run in duplicate). An asterisk (*) indicates $p \leq 0.05$ or ** indicates $p \leq 0.01$ for paired t-test between lipofectamine carriers for a particular NANP type. NC—negative control (untreated PBMC); PC—positive control (5 µg/mL ODN2216); L2K—Lipofectamine™ 2000 reagent; LMM—Lipofectamine™ MessengerMAX™ reagent.

When addressing the specific effect of lipofectamine carrier, we determined that lipofectamine carrier type did not affect NANP-induced IFN production except with the RNA fibers. In the case of RNA fibers, delivery of RNA fibers with LMM led to decreased IFN production for all four IFNs tested as compared to L2K (Figure 3B). This difference in IFN production may reflect the design of the LMM carrier, which was optimized to deliver mRNA.

3. Discussion

The greater degree of NANP uptake in the monocyte population as compared to the lymphocyte population in the presence of a lipofectamine carrier (Figure 2) is consistent with our previous studies [5,7] and data published by other research groups using DNA origami [13]. The uptake of these NANPs in the monocytes was higher than in lymphocytes regardless of tested carriers—L2K, LMM (Figure 2), or dendrimers [7]. Furthermore, while there were differences seen between L2K-mediated and LMM-mediated uptake for a few NANPs in the lymphocyte and monocyte populations, the differences are less than two-fold except for AF488 gMFI in monocytes for RNA cubes (Figure 2B,C). Differences less than two-fold are unlikely to lead to a biologically significant change. Moreover, these differences seen in uptake did not correspond to downstream differences in IFN production (Figure 3). We observed decreased PBMC IFN production after treatment with LMM-delivered RNA fibers compared to L2K-delivered RNA fibers. Unfortunately, we did not have AF488-RNA fibers to test RNA fiber uptake in monocyte and lymphocyte populations. Therefore, we do not have data to indicate whether the decrease in IFN production in PBMC from LMM-delivered RNA fibers is due to a lack of NANP uptake or another downstream process.

Interestingly, the current study and one of our earlier studies [7] observed the association of DNA cubes with ~60% of the monocyte population in the absence of a carrier (Figure 2C, left) [7]. This observation was also similar to the study by Du et al. investigating the uptake of DNA origami [13] but in contrast to the initial report by Hong et al., in which the uptake of DNA NANPs by monocytes was detected only in the presence of L2K [5]. We hypothesize that differences in the type of flow cytometer used in these studies may explain the observed discrepancy in the test results. Our current research and reports by Avila et al. and Du et al. utilized digital flow cytometers, which adjust the instrument settings automatically and, thus, are more sensitive at detecting even low fluorescent signal [7,13]. In contrast, the initial study by Hong et al. used a traditional cytometer which involves manual adjustment of instrument settings and often leads to the relocation of objects with weak fluorescence outside of the data collection gates [5]. Furthermore, even though the use of the NovoCyte flow cytometer in our current study revealed ~60% of the monocyte population was positive for DNA cube in the absence of any carrier (Figure 2C, left), this increase was not accompanied by increased AF488 gMFI (Figure 2C, right) nor was it accompanied by detectable IFN production (Figure 3), further suggesting that the association on the individual cell level was relatively low. This could imply that the NovoCyte 3005 (and possibly other digital cytometers with similar properties) is more sensitive than the previously used FACSCalibur [5] in the ability to detect low NANP quantities associated with the cells. A cross-validation between the two instruments would help verify this hypothesis, but it was not feasible because FACSCalibur is no longer available.

4. Materials and Methods

4.1. Materials

DNA strands (PCR forward and reverse primers and templates for RNA NANPs and individual oligos for DNA NANPs) and fluorescently labeled oligos (3′- Alexa Fluor 488) were obtained from Integrated DNA Technologies (IDT), Inc. MyTaq Mix, was purchased from Bioline. A DNA Clean & Concentrator kit was obtained from Zymo Research. RQ1 RNase-Free DNase was purchased from Promega (3 u/50 μL). Phosphate-buffered saline (PBS), RPMI-1640 medium, penicillin–streptomycin solution, L-glutamine, ficoll-paque premium, fetal bovine serum (FBS), and HyPure cell-culture-grade water were all obtained from Cytiva/GE Heathcare Life Sciences (Marlborough, MA, USA). Opti-MEM™ I reduced serum medium and Hank's balanced salt solution (HBSS) were from Gibco (Gaithersburg, MD, USA). Acridine orange (AO)/propidium iodide (PI) staining solution was from Nexcelom Bioscience (Lawrence, MA, USA). Oligodeoxyribonucleotide, a human TLR9 ligand (ODN2216), was from InvivoGen (San Diego, CA, USA). NovoFlow, NovoRinse, and NovoClean were from Agilent Technologies (Santa Clara, CA, USA). Lipofectamine™ MessengerMAX™ reagent and Lipofectamine™ 2000 reagent were obtained from Invitrogen (Waltham, MA, USA). Paraformaldehyde (PFA) 20% Solution was from Electron Microscopy Science (Hatfield, PA, USA). A custom 4-plex Multiplex (IFNα; IFNβ; IFNλ; IFNω) kit with sample diluent, calibrator 1, calibrator 2, detection solution, streptavidin-HRP, substrate A, substrate B+, and wash buffer was obtained from Quansys BioSciences (Logan, UT, USA).

4.2. NANP Preparation

All sequences used for NANP preparation are provided in the Supporting Information. DNA templates were amplified via PCR using MyTaq Mix. The DNA Clean & Concentrator kit was used to purify the amplified PCR products, followed by in vitro run-off transcription using T7 RNA Polymerase in 80 mM HEPES-KOH (pH 7.5), 2.5 mM spermidine, 50 mM DTT, 25 mM $MgCl_2$, and 5 mM of each rNTP at 37 °C over 3.5 h. Transcription was stopped through adding RQ1 RNase-Free DNase and incubating at 37 °C for 30 min. For the purification of RNA strands, denaturing polyacrylamide gel electrophoresis (PAGE, 8%) in the presence of 8 m urea run in 89 mM tris-borate, 2 mM EDTA (TBE, pH 8.2) was run at 13 W for 2 h. UV was used to visualize the RNA bands; the bands were then excised and eluted overnight in 300 mM NaCl, TBE (pH 8.2) at 4 °C. To precipitate the RNAs, the elution was mixed with 2.5 volumes of 100% EtOH and stored at −20 °C for 3 h. Then, the samples were centrifuged at 10.0× *g* for 30 min at 4 °C. The pellet was washed with 90% EtOH for 10 min via centrifugation at 10.0× *g* at 4 °C; this step was repeated twice. The pelleted samples were vacuum-dried at 55 °C with IR in a CentriVap micro-IR vacuum concentrator (Labconco), then dissolved in HyPure cell-culture-grade water. The concentration of each strand was measured using a NanoDrop 2000 (ThermoFisher) at 260 nm. The fourteen RNA strands were stored at −20 °C until use.

All NANPs were assembled in a one-pot thermal anneal through combining each strand in an equimolar ratio with HyPure cell-culture-grade water. The DNA cubes, RNA cubes, AF488-DNA cubes, and AF488-RNA cubes were heated to 95 °C for 2 min, then mixed with assembly buffer (89 mM tris-borate (pH 8.2), 2 mM $MgCl_2$, 50 mM KCl) and incubated at 45 °C for 30 min, and then stored at 4 °C until use. The RNA rings, AF488-RNA fibers, and RNA rings were heated to 95 °C for 2 min, snap-cooled on ice for 2 min, mixed with the assembly buffer, and incubated at 30 °C for 30 min, then stored at 4 °C until use.

4.3. Characterization of NANPs

Successful assembly of NANPs was confirmed via visualization on 8% native-PAGE (37.5:1 acrylamide:bis-acrylamide). The gel was prepared on a Mini-PROTEAN Tetra Cell system (Bio-Rad), pre-run for 5 min at 150 V with running buffer (89 mM TB (pH 8.2), and 2 mM $MgCl_2$). 2 μL of each sample was mixed with 2 μL loading buffer (Assembly buffer, 30% glycerol, bromophenol blue, xylene cyanol), and loaded per well. The loaded gel was run at 300 V for 30 min in a 4 °C cold room. The gel was stained with ethidium bromide (EtBr, 0.5 μg mL^{-1}) for 5 min, then washed twice with double-deionized water (ddiH2O). Then, the gel was imaged using a ChemiDoc MP (Bio-Rad). The Alexa Fluor 488-labeled NANPs' gel was imaged before EtBr staining via the Alexa Fluor 488 setting on the ChemiDoc MP system.

Atomic force microscopy (AFM) imaging of NANPs was performed on a freshly cleaved 1-(3-aminopropyl) silatrane-modified mica surface as previously described [7,14,15].

The AFM imaging was performed in tapping mode on the MultiMode AFM Nanoscope IV system (Bruker Instruments, Billerica, MA, USA).

For TEM imaging, 10 μL of corresponding 1 μM NANP stock and 2 μL of L2K or LMM were repeatedly mixed through pipetting up and down. The complexed samples were incubated at room temperature for 5–30 min. Stock L2K and LMM complexes were used for imaging except for the LMM + RNA fiber, which was diluted 10-fold in water before imaging. Samples were vortexed and 5 μL of each sample was applied to a glow-discharged carbon-coated 200 mesh Cu grid (EMS, Hatfield, PA, USA) for LMM/LMM-NANPs complexes or carbon-coated 400 mesh Cu/Rh grid (Ted Pella, Redding, CA, USA) for L2K/L2K-NANPs complexes and allowed to dry for 1 min at room temperature. Staining with 5 μL of 1% uranyl acetate (EMS, Hatfield, PA for LMM samples or Polysciences, Warrington, PA, USA for L2K samples) was repeated twice followed by final blotting and air-drying the grid. An FEI Tecnai T20 transmission electron microscope operating at 200 kV with a Gatan 2 k × 2 k Eagle camera was used to image the LMM grids and an FEI Talos L120C TEM with Gatan 4 k × 4 k OneView camera was used to image the L2K grids. A bridging experiment analyzing L2K on 200 mesh Cu grids (EMS, Hatfield, PA, USA) and images from an FEI Tecnai T20 transmission electron microscope operating at 200 kV with a Gatan 2 k × 2 k Eagle camera was conducted to verify that differences in instrumentation do not affect the results; the image is included in the Supplementary Materials (Figure S1).

4.4. PBMC Isolation

Healthy human donor whole blood was collected in li-heparin vacutainers (BD Bio-Sciences) under NCI-Frederick protocol OH9-C-N046. The whole blood was used for PBMC isolation as specified in NCL protocol ITA-10 [16]. In brief, whole blood was diluted with PBS at a 1:1 ratio, layered over ficoll-paque at a ratio of 4:3 (4 mL diluted blood for every 3 mL ficoll-paque), and centrifuged for 30 min at room temperature at 900× *g* with no brake. The mononuclear cell layer containing the PBMCs was then removed, collected, and washed twice with HBSS (centrifuged for 10 min at 400× *g*). The PBMCs were resuspended in complete RPMI-1640 medium (10% heat-inactivated FBS, 100 U/mL penicillin, 100 μg/mL streptomycin, and 2 mM L-glutamine) and counted on a Cellometer using a 1:1 ratio of the cell suspension to AOPI. Once the PBMCs were counted, samples were diluted to 1.25×10^6 cells/mL using a complete RPMI-1640 medium.

4.5. Uptake of Alexa Fluor-488 NANPs in PBMCs

PBMCs were aliquoted into a 96-well round bottom plate with 160 μL cell suspension (1.25×10^6 cells/mL) per well. The AlexaFluor-488 NANPs (AF488-DNA cubes; AF488-RNA cubes; and AF488-RNA rings) and appropriate controls (untreated controls and no-carrier controls) were then prepared in microcentrifuge tubes using Opti-MEM^{TM} I reduced serum medium and lipofectamine reagents. An aliquot of 15 μL of 1 μM stock of appropriate NANPs was combined with 3 μL of lipofectamine reagent (LMM or L2K) or 3 μL of Opti-MEM^{TM} I reduced serum medium for the no-carrier controls and incubated between 5–30 min in the dark at room temperature. The untreated controls consisted of either complete RPMI-1640 media (Complete Media) only or Opti-MEM^{TM} I reduced serum medium (OptiMEM) only. After the incubation, 282 μL of Opti-MEM^{TM} I reduced serum medium was added to each sample (except negative controls) for a total volume of 300 μL and an NANP concentration of 50 nM where applicable. An aliquot of 40 μL prepared sample or control was added to each appropriate well of the prepared 96-well plate with PBMC suspension for a final volume of 200 μL (cells at 1×10^6 cells/mL; NANP at 10 nM final concentration). The 96-well plate was placed in a humidified 37 °C/95% CO_2 incubator for approximately 20 h.

The PBMC samples were then prepared for acquisition on a NovoCyte 3005 flow cytometer (Agilent Technologies, Inc., Santa Clara, CA, USA). The plate was removed from the incubator and centrifuged for 5 min at 400× *g*. The supernatants from each well were then aspirated and discarded, leaving the cell pellet undisturbed. The samples were washed twice with 150 μL 1× PBS (centrifuged 400× *g* for 5 min). The cell pellets were then fixed with 2% PFA for 15 min at room temperature and washed twice more with 1× PBS. Each cell pellet was resuspended in 150 μL 1× PBS for acquisition on the flow cytometer. On the NovoExpress software, side-scatter and forward-scatter area and height parameters were selected along with the area and height parameters for the FITC (488) channel. All other parameters remained unselected. Samples were then acquired with the instrumentation and analyzed using GraphPad Prism 9 (Graph Pad Software, Boston, MA, USA) and NovoExpress software version 1.5.6 (Agilent Technologies, Inc., Santa Clara, CA, USA).

4.6. IFN Production of PBMCs after NANPs Treatment

PBMCs were aliquoted into 96-well round bottom plates with 160 μL cell suspension (1.25×10^6 cells/mL) per well. The NANPs (DNA cubes; RNA cubes; RNA fibers; and RNA rings) and appropriate controls (negative control, positive control (5 μg/mL ODN2216), vehicle controls, no-carrier controls) were then prepared in microcentrifuge tubes. An aliquot of 20 μL of 1 μM stock of appropriate NANPs was combined with 4 μL of lipofectamine reagent (LMM or L2K) or 4 μL of Opti-MEM™ I reduced serum medium for the no-carrier controls and incubated between 5–30 min at room temperature. The vehicle controls consisted of 20 μL Opti-MEM™ I reduced serum medium combined with 4 μL of appropriate lipofectamine reagent. The negative control consisted of Opti-MEM™ I decreased serum medium only. The positive control consisted of 10 μL ODN2216 1 mg/mL stock diluted in 390 μL Opti-MEM™ I reduced serum medium for a 25 μg/mL concentration. After the incubation, 376 μL of Opti-MEM™ I reduced serum medium was added to each sample (except the positive control) for a total volume of 400 μL and a NANPs concentration of 50 nM where applicable. An aliquot of 40 μL prepared sample or control was added to each appropriate well of the prepared 96-well plates with PBMC suspension for a final volume of 200 μL (cells at 1×10^6 cell/mL). NANP samples were at a 10 nM final concentration, and the positive control samples were at a final concentration of 5 μg/mL. The well plates were placed in a humidified 37 °C/95% CO_2 incubator for approximately 20 h. After the incubation, the plates were centrifuged for 10 min at 700× *g*. Supernatant aliquots were then collected in newly labeled 96-well plates and stored at −80 °C.

A custom 4-plex Multiplex (IFNα; IFNβ; IFNλ; IFNω) from Quansys BioSciences was then used to analyze the freeze–thawed aliquots according to the manufacturer's manual and NCL Protocol ITA-27 [17]. All reagents needed were included in the kit except for de-ionized water and cell-culture-grade water and prepared when indicated by the manual. In brief, the supernatants were thawed (partially at room temperature and partially at 37 °C). The calibration standards were prepared using the sample diluent in a 96-well polypropylene plate. The supernatant samples were diluted 2-fold with the sample diluent. 50 μL aliquots of calibration standards and supernatants were loaded into appropriate wells of the provided multiplex plate and incubated at room temperature for 2 h on a shaker (500 rpm). The multiplex plate was washed 3 times with wash buffer using a plate washer. The detection mix was then added to the multiplex plate and incubated for 1 h at room temperature on a shaker. The multiplex plate was then washed 3 times. A 50 μL aliquot of streptavidin-HRP was added to each well, and the plate was incubated for 20 min on the shaker. The multiplex plate was then washed 6 times, and 50 μL of ChemiLum substrate (Substrate A combined with Substrate B+) was added to each well. The plate was then read using the Quansys ImagePro, and the resulting data were analyzed using Microsoft Excel and GraphPad Prism.

5. Conclusions

This is the first study to demonstrate that the lipofectamine type of commercial delivery agents can be used as a simple tool to mediate change in the immunorecognition of different NANPs. LMM decreased IFN production in response to RNA fibers, which may be linked to their linear structure. This sensitivity to lipofectamine carriers could be used to modify PBMC response to NANPs precisely.

Supplementary Materials: The following supporting information can be downloaded at: https://www.mdpi.com/article/10.3390/molecules28114484/s1, Sequences used in this project and Figure S1: Comparison of lipofectamine 2000 (L2K) morphology by TEM performed using different grids.

Author Contributions: H.S.N., Y.R., and J.X. conducted experiments and analyzed the data. J.D.C. analyzed the TEM data. M.A.D. and K.A.A. conceived, designed and supervised the study. All authors wrote and reviewed the manuscript. All authors have read and agreed to the published version of the manuscript.

Funding: The study was funded in part (M.A.D., J.D.C., J.X and H.S.N.) by federal funds from the National Cancer Institute, National Institutes of Health, under contract 75N91019D00024. The content of this publication does not necessarily reflect the views or policies of the Department of Health and Human Services, nor does mention of trade names, commercial products, or organizations imply endorsement by the U.S. Government. Research reported in this publication was supported by the National Institute of General Medical Sciences of the National Institutes of Health under Award Number R35GM139587 (to K.A.A.). The content is solely the responsibility of the authors and does not necessarily represent the official views of the National Institutes of Health.

Institutional Review Board Statement: All experiments involving human whole blood were performed according to the IRB-approved NCI-Frederick protocol OH99-C-N046.

Informed Consent Statement: Not applicable.

Data Availability Statement: All data are provided in manuscript.

Conflicts of Interest: The authors declare no conflict of interest.

Sample Availability: Samples of the compounds (NANPs) are available from the authors upon reasonable request.

References

1. Halman, J.R.; Satterwhite, E.; Roark, B.; Chandler, M.; Viard, M.; Ivanina, A.; Bindewald, E.; Kasprzak, W.K.; Panigaj, M.; Bui, M.N.; et al. Functionally-interdependent shape-switching nanoparticles with controllable properties. *Nucleic Acids Res.* **2017**, *45*, 2210–2220. [CrossRef] [PubMed]
2. Chandler, M.; Afonin, K.A. Smart-Responsive Nucleic Acid Nanoparticles (NANPs) with the Potential to Modulate Immune Behavior. *Nanomaterials* **2019**, *9*, 611. [CrossRef] [PubMed]
3. Johnson, M.B.; Halman, J.R.; Miller, D.K.; Cooper, J.S.; Khisamutdinov, E.F.; Marriott, I.; Afonin, K.A. The immunorecognition, subcellular compartmentalization, and physicochemical properties of nucleic acid nanoparticles can be controlled by composition modification. *Nucleic Acids Res.* **2020**, *48*, 11785–11798. [CrossRef] [PubMed]
4. Panigaj, M.; Skelly, E.; Beasock, D.; Marriott, I.; Johnson, M.B.; Salotti, J.; Afonin, K.A. Therapeutic immunomodulation by rationally designed nucleic acids and nucleic acid nanoparticles. *Front. Immunol.* **2023**, *14*, 1053550. [CrossRef] [PubMed]
5. Hong, E.; Halman, J.R.; Shah, A.B.; Khisamutdinov, E.F.; Dobrovolskaia, M.A.; Afonin, K.A. Structure and Composition Define Immunorecognition of Nucleic Acid Nanoparticles. *Nano Lett.* **2018**, *18*, 4309–4321. [CrossRef] [PubMed]
6. Hong, E.; Halman, J.R.; Shah, A.; Cedrone, E.; Truong, N.; Afonin, K.A.; Dobrovolskaia, M.A. Toll-Like Receptor-Mediated Recognition of Nucleic Acid Nanoparticles (NANPs) in Human Primary Blood Cells. *Molecules* **2019**, *24*, 1094. [CrossRef] [PubMed]
7. Avila, Y.I.; Chandler, M.; Cedrone, E.; Newton, H.S.; Richardson, M.; Xu, J.; Clogston, J.D.; Liptrott, N.J.; Afonin, K.A.; Dobrovolskaia, M.A. Induction of Cytokines by Nucleic Acid Nanoparticles (NANPs) Depends on the Type of Delivery Carrier. *Molecules* **2021**, *26*, 652. [CrossRef] [PubMed]
8. Bila, D.; Radwan, Y.; Dobrovolskaia, M.A.; Panigaj, M.; Afonin, K.A. The Recognition of and Reactions to Nucleic Acid Nanoparticles by Human Immune Cells. *Molecules* **2021**, *26*, 4231. [CrossRef] [PubMed]
9. Nordmeier, S.; Ke, W.; Afonin, K.A.; Portnoy, V. Exosome mediated delivery of functional nucleic acid nanoparticles (NANPs). *Nanomedicine* **2020**, *30*, 102285. [CrossRef] [PubMed]
10. Chu, Y.; Masoud, M.; Gebeyehu, G. Transfection Agents. International Patent WO2000027795A1, May 2000. Available online: https://patents.google.com/patent/WO2000027795A1/en (accessed on 1 March 2023).
11. ThermoFisher Scientific. Lipofectamine™ 2000 Transfection Reagent. Available online: https://www.thermofisher.com/order/catalog/product/11668019 (accessed on 27 February 2023).
12. ThermoFischer Scientific. Lipofectamine™ MessengerMAX™ Transfection Reagent. Available online: https://www.thermofisher.com/order/catalog/product/LMRNA001 (accessed on 27 February 2023).

13. Du, R.R.; Cedrone, E.; Romanov, A.; Falkovich, R.; Dobrovolskaia, M.A.; Bathe, M. Innate Immune Stimulation Using 3D Wireframe DNA Origami. *ACS Nano* **2022**, *16*, 20340–20352. [CrossRef] [PubMed]
14. Ke, W.; Hong, E.; Saito, R.F.; Rangel, M.C.; Wang, J.; Viard, M.; Richardson, M.; Khisamutdinov, E.F.; Panigaj, M.; Dokholyan, N.V.; et al. RNA-DNA fibers and polygons with controlled immunorecognition activate RNAi, FRET and transcriptional regulation of NF-kappaB in human cells. *Nucleic Acids Res.* **2019**, *47*, 1350–1361. [CrossRef] [PubMed]
15. Shlyakhtenko, L.S.; Gall, A.A.; Filonov, A.; Cerovac, Z.; Lushnikov, A.; Lyubchenko, Y.L. Silatrane-based surface chemistry for immobilization of DNA, protein-DNA complexes and other biological materials. *Ultramicroscopy* **2003**, *97*, 279–287. [CrossRef] [PubMed]
16. Potter, T.M.; Cedrone, E.; Neun, B.W.; Dobrovolskaia, M.A. NCL Method-10: Preparation of Human Whole Blood and Peripheral Blood Mononuclear Cell Cultures to Analyze Nanoparticle Potential to Induce Cytokines In Vitro. *NCI Hub.* **2020**. [CrossRef]
17. Cedrone, E.; Potter, T.M.; Neun, B.W.; Tyler, A.; Dobrovolskaia, M.A. NCL Method ITA-27: Multiplex Enzyme-Linked Immunosorbent Assay (ELISA) for Detection of Human Cytokines in Culture Supernatants. *NCI Hub.* **2020**. [CrossRef]

molecules

MDPI

Review

Nanomedicine Reformulation of Chloroquine and Hydroxychloroquine

David M. Stevens, Rachael M. Crist and Stephan T. Stern *

Nanotechnology Characterization Laboratory, Cancer Research Technology Program, Leidos Biomedical Research, Inc., Frederick National Laboratory for Cancer Research sponsored by the National Cancer Institute, Frederick, MD 21702, USA; dstevens5654@gmail.com (D.M.S.); cristr@mail.nih.gov (R.M.C.)
* Correspondence: sternstephan@mail.nih.gov

Abstract: The chloroquine family of antimalarials has a long history of use, spanning many decades. Despite this extensive clinical experience, novel applications, including use in autoimmune disorders, infectious disease, and cancer, have only recently been identified. While short term use of chloroquine or hydroxychloroquine is safe at traditional therapeutic doses in patients without predisposing conditions, administration of higher doses and for longer durations are associated with toxicity, including retinotoxicity. Additional liabilities of these medications include pharmacokinetic profiles that require extended dosing to achieve therapeutic tissue concentrations. To improve chloroquine therapy, researchers have turned toward nanomedicine reformulation of chloroquine and hydroxychloroquine to increase exposure of target tissues relative to off-target tissues, thereby improving the therapeutic index. This review highlights these reformulation efforts to date, identifying issues in experimental designs leading to ambiguity regarding the nanoformulation improvements and lack of thorough pharmacokinetics and safety evaluation. Gaps in our current understanding of these formulations, as well as recommendations for future formulation efforts, are presented.

Keywords: chloroquine; hydroxychloroquine; nanomedicine; nanoformulation

Citation: Stevens, D.M.; Crist, R.M.; Stern, S.T. Nanomedicine Reformulation of Chloroquine and Hydroxychloroquine. *Molecules* **2021**, *26*, 175. https://doi.org/10.3390/molecules26010175

Academic Editor: Derek J. McPhee
Received: 8 December 2020
Accepted: 29 December 2020
Published: 31 December 2020

1. Introduction

Chloroquine (CQ) and hydroxychloroquine (HCQ) have been used for decades in the prevention and treatment of malaria and in the treatment of some autoimmune diseases such as lupus erythematosus and rheumatoid arthritis due to their immunomodulatory properties [1–3]. Despite being considered old drugs, CQ and HCQ have generated new interest due to their anticancer activity both in preclinical and clinical studies [4,5]. Researchers have shown these drugs act through a variety of antineoplastic mechanisms such as autophagy disruption, tumor vessel normalization, immunomodulation, and inhibition of metastasis, acting both directly on the tumor parenchyma and tumor microenvironment [6,7]. Chloroquines have been shown effective either as monotherapies or as adjunct therapies, sensitizing cancer cells to existing cytostatic agents as well as targeted therapies [7]. For example, HCQ has been shown to synergize with MEK pathway inhibitors for effective treatment of RAS-driven cancers, and CQ has been shown to inhibit melanoma growth through modifying tumor-associated macrophage (TAM) from the M2 immunosuppressive/pro-tumor phenotype to M1 immunostimulatory/antitumor phenotype [8,9].

CQ and HCQ have also recently received worldwide attention due to their potential use in treating coronavirus disease 2019 (COVID-19) caused by severe acute respiratory syndrome coronavirus 2 (SARS-CoV-2) infection. Previous studies showed in vitro efficacy of these drugs against Middle East respiratory syndrome coronavirus (MERS-CoV) and severe acute respiratory coronavirus (SARS-CoV), and a recent study demonstrated CQ could effectively inhibit viral infection of SARS-CoV-2 in vitro [10–12]. As a result, scientists suggested their assessment in patients, leading to emergency use authorization for HCQ and the initiation of several clinical trials. However, interest in these drugs sharply declined following a retrospective analysis claiming COVID-19 patients were more likely to die of irregular heart rhythms when taking CQ or HCQ, resulting in revocation of the FDA's emergency use authorization [13,14]. This report was later retracted due to data validity concerns; however, many clinical trials had already been terminated. It should also be noted that recent data have questioned the original in vitro findings supporting inhibition of viral replication by CQ, demonstrating that the CQ-sensitive viral activation mechanism in the Vero cell line utilized was not relevant to human lung cells [15]. For these reasons, the use of these drugs for the prevention or treatment of COVID-19 remains extremely controversial.

CQ and HCQ are both basic amphiphiles that concentrate in the lysosome and inhibit lysosomal function as their primary mechanism of action [16]. While CQ and HCQ also have similar toxicity profiles and are equipotent, chloroquine is much more toxic (2-fold) [16]. Although short-term administration of either drug is generally well-tolerated, except in patients predisposed to arrhythmia, chronic dosing and high-dose regimens can cause severe side effects such as irreversible retinal toxicity [17–19]. CQ and HCQ have similar pharmacokinetic (PK) properties, including high volume of distribution and prolonged plasma half-lives between 40 and 50 days, which requires weeks of dosing to achieve steady-state therapeutic concentrations [20]. Reformulation of CQ and HCQ to improve their PK and safety profile may support the use of these drugs for applications such as cancer and infectious diseases.

Nanoparticle drug delivery is one promising strategy to overcome drug liabilities such as poor PK and toxicity while improving site-specific drug delivery. Nanomedicines can provide a variety of benefits, such as improving the solubility of hydrophobic drugs, protecting drugs from degradation, and altering tissue distribution through passive or active targeting mechanisms [21]. Indeed, various nanomedicines have been developed and clinically approved that enhance the safety and/or efficacy of drugs and legacy formulations [22]. Overall, CQ and HCQ therapy may benefit from reformulation, and this review will discuss the efforts to formulate these drugs through nanomedicine approaches (Figure 1).

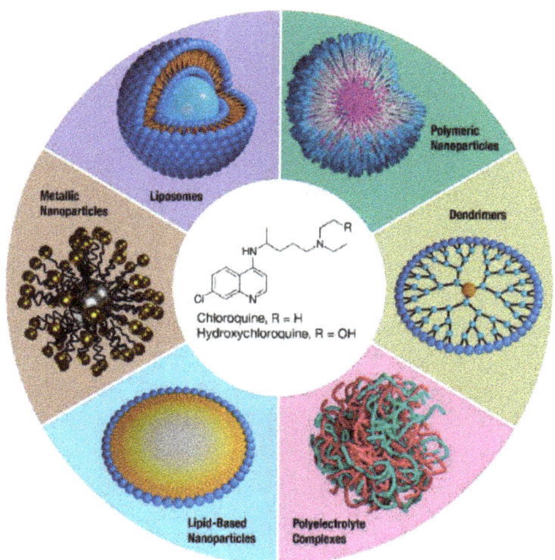

Figure 1. Nanomedicine formulations of CQ and HCQ. A variety of nanotechnology platforms are being explored in the reformulation efforts of improving the overall safety and efficacy of CQ and HCQ.

2. Liposomes

Liposomes are spherical vesicles consisting of one or more phospholipid bilayers and are capable of loading drugs within their aqueous core or lipid bilayer. Liposomes are generally very stable with long circulatory half-lives, and changes to their surface chemistry, such as hydrophilic coating (e.g., polyethylene glycol; PEG) or targeting moieties (e.g., antibodies), can result in decreased uptake by the mononuclear phagocytic system (MPS) and site-specific delivery, respectively [23,24]. Liposomal formulations of CQ, with and without erythrocyte-specific antibody targeting fragments, were first developed during the 1980s and provided better suppression of parasitemia compared to unformulated CQ in malaria parasite *P. berghei*-infected animals (Table 1) [25–30]. Despite these early successes, liposomal CQ did not progress toward clinical applications, and only a few liposomal CQ formulations have been published since. For example, Fotoran et al. developed micron-sized, multilamellar liposomes for loading CQ through interlayer hydrogen bonding [31]. In comparison to unformulated CQ, this formulation only provided a significant reduction in parasitemia for two of the thirteen-day efficacy study, suggesting only a modest improvement in therapy.

It is worth noting that these studies utilized non-PEGylated liposomes, which are known to be rapidly cleared by resident macrophages in MPS organs such as the liver and spleen [32]. Although this is unfavorable for many applications, since it lowers drug exposure to non-MPS tissues, some researchers have utilized non-PEGylated liposomes as a strategy to increase drug exposure to macrophages and improve treatment of macrophage-based infections. For example, in a *C. neoformans* murine model, liposomal CQ in combination with fluconazole provided better antifungal prophylaxis and treatment compared to free drug controls due to enhanced liposomal drug uptake by macrophages [33,34]. Most modern liposomal formulations contain a PEG surface coating that reduces macrophage clearance and increases circulatory time, which may be desirable for malaria and cancer indications. In one recent example, a CQ formulation using PEGylated liposomes with antibody targeting to the erythrocyte surface protein glycophorin A provided robust CQ delivery to uninfected and Plasmodium-infected red blood cells, resulting in superior efficacy compared to unformulated CQ in *P. falciparum*-infected mice [35]. Overall, these studies support the use of liposomal formulations for delivering CQ to erythrocytes and macrophages for malaria and antifungal applications, but additional PK and toxicology studies would be informative to evaluate their safety profile moving forward.

Liposomes initially found clinical success as drug carriers in cancer treatment with the development of Doxil® (liposomal doxorubicin), which reduced the drug's dose-limiting cardiotoxicity and increased tumor exposure due to the enhanced permeability and retention (EPR) effect [36]. The EPR effect concept was first introduced by Matsumura and Maeda et al. in 1986; this ability of nanoparticle-based formulations to accumulate in tumor tissue is now widely recognized and was recently reviewed by Price et al. [37,38]. In particular, liposomes have become a commonly used formulation to passively target one or multiple drugs to tumors. Due to CQ's anticancer activity, researchers have developed liposomal formulations combining CQ and other chemotherapeutics for enhanced anticancer efficacy. For example, liposomes co-loaded with CQ and paclitaxel (PTX) or doxorubicin (DXR) resulted in tumor growth suppression in A549/T-tumor-bearing mice and MCF-7/ADR-tumor-bearing zebrafish, respectively [39,40]. However, the authors did not compare to unformulated drug controls in the efficacy or drug distribution studies, and therefore, it is unclear if the liposomal formulations provided any benefits to CQ delivery, a major shortcoming of these studies.

Table 1. Nanomedicine formulations of CQ and HCQ tested in vivo.

Formulation	Drug Loading Method	Drug	Co-drug	Indication	Reference
Liposome (Anti-mouse erythrocyte F(ab′)$_2$ targeted- egg PC/Chol/gangliosides)	Passive loading by thin film sonication	CQ	-	Malaria	[25]
Liposome (Mouse monoclonal antibody F$_{10}$ targeted- egg PC/chol/gangliosides)	Passive loading by thin film sonication	CQ	-	Malaria	[26]
Liposome (Soybean PC/egg PG/Chol)	Passive loading by reverse-phase evaporation, sonication, and extrusion	CQ	-	Malaria	[28–30]
Liposome (Anti-mouse erythrocyte Fab′ targeted-MPC-PE, Chol/PC/PS)	Passive loading by reverse-phase evaporation, sonication, and extrusion	CQ	-	Malaria	[27]
Liposome (Multilamellar- DOPC/DPGG/amine-N-[4-(p-maleimidophenyl) butyramide)])	Passive loading by thin film sonication	CQ	-	Malaria	[31]
Liposome (Glycophorin A targeted-DOPC/DSPC/DSPE-PEG2000-Mal)	Passive loading by thin film hydration, sonication, and extrusion	CQ	-	Malaria	[35]
Liposome (Egg PC/Chol)	Passive loading by thin film hydration and sonication	CQ	Fluconazole	*C. neoformans*	[33,34]
Liposome (Soybean PC/Chol)	Thin film hydration, sonication, and active loading using a citrate pH-gradient	CQ	PTX	Lung cancer	[39]
Liposome (Soybean PC/Chol)	Thin film hydration, sonication, and extrusion, followed by active loading using a citrate pH-gradient	CQ	DXR	Breast cancer	[40]
Liposome (Soybean PC/Chol/DSPE-mPEG2000)	Thin film hydration, sonication, and active loading using a citrate pH-gradient	HCQ	Tat-Beclin1 peptide	Breast cancer	[41]
Liposome (Soybean PC/Chol/DSPE-mPEG2000)	Thin film hydration, sonication, and active loading using a citrate pH-gradient	HCQ	VNP20009	Melanoma	[42]
Liposome (ITGAV-ITGB3/integrin $\alpha_v\beta_3$ receptor-targeted- soybean PC/Chol/DSPE-PEG2000-Mal)	Thin film hydration, sonication, and active loading using a citrate pH-gradient	HCQ	DXR	Melanoma	[43]
Liposome (ITGAV-ITGB3/integrin $\alpha_v\beta_3$ receptor-targeted- soybean PC/Chol/DSPE-PEG2000-Mal)	Thin film hydration, sonication, and active loading using a sulfate pH-gradient	HCQ	PTX	Pancreatic cancer	[44]
Liposome (Neuropilin-1/integrin $\alpha_v\beta_3$ receptor-targeted-soybean PC/Chol/DSPE-PEG2000-Mal)	Thin film hydration, sonication, and active loading using a sulfate pH-gradient	HCQ	PTX	Melanoma	[45]
Liposome (Neuropilin-1/integrin $\alpha_v\beta_3$ receptor-targeted-soybean PC/Chol/DSPE-PEG2000-Mal)	Thin film hydration, sonication, and active loading using a sulfate pH-gradient	HCQ	ZD6473	Glioma	[46]
Liposome (LRP1-targeted-DSPC/DOPC/DSPE-PEG2000-Mal)	Thin film hydration, sonication, and extrusion, followed by active loading using a citrate pH-gradient	HCQ	Chlorin e6	Glioma	[47]

Table 1. *Cont.*

Formulation	Drug Loading Method	Drug	Co-drug	Indication	Reference
Liposome (Chol-HCQ/PC)	Passive loading by thin film hydration and sonication	HCQ (cholesterol modified)	-	Pulmonary fibrosis	[48]
Polymeric micelle (mPEG-PLA)	Thin film hydration	CQ	DXR, PTX, cis-platin	Ovarian cancer	[49]
Polymeric NP (PLGA)	Water-in-oil-in-water (w/o/w) double emulsion solvent evaporation method	CQ	pDNA	Colon cancer	[50]
Polymeric NP (PLGA)	Water-in-oil-in-water (w/o/w) double emulsion solvent evaporation method	HCQ	OVA	Vaccine	[51]
Polymeric NP (CD-20 antibody-targeted- PCL/PLA)	Not described	HCQ	Chlorambucil		[52]
Acrylamide nanogel	Electrostatic complexation	CQ	DXR	Breast cancer	[53]
Dendrimer (Bis-MPA/glycine)	Emulsion evaporation method	CQ	-	Malaria	[54]
Dendrimer (PEG-PLL/galactose)	Equilibrium dialysis	CQ	-	Malaria	[55]
Dendrimer (PEG-PLL/chondroitin A sulfate)	Equilibrium dialysis	CQ	-	Malaria	[56]
Dendrimer (PEI/triphenylphosphate)	Precipitation	CQ	DXR	Prostate cancer	[57]
Polyelectrolyte complex (Poly(amidoamine))	Electrostatic interaction	CQ	-	Malaria	[58,59]
Polyelectrolyte complex (Chitosan/tripolyphosphate)	Electrostatic interaction	CQ	-	Malaria	[60–63]
SLN (Compritol® proprietary lipid)	Melt homogenization method	CQ	-	Arthritis	[64]
Lipid nanoemulsion	Microemulsion method	CQ	-	Malaria	[65]
Niosome gel	Emulsion evaporation method	HCQ	-	Oral lichen planus	[66]
Gold NP	Conjugated to gold np via HCQ thiol prodrug	HCQ	DXR	Glioma	[67]
Titanium dioxide NP	Inclusion complex	HCQ	-	Breast cancer	[68]
Polymer prodrug(poly(N-(2-hydroxypropyl) methacrylamide-co-methacryloylated HCQ))	Polymer ester prodrug	HCQ	-	Inflammatory bowel disease	[69]

Bis-MPA: 2,2′-bis(hydroxymethyl)propionic acid; Chol: cholesterol; DOPC: 2-dioleoyl-sn-glycero-3-phosphocholine; DPGG: 1,2-dipalmitoyl-galloylglycerol; DSPC: 1,2-distearoyl-sn-glycero-3-phosphocholine; DSPE-PEG2000: 1,2-distearoyl-sn-glycero-3-phosphoethanolamine-N-[methoxy(polyethylene glycol)-2000]; DSPE-PEG2000-Mal: 1,2-distearoyl-sn-glycero-3-phosphoethanolamine-N-[maleimide(polyethylene glycol)-2000]); MPB-PE:maleimido-4-(p-phenylbutyrate)-phosphatidylethanolamine; mPEG: methoxypoly(ethylene glycol)-b-poly(lactic acid); PC: phosphatidylcholine; PCL: polycaprolactone; PEG-PLL: poly(ethylene glycol)-block-poly(L-lysine); PEI: polyethylenimine; PG: phosphatidylglycerol; PLA: polylactic acid; PLGA: poly(lactic-co-glycolic acid); PS: phosphatidylserine.

Several HCQ-loaded liposomes have also been developed for cancer indications. For example, Wang et al. combined HCQ-loaded liposomes with TAT-Beclin 1 peptide to induce autophagy catastrophe in a 4T1 breast cancer model [41]. The combination treatment significantly reduced tumor growth compared to unformulated drug controls and HCQ liposomes alone, suggesting improved liposomal HCQ tumor exposure and supporting the strategy of inducing autophagy catastrophe in tumor cells to treat cancer. A similar strategy was used to combine HCQ-loaded liposomes with Salmonella VNP20009 antitumor peptide in a B16F10 melanoma xenograft model [42]. The liposomes increased HCQ concentrations 4-fold within the tumor compared to the free drug control 24 h following injection, with no difference in liver or spleen concentrations at this same time point. This improvement in tumor drug exposure resulted in 90% survival in comparison to 20% survival for the free drug HCQ + VNP20009 combination control group, and no survival in the HCQ, VNP20009, or HCQ-liposome only groups. These studies strongly support the use of liposome formulations to increase HCQ delivery to the tumor site and improve efficacy when combined with other anticancer drugs.

To further improve the delivery of HCQ to tumors, liposomes have been modified with various targeting ligands to enable tumor-specific drug delivery. For example, liposomes decorated with pH-sensitive RGD peptides for targeting ITGAV-ITGB3/integrin $\alpha v\beta 3$ receptors were used for HCQ delivery to melanoma tumors [43]. Both untargeted and targeted versions of the liposomes significantly decreased drug exposure to the heart, spleen, lung, and kidney compared to unformulated drug control, while only the targeted liposome version significantly increased HCQ concentrations within the tumor 24 h post-injection. As a monotherapy, the formulation achieved a median survival of 30 days compared to 25 days from the untargeted liposome treatment group and 15 days from HCQ free drug control. However, when combined with liposomes containing DXR, the median survival improved to >60 days, and tumor growth was significantly inhibited compared to free drug controls or DXR liposomes only. This same ITGAV-ITGB3/integrin $\alpha v\beta 3$ receptor-targeted liposome formulation was also used to co-deliver HCQ and PTX for pancreatic cancer therapy [44]. This formulation achieved significantly better tumor growth inhibition and reduction of metastatic tumor nodules in a BxPC-3-luc orthotopic tumor model compared to targeted liposomes containing either HCQ or PTX, untargeted liposomes containing both drugs, and PTX + HCQ free drug control while not affecting body weight. This ITGAV-ITGB3/integrin $\alpha v\beta 3$ receptor-targeted liposomal formulation not only significantly changed HCQ distribution toward the tumor, but also provided excellent anticancer efficacy when combined with chemotherapeutics.

In addition to integrin $\alpha v\beta 3$ receptors, Yin et al. also targeted neuropilin-1 receptors on melanoma cells for co-delivery of HCQ and PTX [45]. This targeted liposomal formulation significantly inhibited tumor growth and effectively inhibited metastasis in a B16F10 melanoma model compared to an untargeted liposome version and unformulated drug controls. Another liposomal formulation, also targeting the integrin $\alpha v\beta 3$ and neuropilin-1 receptors was co-loaded with HCQ and tyrosine kinase inhibitor ZD6474 and evaluated for efficacy in a C6 glioma model [46]. Interestingly, these liposomes did not significantly change drug exposure to the heart, liver, spleen, lung, or kidney compared to untargeted liposome version or free drug controls, but they did achieve a 4.9-fold increase in drug exposure to the brain in C6 intracranial tumor-bearing mice. This improvement in drug delivery across the blood-brain barrier (BBB) resulted in significantly prolonged median survival time with the targeted, co-loaded liposomes (41 days) compared to the untargeted liposome version (35 days) and unformulated, free drug controls (28 days).

Ultrasound (US) is another method that has been investigated as a means to improve nanoparticle delivery across the BBB and is also involved in sonodynamic therapy [70,71]. This strategy was used to improve the delivery of HCQ and sonoactive chlorin e6 to glioma tumors using angiopep-2 peptide-modified liposomes that target low-density lipoprotein receptor-related protein 1 (LRP1) [47]. Combined with ultrasonic pulse, the targeted liposome containing both drugs achieved the greatest median survival time of 52 days compared to 40 days from the untargeted liposome version and 33 days from chlorin e6 + HCQ unformulated drug controls. Combining autophagy inhibitors with sonodynamic therapy through targeted drug delivery to brain tumors may offer a novel therapeutic strategy for glioma.

Drugs that are not suitable for remote loading into the liposomal aqueous core or are not sufficiently lipophilic to associate with the lipid bilayer can be conjugated to a lipid anchor to facilitate loading within the lipid bilayer [72]. Although researchers have shown HCQ can be successfully incorporated in the liposomal aqueous core with high drug loading, Liu et al. developed a liposome bilayer-loaded cholesterol-modified version of HCQ for the treatment of pulmonary fibrosis [48]. Both cholesterol-modified HCQ liposomes and core-loaded HCQ liposomes inhibited the development of bleomycin-induced pulmonary fibrosis in Sprague-Dawley rats; however, the authors did not compare to unformulated HCQ, so the benefits of using a liposomal bilayer-loaded cholesterol-modified HCQ formulation remain unclear.

3. Polymeric Nanoparticles

Polymer-based nanoparticles have been used to improve the solubility of hydrophobic drugs and facilitate enhanced tumor distribution through the EPR effect. Polymeric micelles, one of several different types of polymer-based nanoparticles, generally consist of amphipathic polymers that co-precipitate with drugs to form a hydrophobic core surrounded by a hydrophilic shell. These formulations have been shown to have low critical micelle concentrations (CMC) and have better stability than traditional surfactant micellar systems due to hydrophobic interactions between the drug and polymer [73]. CQ is a hydrophobic drug with a high logP of 4.72 and is predicted to be suitable for polymeric micelle formulations based on previous analysis of how drug properties influence nanomedicine compatibility [74,75]. Despite this, few examples of CQ-polymeric micelles have been reported. In one study, micelles composed of methoxy PEG-b-poly(L-lactic acid) (mPEG-PLA) were used to co-load CQ with either DXR, PTX, or cis-platin [49]. In all cases, the micellar formulations provided superior efficacy in ovarian cancer models compared to unformulated drug combinations, indicating improved tumor distribution.

In addition to micelles, polymeric nanoparticles can be formed through emulsion techniques. This approach can be used to encapsulate hydrophilic drugs and biologics within the polymer matrix and do not require amphipathic polymers. For example, Yang et al. developed a nanoparticle composed of poly(lactic-co-glycolic acid) (PLGA) for co-delivery of CQ and pDNA expressing the mSurvivin-T34A protein [50]. In this case, CQ was used for pDNA compaction through electrostatic interactions as well as for improving lysosome escape of the pDNA following cell uptake. This formulation provided better tumor growth inhibition compared to pDNA/PLGA nanoparticle without CQ in a CT26 tumor model. However, the authors did not compare to a CQ free drug control or to pDNA/PLGA + CQ administered separately, so it is unclear if the improvement in efficacy is due simply to the addition of CQ or to an improvement in CQ drug delivery.

HCQ has also been formulated with biologics to aid in lysosome escape. For example, Liu et al. developed a PLGA nanoparticle co-loaded with HCQ and ovalbumin (OVA) as a model antigen for a proof-of-concept vaccine delivery formulation [51]. This formulation provided statistically significant tumor growth inhibition in an OVA-sensitive E.G7-OVA xenograft tumor model compared to free OVA or OVA-nanoparticles alone, but the authors did not include controls for unformulated HCQ administered alone or in combination with OVA-nanoparticles. Further studies are required to determine if there is a benefit to formulating CQ or HCQ to facilitate cytosolic delivery of biologics, or if the same effects can be achieved by simply administering the drugs separately.

Similar to liposomes, polymeric nanoparticles can be coated with antibodies to enable tumor-specific drug delivery, but few have been developed for CQ or HCQ. In one example, a cd20-antibody-targeted poly(caprolactone)/PLA nanoparticle was co-loaded with HCQ and chlorambucil and evaluated for efficacy in a Burkitt lymphoma animal model [52].

The targeted nanoparticle provided 90% survival after 120 days compared to 40% survival in animals treated with the antibody alone and 0% survival in animals treated with untargeted, drug-loaded nanoparticles or free drug combination controls. Interestingly, at non-toxic doses, the untargeted version of the nanoparticle provided worse survival (0%) compared to the free drug combination control (33%), indicating an untargeted polymeric nanoparticle may unfavorably change tissue distribution of these drugs.

Although most polymers used in drug delivery are biodegradable, some non-biodegradable polymers such as acrylamide-based polymers have shown success for small molecule and oligonucleotide delivery [76,77]. One major advantage of acrylic polymers is the wide selection of functionalized monomers available to form polymers with different physicochemical properties. For example, poly(N-isopropylacrylamide-co-acrylic acid) (PNIPAM-Aac) is a negatively charged polymer that can undergo electrostatic complexation with positively charged molecules. This approach was used to co-load CQ and DXR within PNIPAM-Aac nanogels to induce autophagy catastrophe within tumor cells [53]. Despite successful drug loading, CQ release in PBS was rapid, with more than 50% in the first two hours and more than 95% over 12 h. This rapid drug release is likely too fast to benefit from any passive tumor targeting of the nanoparticle. Indeed, in an efficacy study in an MCF-7 breast cancer model, the nanoparticles containing both drugs did not achieve a statistically significant decrease in tumor weight compared to CQ-only nanogels. The authors also did not compare to a free drug DXR + CQ control to prove the benefit of nanoparticle delivery.

Overall, due to a lack of appropriate controls, there is limited data to support the utilization of polymeric nanoparticles for improving the delivery of either CQ or HCQ.

4. Dendrimers

Dendrimers are repetitively branched molecules generally constructed as macromolecular polymers with variable cores and terminal groups to facilitate drug encapsulation and drug delivery [78]. Properties such as size, morphology, and surface chemistry can be controlled through synthetic chemistry steps and designed for specific drug delivery needs. To improve CQ delivery to Plasmodium-infected red blood cells, Marti Coma-Cros et al. designed cationic dendrimers based on Pluronic F127 and 2,2′-bis(glycyloxymethyl)propionic acid as well as a hyperbranched dendrimer derived from 2,2′-bis(hydroxymethyl)propionic acid [54]. Although both dendrimer formulations were capable of loading CQ and demonstrated parasite growth inhibition in vitro, they provided worse survival outcomes (20%) in *P. yoelii*-infected mice compared to CQ control (80%), indicating the formulations significantly reduced the antimalarial efficacy of CQ. One possible explanation for this decrease in efficacy could be due to a reduction in systemic drug exposure. Previously, dendrimers composed of PEG and poly(lysine) with and without galactose terminal groups significantly reduced the maximum concentration (Cmax) and area under the concentration-time curve (AUC) of CQ in comparison to unformulated CQ [55]. A similar CQ-loaded PEG-poly(lysine dendrimer) with a chondroitin sulfate A coating also significantly reduced Cmax compared to free drug (13.85 and 50.23 μg/mL, respectively), but increased AUC from 74.72 to 120.58 μg*h/mL; however, in this case the differences in PK were likely due to the routes of administration, since the unformulated drug was administered intravenously and the dendrimer formulation was administered intramuscularly [56].

Alternatively, Panagiotaki et al. designed dendrimers composed of poly(ethylenimine) with triphenylphosphate terminal groups to facilitate mitochondrial delivery of DXR and CQ for improved cancer therapy [57]. Dendrimer formulations were developed for each drug and, when administered together, significantly reduced tumor volume in DU145 tumor-bearing mice. However, the efficacy was only slightly better than the DXR-only dendrimer, and the authors did not compare to a CQ-only dendrimer formulation or DXR + CQ free drug control. Therefore, it is unclear whether this formulation provided any benefit to the delivery or anticancer efficacy of CQ.

5. Polyelectrolyte Complexes

Polyelectrolyte complexes, also sometimes referred to as polyplexes and coacervates, are formed by mixing oppositely charged polyionic species in an aqueous medium, and various ionic polymers have been investigated extensively for their ability to complex with nucleic acids [79]. However, their use for delivering small molecule drugs has been limited, likely due to the necessity of multiple charge sites per drug molecule to allow stable complexation with the polymer.

CQ is positively charged at physiological pH due to its two ionizable amine groups, and because of this, researchers have attempted to load the drug into complexes containing ionic polymers. In one example, Urban et al. developed poly(amidoamine) polymers that formed ~10 nm complexes when mixed with CQ [59]. Drug release from the formulations in PBS was nearly identical to unformulated CQ, indicating formulation instability. Surprisingly, *P. yoelii*-infected mice treated with the polymer/CQ complexes achieved 100% survival 30-days post-infection compared to 0% survival in the unformulated CQ control group. Although the polymers alone were shown to reduce parasitemia in vitro, polymer-only controls were not included in the in vivo efficacy study. Therefore, it is unclear whether the improved survival is due to an improvement in CQ delivery or rather due to additive or synergistic effects of the drug and polymers. Furthermore, the formulations provided no statistically significant improvement in survival compared to CQ alone in *P. yoelii*-infected mice when administered orally [58].

Another CQ-polyelectrolyte complex, composed of chitosan and tripolyphosphate, was shown to reduce parasitemia to a greater extent than unformulated CQ in several efficacy studies in *P. berghei*-infected mice [60–63]. However, the authors did not use vehicle-only controls in any of the studies to rule out the possible antimalarial activity of the polymer complex itself. Overall, these studies support the use of combining ionic polymers with CQ to improve malaria treatment since there is evidence of better survival outcomes and reduced parasitemia, possibly due to additive effects between CQ and the ionic polymers, rather than improved delivery to target cells.

6. Non-Liposomal Lipid-Based Nanoparticles

In addition to liposomes, there are a variety of other lipid-based nanoparticles including solid lipid nanoparticles (SLN), nanoemulsions, and niosomes. These formulations are generally used for improving the solubility and delivery of hydrophobic drugs and are highly biocompatible and biodegradable due to their physiological lipid compositions.

Unlike other lipid-based carriers, SLN contain a solid lipid core and are often utilized as oral formulations to improve solubility and intestinal absorption of hydrophobic drugs [80,81]. CQ is typically administered orally and has highly variable bioavailability ranging from 52% to 102% as an oral solution and 67–114% as a tablet [82]. It has also been shown that taking CQ with food results in significantly higher Cmax and AUC, and it is recommended to avoid an upset stomach during CQ dosing [83]. Despite having high oral bioavailability, Bhalekar et al. attempted to improve CQ oral delivery and intestinal lymphatic uptake using a SLN formulation for arthritis therapy [64]. The SLN formulation achieved 2-fold increases in Cmax, time of maximum concentration (Tmax), and AUC in comparison to standard CQ suspension, reportedly due to intestinal lymphatic uptake and bypassing first-pass metabolism. Consequently, the SLN formulation achieved greater paw volume reduction compared to the standard CQ suspension in the arthritis mouse model.

In addition to loading drugs, lipid-based carriers have been shown to inhibit malarial parasitemia in erythrocytes [84]. Due to these properties, Baruah et al. developed CQ-loaded, cationic nanoemulsions to improve antimalarial efficacy [65]. The formulation suppressed parasitemia by 99.68% compared to only 76.5% by unformulated CQ in *P. berghei*-infected mice 5 days post-infection. However, the blank lipid emulsion reduced parasitemia by 35.35%, indicating the lipid emulsion alone inhibited malarial infection. Therefore, it is unclear if the efficacy from the CQ nanoemulsion is due to an improvement in drug delivery or simply additive or synergistic effects with the lipid emulsion and drug.

Niosomes are another class of drug delivery vehicle capable of loading both hydrophobic and hydrophilic drugs. Niosomes are similar to liposomes in that they also contain a bilayer and an aqueous core. Unlike liposomes, which typically utilize phospholipids, niosomes are formed from mixtures of non-ionic surfactant molecules and cholesterol. Niosomes have been used for transdermal drug delivery due to their ability to improve drug penetration through the skin and provide local and sustained drug release [85]. This strategy was used to develop a HCQ-loaded niosome formulation dispersed in a Pluronic F-127 gel for the treatment of oral lichen planus [66]. Human patients applied the niosome gel with or without the drug (placebo group) to their lesion every day for four months. Patients receiving the HCQ-containing gel observed an average lesion size reduction of 64.28% compared to only 3.94% reduction in the placebo group. On a pain score from 0 to 10, where 0 is no pain and 10 is the worst pain, patients in the gel and placebo groups reported pain scores of 4 and 3 pre-treatment and 1 and 3 post-treatment, respectively. Although these data support the benefits of this HCQ niosome gel in human patients, the authors did not compare to HCQ gel control, HCQ free drug control, or standard of care (corticosteroids). Therefore, it is unclear whether encapsulation within niosome provided any benefits to the delivery of HCQ.

7. Metal Nanoparticles

Metallic nanoparticles have been successfully implemented as contrast agents and many are being investigated as therapeutic agents and drug delivery vehicles [86,87]. One of their limitations for drug delivery is the requirement of functional groups on the drug that can undergo chelation with metals. For example, thiol-containing drugs can be conjugated to the surface of gold nanoparticles through Au-thiol bonding. Upon cell entry, thiol-exchange with intracellular glutathione releases the drug. Drugs without thiol groups must be chemically modified as prodrugs in order to conjugate to gold nanoparticles and allow the release of the parent drug. Ruan et al. used this strategy to modify DXR and HCQ as ester prodrugs containing terminal thiol groups to enable coupling to gold nanoparticles and evaluated these nanoparticles for antiglioma efficacy [67]. The nanoparticles containing both drugs resulted in a 56-day median survival in C6 glioma-bearing mice compared to 44 days from nanoparticles containing only DXR; however, the results were not statistically significant. The nanoparticles containing only HCQ resulted in a 38-day median survival compared to 30 days from the free HCQ treatment group, though a better control would have been the modified version of HCQ since this is the molecule that is released from the gold nanoparticle. The authors described in vitro DXR release in PBS at acidic pH, but they did not investigate HCQ release, and drug release in plasma would be a better predictor of nanoparticle stability in vivo since plasma contains both glutathione and esterase enzymes. Therefore, the stability of the HCQ prodrug and its chelation with the nanoparticle surface are unclear.

HCQ has also been used to enhance sonodynamic therapy of metallic nanoparticles through autophagy disruption. For example, Feng et al. designed HCQ-loaded hollow mesoporous titanium dioxide nanoparticles that are coated with a cancer cell membrane to allow homologous targeting to the tumor [68]. HCQ release in PBS from coated nanoparticles was much slower than that of uncoated nanoparticles, but the release became equivalent to the uncoated particles when exposed to US irradiation, suggesting a US responsive drug release mechanism. In MCF-7 tumor-bearing mice, the cancer cell membrane coated nanoparticles extended the systemic half-life of HCQ to 12.3 ± 1.7 h, which was higher than that of uncoated nanoparticles (8.7 ± 1.3 h) and free HCQ (3.4 ± 0.4 h). However, it is unclear if the authors measured the total drug fraction in the blood or the released (pharmacologically active) fraction. The PK of nanomedicines is very complex since total drug concentration in the plasma and blood, as well as tissues, is comprised of encapsulated and unencapsulated drug fractions, and both fractions can contribute to drug efficacy and toxicity [88]. Nevertheless, the cancer cell membrane coated nanoparticles containing HCQ combined with tumor US irradiation significantly reduced tumor growth compared to empty nanoparticles + US and free HCQ controls, supporting the strategy of combining US with autophagy disruption. However, the degree to which the nanoparticle improved HCQ exposure of the tumor site remains unknown, and treatment of HCQ + nanoparticle + US may have been just as effective.

8. Conclusions and Perspectives

There is new interest in repurposing CQ and HCQ for novel applications such as cancers, as well as improving therapy for their traditional indications such as infectious and inflammatory diseases. Nanomedicines have been evaluated for their ability to improve the safety and efficacy of chloroquines. There are a variety of nanoparticle types, with each having their own advantages and disadvantages, and it is important to understand the liabilities and physicochemical properties of the drug being formulated in order to select the most appropriate platform. In the case of CQ and HCQ, off-target toxicities can be reduced, and efficacy enhanced using a combination of site-specific drug delivery and controlled release; the balance between delivery and release kinetics being a crucial factor in improving therapeutic index [89]. In order to achieve this, researchers have tested nearly every type of nanomedicine available, with many failing to conclusively demonstrate benefits to CQ or HCQ therapy.

Polymeric nanoparticles, which have been successful in formulating hydrophobic drugs in preclinical and clinical studies, are typically unstable formulations that release their drug immediately after injection, thereby eliminating any potential benefits of nanoparticle distribution and essentially acting as solubilizing formulations. For example, Genexol® PM, a polymeric nanoparticle formulation of PTX that is approved as a cancer therapy in South Korea, has been shown to completely release its drug within 10 min after exposure to plasma [90]. Dendrimers and polyelectrolyte complexes have shown promising preclinical results for gene delivery but have been less successful in formulating small molecule drugs. Dendrimer-drug conjugates of chemotherapeutics are currently undergoing clinical trials, and this may prove to be a more useful strategy since drug release stability is controlled through the linker chemistry [91,92]. Metallic nanoparticles have been approved as contrast and therapeutic agents, but none have proven useful for improving the delivery of small molecule drugs, likely due to insufficiently stable drug-metal interactions. All of these nanoparticle types have been used to reformulate CQ and HCQ, but most have not provided sufficient evidence of improving their efficacy and safety profile. In many cases, appropriate controls were missing, and it was unclear if the efficacy of CQ and HCQ was due to an improvement in drug delivery or if the same results could be achieved using the unformulated drugs. Therefore, additional PK and efficacy studies with appropriate controls are needed to support the use of these nanomedicine formulations for CQ or HCQ delivery. Further, toxicity studies are also rarely performed on these formulations and are necessary for evaluation of improvements to the therapeutic index overall.

On the other hand, liposomal formulations appeared to provide a clear benefit to the delivery of CQ and HCQ in various malaria and tumor models, respectively. With its ionizable amine groups, CQ can be actively loaded into the aqueous liposomal core, and erythrocyte-specific targeting ligands on the surface of the liposomes improve drug uptake within red blood cells, a target for malaria. Since lipids have been shown to inhibit Plasmodium infection, combining CQ with lipid-based carriers may provide not only better drug delivery to uninfected and infected red blood cells, but also synergistic efficacy. Liposomes also make a good choice for improving the delivery of these drugs to tumors. With their ~100 nm size and good stability, liposomes are able to accumulate within the tumor microenvironment via the EPR effect and deliver their therapeutic cargo [93]. With the help of targeting peptides on their surface, liposomes were able to co-deliver HCQ and other chemotherapeutics to significantly improve efficacy and survival outcomes and appear to be a promising strategy for cancer therapy moving forward. However, one disadvantage of liposomes is that they are generally very stable with extremely long drug release half-lives. For example, Doxil has a drug release half-life greater than 100 h, and there are currently efforts to design less stable liposomes that provide faster drug release rates at the site of interest [94–96].

One notable absence in the above nanotechnology formulation discussion of chloroquines is polymer prodrug systems, a major drug delivery class that has scarcely been evaluated for these drugs and may offer an ideal balance of targeting and stability. Polymer prodrugs can be designed to be biodegradable, provide site-specific targeting, and enable controlled drug release through the polymer-drug linker chemistry [97]. This strategy has proven useful for the delivery of small molecule drugs for cancer and neurological diseases, and there are several candidates in clinical trials [98]. To our knowledge, only a single example of a polymer prodrug of HCQ evaluated in vivo has been published, and it demonstrated substantially better efficacy and lower toxicity compared to unformulated HCQ in a mouse model of colitis [69].

It should be emphasized that despite the promising preclinical data for some of the formulations presented in this review, none of the formulations have made it to the clinical stage. The lack of clinical development is likely due to poor intellectual property protections and uncertain commercial promise for the formulation platforms presented, many of which rely on generic formulation strategies. However, it is expected that the recent commercial success of novel nanotechnology-based delivery platforms and renewed interest in chloroquine drugs for novel indications, such as cancer, will fuel future clinical development of chloroquine nanoformulations [9,99]. Overall, reformulation efforts of CQ and HCQ through nanomedicine approaches have shown some promising improvements in efficacy and safety, but further developments are warranted.

Author Contributions: Writing—original draft preparation, D.M.S. and S.T.S.; writing—review and editing, D.M.S., R.M.C. and S.T.S. All authors have read and agreed to the published version of the manuscript.

Funding: This project has been funded in whole or in part with Federal funds from the National Cancer Institute, National Institutes of Health, under Contract No. 75N91019D00024. The content of this publication does not necessarily reflect the views or policies of the Department of Health and Human Services, nor does mention of trade names, commercial products, or organizations imply endorsement by the U.S. Government.

Institutional Review Board Statement: Not applicable.

Informed Consent Statement: Not applicable.

Data Availability Statement: No new data were created or analyzed in this study. Data sharing is not applicable to this article.

Acknowledgments: The authors thank Joseph Meyer (Leidos Biomedical Research, Inc.) for the figure.

Conflicts of Interest: The authors declare no conflict of interest.

References

1. Chew, C.Y.; Mar, A.; Nikpour, M.; Saracino, A.M. Hydroxychloroquine in dermatology: New perspectives on an old drug. *Australas J. Dermatol.* **2020**, *61*, e150–e157. [CrossRef]
2. Mian, A.; Ibrahim, F.; Scott, D.L. A systematic review of guidelines for managing rheumatoid arthritis. *BMC Rheumatol.* **2019**, *3*, 42. [CrossRef] [PubMed]
3. Taherian, E.; Rao, A.; Malemud, C.J.; Askari, A.D. The biological and clinical activity of anti-malarial drugs in autoimmune disorders. *Curr. Rheumatol. Rev.* **2013**, *9*, 45–62. [CrossRef]
4. Dolgin, E. Anticancer autophagy inhibitors attract 'resurgent' interest. *Nat. Rev. Drug Discov.* **2019**, *18*, 408–410. [CrossRef]
5. Shi, T.T.; Yu, X.X.; Yan, L.J.; Xiao, H.T. Research progress of hydroxychloroquine and autophagy inhibitors on cancer. *Cancer Chemother. Pharmacol.* **2017**, *79*, 287–294. [CrossRef] [PubMed]
6. Maes, H.; Kuchnio, A.; Peric, A.; Moens, S.; Nys, K.; De Bock, K.; Quaegebeur, A.; Schoors, S.; Georgiadou, M.; Wouters, J.; et al. Tumor vessel normalization by chloroquine independent of autophagy. *Cancer Cell* **2014**, *26*, 190–206. [CrossRef]
7. Verbaanderd, C.; Maes, H.; Schaaf, M.B.; Sukhatme, V.P.; Pantziarka, P.; Sukhatme, V.; Agostinis, P.; Bouche, G. Repurposing Drugs in Oncology (ReDO)-chloroquine and hydroxychloroquine as anti-cancer agents. *Ecancermedicalscience* **2017**, *11*, 781. [CrossRef] [PubMed]

8. Chen, D.; Xie, J.; Fiskesund, R.; Dong, W.; Liang, X.; Lv, J.; Jin, X.; Liu, J.; Mo, S.; Zhang, T.; et al. Chloroquine modulates antitumor immune response by resetting tumor-associated macrophages toward M1 phenotype. *Nat. Commun.* **2018**, *9*, 873. [CrossRef] [PubMed]
9. Kinsey, C.G.; Camolotto, S.A.; Boespflug, A.M.; Guillen, K.P.; Foth, M.; Truong, A.; Schuman, S.S.; Shea, J.E.; Seipp, M.T.; Yap, J.T.; et al. Protective autophagy elicited by RAF–>MEK–>ERK inhibition suggests a treatment strategy for RAS-driven cancers. *Nat. Med.* **2019**, *25*, 620–627. [CrossRef]
10. Dyall, J.; Coleman, C.M.; Hart, B.J.; Venkataraman, T.; Holbrook, M.R.; Kindrachuk, J.; Johnson, R.F.; Olinger, G.G., Jr.; Jahrling, P.B.; Laidlaw, M.; et al. Repurposing of clinically developed drugs for treatment of Middle East respiratory syndrome coronavirus infection. *Antimicrob. Agents Chemother.* **2014**, *58*, 4885–4893. [CrossRef]
11. Keyaerts, E.; Vijgen, L.; Maes, P.; Neyts, J.; Van Ranst, M. In vitro inhibition of severe acute respiratory syndrome coronavirus by chloroquine. *Biochem. Biophys. Res. Commun.* **2004**, *323*, 264–268. [CrossRef] [PubMed]
12. Wang, M.; Cao, R.; Zhang, L.; Yang, X.; Liu, J.; Xu, M.; Shi, Z.; Hu, Z.; Zhong, W.; Xiao, G. Remdesivir and chloroquine effectively inhibit the recently emerged novel coronavirus (2019-nCoV) in vitro. *Cell Res.* **2020**, *30*, 269–271. [CrossRef] [PubMed]
13. Coronavirus (COVID-19) Update: FDA Revokes Emergency Use Authorization for Chloroquine and Hydroxychloroquine. Available online: https://www.fda.gov/news-events/press-announcements/coronavirus-covid-19-update-fda-revokes-emergency-use-authorization-chloroquine-and (accessed on 18 November 2020).
14. Mehra, M.R.; Ruschitzka, F.; Patel, A.N. Retraction-Hydroxychloroquine or chloroquine with or without a macrolide for treatment of COVID-19: A multinational registry analysis. *Lancet* **2020**, *395*, 1820. [CrossRef]
15. Hoffmann, M.; Mosbauer, K.; Hofmann-Winkler, H.; Kaul, A.; Kleine-Weber, H.; Kruger, N.; Gassen, N.C.; Muller, M.A.; Drosten, C.; Pohlmann, S. Chloroquine does not inhibit infection of human lung cells with SARS-CoV-2. *Nature* **2020**, *585*, 588–590. [CrossRef]
16. Browning, D.J. Pharmacology of Chloroquine and Hydroxychloroquine. In *Hydroxychloroquine and Chloroquine Retinopathy*; Springer: New York, NY, USA, 2014.
17. Leung, L.S.; Neal, J.W.; Wakelee, H.A.; Sequist, L.V.; Marmor, M.F. Rapid Onset of Retinal Toxicity From High-Dose Hydroxychloroquine Given for Cancer Therapy. *Am. J. Ophthalmol.* **2015**, *160*, 799–805 e1. [CrossRef]
18. Michaelides, M.; Stover, N.B.; Francis, P.J.; Weleber, R.G. Retinal toxicity associated with hydroxychloroquine and chloroquine: Risk factors, screening, and progression despite cessation of therapy. *Arch. Ophthalmol.* **2011**, *129*, 30–39. [CrossRef]
19. Offerhaus, J.A.; Wilde, A.A.M.; Remme, C.A. Prophylactic (hydroxy)chloroquine in COVID-19: Potential relevance for cardiac arrhythmia risk. *Heart Rhythm.* **2020**, *17*, 1480–1486. [CrossRef]
20. Furst, D.E. Pharmacokinetics of hydroxychloroquine and chloroquine during treatment of rheumatic diseases. *Lupus* **1996**, *5* (Suppl. 1), S11–S15. [CrossRef]
21. Hua, S.; de Matos, M.B.C.; Metselaar, J.M.; Storm, G. Current Trends and Challenges in the Clinical Translation of Nanoparticulate Nanomedicines: Pathways for Translational Development and Commercialization. *Front. Pharmacol.* **2018**, *9*, 790. [CrossRef]
22. Anselmo, A.C.; Mitragotri, S. Nanoparticles in the clinic: An update. *Bioeng. Transl. Med.* **2019**, *4*, e10143. [CrossRef]
23. Abra, R.M.; Bankert, R.B.; Chen, F.; Egilmez, N.K.; Huang, K.; Saville, R.; Slater, J.L.; Sugano, M.; Yokota, S.J. The next generation of liposome delivery systems: Recent experience with tumor-targeted, sterically-stabilized immunoliposomes and active-loading gradients. *J. Liposome Res.* **2002**, *12*, 1–3. [CrossRef] [PubMed]
24. Woodle, M.C. Surface-modified liposomes: Assessment and characterization for increased stability and prolonged blood circulation. *Chem. Phys. Lipids* **1993**, *64*, 249–262. [CrossRef]
25. Agrawal, A.K.; Singhal, A.; Gupta, C.M. Functional drug targeting to erythrocytes in vivo using antibody bearing liposomes as drug vehicles. *Biochem. Biophys. Res. Commun.* **1987**, *148*, 357–361. [CrossRef]
26. Owais, M.; Varshney, G.C.; Choudhury, A.; Chandra, S.; Gupta, C.M. Chloroquine encapsulated in malaria-infected erythrocyte-specific antibody-bearing liposomes effectively controls chloroquine-resistant Plasmodium berghei infections in mice. *Antimicrob. Agents Chemother.* **1995**, *39*, 180–184. [CrossRef] [PubMed]
27. Peeters, P.A.; Brunink, B.G.; Eling, W.M.; Crommelin, D.J. Therapeutic effect of chloroquine(CQ)-containing immunoliposomes in rats infected with Plasmodium berghei parasitized mouse red blood cells: Comparison with combinations of antibodies and CQ or liposomal CQ. *Biochim. Biophys. Acta* **1989**, *981*, 269–276. [CrossRef]
28. Peeters, P.A.; de Leest, K.; Eling, W.M.; Crommelin, D.J. Chloroquine blood levels after administration of the liposome-encapsulated drug in relation to therapy of murine malaria. *Pharm. Res.* **1989**, *6*, 787–793. [CrossRef]
29. Peeters, P.A.; Huiskamp, C.W.; Eling, W.M.; Crommelin, D.J. Chloroquine containing liposomes in the chemotherapy of murine malaria. *Parasitology* **1989**, *98 Pt 3*, 381–386. [CrossRef]
30. Titulaer, H.A.; Eling, W.M.; Crommelin, D.J.; Peeters, P.A.; Zuidema, J. The parenteral controlled release of liposome encapsulated chloroquine in mice. *J. Pharm. Pharmacol.* **1990**, *42*, 529–532. [CrossRef]
31. Fotoran, W.L.; Muntefering, T.; Kleiber, N.; Miranda, B.N.M.; Liebau, E.; Irvine, D.J.; Wunderlich, G. A multilamellar nanoliposome stabilized by interlayer hydrogen bonds increases antimalarial drug efficacy. *Nanomedicine* **2019**, *22*, 102099. [CrossRef]
32. Gabizon, A.; Chisin, R.; Amselem, S.; Druckmann, S.; Cohen, R.; Goren, D.; Fromer, I.; Peretz, T.; Sulkes, A.; Barenholz, Y. Pharmacokinetic and imaging studies in patients receiving a formulation of liposome-associated adriamycin. *Br. J. Cancer* **1991**, *64*, 1125–1132. [CrossRef]

33. Khan, M.A.; Jabeen, R.; Mohammad, O. Prophylactic role of liposomized chloroquine against murine cryptococcosis less susceptible to fluconazole. *Pharm. Res.* **2004**, *21*, 2207–2212. [CrossRef] [PubMed]
34. Khan, M.A.; Jabeen, R.; Nasti, T.H.; Mohammad, O. Enhanced anticryptococcal activity of chloroquine in phosphatidylserine-containing liposomes in a murine model. *J. Antimicrob. Chemother.* **2005**, *55*, 223–228. [CrossRef] [PubMed]
35. Moles, E.; Urban, P.; Jimenez-Diaz, M.B.; Viera-Morilla, S.; Angulo-Barturen, I.; Busquets, M.A.; Fernandez-Busquets, X. Immunoliposome-mediated drug delivery to Plasmodium-infected and non-infected red blood cells as a dual therapeutic/prophylactic antimalarial strategy. *J. Control. Release* **2015**, *210*, 217–229. [CrossRef] [PubMed]
36. Gabizon, A.; Catane, R.; Uziely, B.; Kaufman, B.; Safra, T.; Cohen, R.; Martin, F.; Huang, A.; Barenholz, Y. Prolonged circulation time and enhanced accumulation in malignant exudates of doxorubicin encapsulated in polyethylene-glycol coated liposomes. *Cancer Res.* **1994**, *54*, 987–992. [PubMed]
37. Matsumura, Y.; Maeda, H. A new concept for macromolecular therapeutics in cancer chemotherapy: Mechanism of tumoritropic accumulation of proteins and the antitumor agent smancs. *Cancer Res.* **1986**, *46 Pt 1*, 6387–6392.
38. Price, L.S.L.; Stern, S.T.; Deal, A.M.; Kabanov, A.V.; Zamboni, W.C. A reanalysis of nanoparticle tumor delivery using classical pharmacokinetic metrics. *Sci. Adv.* **2020**, *6*, eaay9249. [CrossRef] [PubMed]
39. Gao, M.; Xu, Y.; Qiu, L. Enhanced combination therapy effect on paclitaxel-resistant carcinoma by chloroquine co-delivery via liposomes. *Int. J. Nanomed.* **2015**, *10*, 6615–6632.
40. Gao, M.; Xu, Y.; Qiu, L. Sensitization of multidrug-resistant malignant cells by liposomes co-encapsulating doxorubicin and chloroquine through autophagic inhibition. *J. Liposome Res.* **2017**, *27*, 151–160. [CrossRef]
41. Wang, Y.; Tai, X.; Zhang, L.; Liu, Y.; Gao, H.; Chen, J.; Shi, K.; Zhang, Q.; Zhang, Z.; He, Q. A novel antitumour strategy using bidirectional autophagic vesicles accumulation via initiative induction and the terminal restraint of autophagic flux. *J. Control. Release* **2015**, *199*, 17–28. [CrossRef]
42. Wang, Y.; Zhou, Z.; Chen, W.; Qin, M.; Zhang, Z.; Gong, T.; Sun, X. Potentiating bacterial cancer therapy using hydroxychloroquine liposomes. *J. Control. Release* **2018**, *280*, 39–50. [CrossRef]
43. Wang, Y.; Shi, K.; Zhang, L.; Hu, G.; Wan, J.; Tang, J.; Yin, S.; Duan, J.; Qin, M.; Wang, N.; et al. Significantly enhanced tumor cellular and lysosomal hydroxychloroquine delivery by smart liposomes for optimal autophagy inhibition and improved antitumor efficiency with liposomal doxorubicin. *Autophagy* **2016**, *12*, 949–962. [CrossRef] [PubMed]
44. Chen, X.; Yu, Q.; Liu, Y.; Sheng, Q.; Shi, K.; Wang, Y.; Li, M.; Zhang, Z.; He, Q. Synergistic cytotoxicity and co-autophagy inhibition in pancreatic tumor cells and cancer-associated fibroblasts by dual functional peptide-modified liposomes. *Acta Biomater.* **2019**, *99*, 339–349. [CrossRef] [PubMed]
45. Yin, S.; Xia, C.; Wang, Y.; Wan, D.; Rao, J.; Tang, X.; Wei, J.; Wang, X.; Li, M.; Zhang, Z.; et al. Dual receptor recognizing liposomes containing paclitaxel and hydroxychloroquine for primary and metastatic melanoma treatment via autophagy-dependent and independent pathways. *J. Control. Release* **2018**, *288*, 148–160. [CrossRef] [PubMed]
46. Wang, X.; Qiu, Y.; Yu, Q.; Li, H.; Chen, X.; Li, M.; Long, Y.; Liu, Y.; Lu, L.; Tang, J.; et al. Enhanced glioma therapy by synergistic inhibition of autophagy and tyrosine kinase activity. *Int. J. Pharm.* **2018**, *536*, 1–10. [CrossRef] [PubMed]
47. Qu, F.; Wang, P.; Zhang, K.; Shi, Y.; Li, Y.; Li, C.; Lu, J.; Liu, Q.; Wang, X. Manipulation of Mitophagy by "All-in-One" nanosensitizer augments sonodynamic glioma therapy. *Autophagy* **2020**, *16*, 1413–1435. [CrossRef] [PubMed]
48. Liu, L.; Ren, J.; He, Z.; Men, K.; Mao, Y.; Ye, T.; Chen, H.; Li, L.; Xu, B.; Wei, Y.; et al. Cholesterol-modified Hydroxychloroquine-loaded Nanocarriers in Bleomycin-induced Pulmonary Fibrosis. *Sci. Rep.* **2017**, *7*, 10737. [CrossRef] [PubMed]
49. Shao, M.; Zhu, W.; Lv, X.; Yang, Q.; Liu, X.; Xie, Y.; Tang, P.; Sun, L. Encapsulation of chloroquine and doxorubicin by MPEG-PLA to enhance anticancer effects by lysosomes inhibition in ovarian cancer. *Int. J. Nanomed.* **2018**, *13*, 8231–8245. [CrossRef]
50. Yang, C.; Hu, T.; Cao, H.; Zhang, L.; Zhou, P.; He, G.; Song, X.; Tong, A.; Guo, G.; Yang, F.; et al. Facile Construction of Chloroquine Containing PLGA-Based pDNA Delivery System for Efficient Tumor and Pancreatitis Targeting in Vitro and in Vivo. *Mol. Pharm.* **2015**, *12*, 2167–2179. [CrossRef]
51. Liu, J.; Liu, X.; Han, Y.; Zhang, J.; Liu, D.; Ma, G.; Li, C.; Liu, L.; Kong, D. Nanovaccine Incorporated with Hydroxychloroquine Enhances Antigen Cross-Presentation and Promotes Antitumor Immune Responses. *ACS Appl. Mater. Interfaces* **2018**, *10*, 30983–30993. [CrossRef]
52. Mezzaroba, N.; Zorzet, S.; Secco, E.; Biffi, S.; Tripodo, C.; Calvaruso, M.; Mendoza-Maldonado, R.; Capolla, S.; Granzotto, M.; Spretz, R.; et al. New potential therapeutic approach for the treatment of B-Cell malignancies using chlorambucil/hydroxychloroquine-loaded anti-CD20 nanoparticles. *PLoS ONE* **2013**, *8*, e74216. [CrossRef]
53. Zhang, X.; Liang, X.; Gu, J.; Chang, D.; Zhang, J.; Chen, Z.; Ye, Y.; Wang, C.; Tao, W.; Zeng, X.; et al. Investigation and intervention of autophagy to guide cancer treatment with nanogels. *Nanoscale* **2017**, *9*, 150–163. [CrossRef] [PubMed]
54. Marti Coma-Cros, E.; Lancelot, A.; San Anselmo, M.; Neves Borgheti-Cardoso, L.; Valle-Delgado, J.J.; Serrano, J.L.; Fernandez-Busquets, X.; Sierra, T. Micelle carriers based on dendritic macromolecules containing bis-MPA and glycine for antimalarial drug delivery. *Biomater. Sci.* **2019**, *7*, 1661–1674. [CrossRef] [PubMed]
55. Agrawal, P.; Gupta, U.; Jain, N.K. Glycoconjugated peptide dendrimers-based nanoparticulate system for the delivery of chloroquine phosphate. *Biomaterials* **2007**, *28*, 3349–3359. [CrossRef] [PubMed]
56. Bhadra, D.; Bhadra, S.; Jain, N.K. PEGylated peptide dendrimeric carriers for the delivery of antimalarial drug chloroquine phosphate. *Pharm. Res.* **2006**, *23*, 623–633. [CrossRef]

57. Panagiotaki, K.N.; Sideratou, Z.; Vlahopoulos, S.A.; Paravatou-Petsotas, M.; Zachariadis, M.; Khoury, N.; Zoumpourlis, V.; Tsiourvas, D. A Triphenylphosphonium-Functionalized Mitochondriotropic Nanocarrier for Efficient Co-Delivery of Doxorubicin and Chloroquine and Enhanced Antineoplastic Activity. *Pharmaceuticals* **2017**, *10*, 91. [CrossRef]
58. Marti Coma-Cros, E.; Biosca, A.; Marques, J.; Carol, L.; Urban, P.; Berenguer, D.; Riera, M.C.; Delves, M.; Sinden, R.E.; Valle-Delgado, J.J.; et al. Polyamidoamine Nanoparticles for the Oral Administration of Antimalarial Drugs. *Pharmaceutics* **2018**, *10*, 225. [CrossRef]
59. Urban, P.; Valle-Delgado, J.J.; Mauro, N.; Marques, J.; Manfredi, A.; Rottmann, M.; Ranucci, E.; Ferruti, P.; Fernandez-Busquets, X. Use of poly(amidoamine) drug conjugates for the delivery of antimalarials to Plasmodium. *J. Control. Release* **2014**, *177*, 84–95. [CrossRef]
60. Tripathy, S.; Das, S.; Chakraborty, S.P.; Sahu, S.K.; Pramanik, P.; Roy, S. Synthesis, characterization of chitosan-tripolyphosphate conjugated chloroquine nanoparticle and its in vivo anti-malarial efficacy against rodent parasite: A dose and duration dependent approach. *Int. J. Pharm.* **2012**, *434*, 292–305. [CrossRef]
61. Tripathy, S.; Das, S.; Dash, S.K.; Chattopadhyay, S.; Roy, S. The Impact of Nanochloroquine on Restoration of Hepatic and Splenic Mitochondrial Damage against Rodent Malaria. *J. Nanoparticls* **2013**, *2013*, 106152. [CrossRef]
62. Tripathy, S.; Das, S.; Dash, S.K.; Mahapatra, S.K.; Chattopadhyay, S.; Majumdar, S.; Roy, S. A prospective strategy to restore the tissue damage in malaria infection: Approach with chitosan-trypolyphosphate conjugated nanochloroquine in Swiss mice. *Eur. J. Pharmacol.* **2014**, *737*, 11–21. [CrossRef]
63. Tripathy, S.; Mahapatra, S.K.; Chattopadhyay, S.; Das, S.; Dash, S.K.; Majumder, S.; Pramanik, P.; Roy, S. A novel chitosan based antimalarial drug delivery against Plasmodium berghei infection. *Acta Trop.* **2013**, *128*, 494–503. [CrossRef] [PubMed]
64. Bhalekar, M.R.; Upadhaya, P.G.; Madgulkar, A.R. Fabrication and efficacy evaluation of chloroquine nanoparticles in CFA-induced arthritic rats using TNF-alpha ELISA. *Eur. J. Pharm. Sci.* **2016**, *84*, 1–8. [CrossRef] [PubMed]
65. Baruah, U.K.; Gowthamarajan, K.; Ravisankar, V.; Karri, V.; Simhadri, P.K.; Singh, V. Optimisation of chloroquine phosphate loaded nanostructured lipid carriers using Box-Behnken design and its antimalarial efficacy. *J. Drug Target.* **2018**, *26*, 576–591. [CrossRef] [PubMed]
66. Bendas, E.R.; Abdullah, H.; El-Komy, M.H.; Kassem, M.A. Hydroxychloroquine niosomes: A new trend in topical management of oral lichen planus. *Int. J. Pharm.* **2013**, *458*, 287–295. [CrossRef]
67. Ruan, S.; Xie, R.; Qin, L.; Yu, M.; Xiao, W.; Hu, C.; Yu, W.; Qian, Z.; Ouyang, L.; He, Q.; et al. Aggregable Nanoparticles-Enabled Chemotherapy and Autophagy Inhibition Combined with Anti-PD-L1 Antibody for Improved Glioma Treatment. *Nano Lett.* **2019**, *19*, 8318–8332. [CrossRef]
68. Feng, Q.; Yang, X.; Hao, Y.; Wang, N.; Feng, X.; Hou, L.; Zhang, Z. Cancer Cell Membrane-Biomimetic Nanoplatform for Enhanced Sonodynamic Therapy on Breast Cancer via Autophagy Regulation Strategy. *ACS Appl. Mater. Interfaces* **2019**, *11*, 32729–32738. [CrossRef]
69. Kanvinde, S.; Chhonker, Y.S.; Ahmad, R.; Yu, F.; Sleightholm, R.; Tang, W.; Jaramillo, L.; Chen, Y.; Sheinin, Y.; Li, J.; et al. Pharmacokinetics and efficacy of orally administered polymeric chloroquine as macromolecular drug in the treatment of inflammatory bowel disease. *Acta Biomater.* **2018**, *82*, 158–170. [CrossRef]
70 Liu, H.; Zhou, M.; Sheng, Z.; Chen, Y.; Yeh, C.K.; Chen, W.; Liu, J.; Liu, X.; Yan, F.; Zheng, H. Theranostic nanosensitizers for highly efficient MR/fluorescence imaging-guided sonodynamic therapy of gliomas. *J. Cell Mol. Med.* **2018**, *22*, 5394–5405. [CrossRef]
71 Sun, Y.; Wang, H.; Wang, P.; Zhang, K.; Geng, X.; Liu, Q.; Wang, X. Tumor targeting DVDMS-nanoliposomes for an enhanced sonodynamic therapy of gliomas. *Biomater. Sci.* **2019**, *7*, 985–994. [CrossRef]
72 Signorell, R.D.; Luciani, P.; Brambilla, D.; Leroux, J.C. Pharmacokinetics of lipid-drug conjugates loaded into liposomes. *Eur. J. Pharm. Biopharm.* **2018**, *128*, 188–199. [CrossRef]
73 Lavasanifar, A.; Samuel, J.; Kwon, G.S. Poly(ethylene oxide)-block-poly(L-amino acid) micelles for drug delivery. *Adv. Drug Deliv. Rev.* **2002**, *54*, 169–190. [CrossRef]
74 Norvaisas, P.; Ziemys, A. The role of payload hydrophobicity in nanotherapeutic pharmacokinetics. *J. Pharm. Sci.* **2014**, *103*, 2147–2156. [CrossRef] [PubMed]
75 Warhurst, D.C.; Steele, J.C.; Adagu, I.S.; Craig, J.C.; Cullander, C. Hydroxychloroquine is much less active than chloroquine against chloroquine-resistant Plasmodium falciparum, in agreement with its physicochemical properties. *J. Antimicrob. Chemother.* **2003**, *52*, 188–193. [CrossRef] [PubMed]
76. Li, J.; Mo, L.; Lu, C.H.; Fu, T.; Yang, H.H.; Tan, W. Functional nucleic acid-based hydrogels for bioanalytical and biomedical applications. *Chem. Soc. Rev.* **2016**, *45*, 1410–1431. [CrossRef]
77. Seymour, L.W.; Ferry, D.R.; Anderson, D.; Hesslewood, S.; Julyan, P.J.; Poyner, R.; Doran, J.; Young, A.M.; Burtles, S.; Kerr, D.J.; et al. Hepatic drug targeting: Phase I evaluation of polymer-bound doxorubicin. *J. Clin. Oncol.* **2002**, *20*, 1668–1676. [CrossRef]
78. Astruc, D.; Boisselier, E.; Ornelas, C. Dendrimers designed for functions: From physical, photophysical, and supramolecular properties to applications in sensing, catalysis, molecular electronics, photonics, and nanomedicine. *Chem. Rev.* **2010**, *110*, 1857–1959. [CrossRef]
79. Wahane, A.; Waghmode, A.; Kapphahn, A.; Dhuri, K.; Gupta, A.; Bahal, R. Role of Lipid-Based and Polymer-Based Non-Viral Vectors in Nucleic Acid Delivery for Next-Generation Gene Therapy. *Molecules* **2020**, *25*, 2866. [CrossRef]

80. Hu, L.; Tang, X.; Cui, F. Solid lipid nanoparticles (SLNs) to improve oral bioavailability of poorly soluble drugs. *J. Pharm. Pharmacol.* **2004**, *56*, 1527–1535. [CrossRef]
81. Wang, H.; Wang, H.; Yang, W.; Yu, M.; Sun, S.; Xie, B. Improved Oral Bioavailability and Liver Targeting of Sorafenib Solid Lipid Nanoparticles in Rats. *AAPS PharmSciTech* **2018**, *19*, 761–768. [CrossRef]
82. Krishna, S.; White, N.J. Pharmacokinetics of quinine, chloroquine and amodiaquine. Clinical implications. *Clin. Pharm.* **1996**, *30*, 263–299. [CrossRef]
83. Tulpule, A.; Krishnaswamy, K. Effect of food on bioavailability of chloroquine. *Eur. J. Clin. Pharmacol.* **1982**, *23*, 271–273. [CrossRef] [PubMed]
84. Hasan, G.M.; Garg, N.; Dogra, E.; Surolia, R.; Ghosh, P.C. Inhibition of the Growth of Plasmodium falciparum in Culture by Stearylamine-Phosphatidylcholine Liposomes. *J. Parasitol. Res.* **2011**, *2011*, 120462. [CrossRef] [PubMed]
85. Muzzalupo, R.; Tavano, L. Niosomal drug delivery for transdermal targeting: Recent advances. *Res. Rep. Transdermal Drug Deliv.* **2015**, *4*, 23–33. [CrossRef]
86. Fan, M.; Han, Y.; Gao, S.; Yan, H.; Cao, L.; Li, Z.; Liang, X.J.; Zhang, J. Ultrasmall gold nanoparticles in cancer diagnosis and therapy. *Theranostics* **2020**, *10*, 4944–4957. [CrossRef]
87. Israel, L.L.; Galstyan, A.; Holler, E.; Ljubimova, J.Y. Magnetic iron oxide nanoparticles for imaging, targeting and treatment of primary and metastatic tumors of the brain. *J. Control. Release* **2020**, *320*, 45–62. [CrossRef] [PubMed]
88. Ambardekar, V.V.; Stern, S.T. NBCD Pharmacokinetics and Drug Release Methods. In *Non-Biological Complex Drugs; The Science and the Regulatory Landscape*, 1st ed.; Crommelin, D.J.A., de Vlieger, J.S.B., Eds.; Springer International Publishing: Cham, Switzerland, 2015; pp. 261–287.
89. Stern, S.T.; Hall, J.B.; Yu, L.L.; Wood, L.J.; Paciotti, G.F.; Tamarkin, L.; Long, S.E.; McNeil, S.E. Translational considerations for cancer nanomedicine. *J. Control. Release* **2010**, *146*, 164–174. [CrossRef] [PubMed]
90. Skoczen, S.L.; Snapp, K.S.; Crist, R.M.; Kozak, D.; Jiang, X.; Liu, H.; Stern, S.T. Distinguishing Pharmacokinetics of Marketed Nanomedicine Formulations Using a Stable Isotope Tracer Assay. *ACS Pharmacol. Transl. Sci.* **2020**, *3*, 547–558. [CrossRef]
91. Starpharma. DEP®Docetaxel Positive Phase 1 Results; Phase 2 Commences. Available online: https://www.starpharma.com/news/339 (accessed on 18 November 2020).
92. Starpharma. Starpharma to Commence DEP®Cabazitaxel Phase 1/2 Trial. Available online: http://www.starpharma.com/news/356 (accessed on 18 November 2020).
93. El Maghraby, G.M.; Arafa, M.F. Liposomes for Enhanced Cellular Uptake of Anticancer Agents. *Curr. Drug Deliv.* **2020**, *17*, 861–873. [CrossRef]
94. Charrois, G.J.; Allen, T.M. Drug release rate influences the pharmacokinetics, biodistribution, therapeutic activity, and toxicity of pegylated liposomal doxorubicin formulations in murine breast cancer. *Biochim. Biophys. Acta* **2004**, *1663*, 167–177. [CrossRef]
95. Shmeeda, H.; Amitay, Y.; Gorin, J.; Tzemach, D.; Mak, L.; Stern, S.T.; Barenholz, Y.; Gabizon, A. Coencapsulation of alendronate and doxorubicin in pegylated liposomes: A novel formulation for chemoimmunotherapy of cancer. *J. Drug Target.* **2016**, *24*, 878–889. [CrossRef]
96. Silverman, L.; Barenholz, Y. In vitro experiments showing enhanced release of doxorubicin from Doxil(R) in the presence of ammonia may explain drug release at tumor site. *Nanomedicine* **2015**, *11*, 1841–1850. [CrossRef] [PubMed]
97. Vicent, M.J.; Duncan, R. Polymer conjugates: Nanosized medicines for treating cancer. *Trends Biotechnol.* **2006**, *24*, 39–47. [CrossRef] [PubMed]
98. Ekladious, I.; Colson, Y.L.; Grinstaff, M.W. Polymer-drug conjugate therapeutics: Advances, insights and prospects. *Nat. Rev. Drug Discov.* **2019**, *18*, 273–294. [CrossRef]
99. Chung, Y.H.; Beiss, V.; Fiering, S.N.; Steinmetz, N.F. COVID-19 Vaccine Frontrunners and Their Nanotechnology Design. *ACS Nano* **2020**, *14*, 12522–12537. [CrossRef] [PubMed]

molecules

MDPI

Review

Dissecting Functional Biological Interactions Using Modular RNA Nanoparticles

Kaitlin Klotz [1], Yasmine Radwan [2] and Kausik Chakrabarti [1,*]

1 Department of Biological Sciences, University of North Carolina at Charlotte, 9201 University City Blvd., Charlotte, NC 28223, USA
2 Nanoscale Science Program, Department of Chemistry, University of North Carolina at Charlotte, Charlotte, NC 28223, USA
* Correspondence: kchakrabarti@uncc.edu

Abstract: Nucleic acid nanoparticles (NANPs) are an exciting and innovative technology in the context of both basic and biomedical research. Made of DNA, RNA, or their chemical analogs, NANPs are programmed for carrying out specific functions within human cells. NANPs are at the forefront of preventing, detecting, and treating disease. Their nucleic acid composition lends them biocompatibility that provides their cargo with enhanced opportunity for coordinated delivery. Of course, the NANP system of targeting specific cells and tissues is not without its disadvantages. Accumulation of NANPs outside of the target tissue and the potential for off-target effects of NANP-mediated cargo delivery present challenges to research and medical professionals and these challenges must be effectively addressed to provide safe treatment to patients. Importantly, development of NANPs with regulated biological activities and immunorecognition becomes a promising route for developing versatile nucleic acid therapeutics. In a basic research context, NANPs can assist investigators in fine-tuning the structure-function relationship of final formulations and in this review, we explore the practical applications of NANPs in laboratory and clinical settings and discuss how we can use established nucleic acid research techniques to design effective NANPs.

Keywords: nucleic acid nanoparticle; RNA motif; RNA domain; SHAPE analysis

Citation: Klotz, K.; Radwan, Y.; Chakrabarti, K. Dissecting Functional Biological Interactions Using Modular RNA Nanoparticles. *Molecules* **2023**, *28*, 228. https://doi.org/10.3390/molecules28010228

Academic Editor: Maria Camilla Bergonzi

Received: 8 December 2022
Revised: 22 December 2022
Accepted: 25 December 2022
Published: 27 December 2022

1. Introduction

Nucleic Acid Nanoparticles (NANPs) are a subtype of therapeutic nucleic acids (TNAs) exclusively composed of specialized oligonucleotides designed to carry out defined architectures and functions such as delivery of therapeutic agents, biosensing, and immunostimulation [1]. NANPs can be specified by the designer to deliver functional groups capable of modulating their biological activities while adding regulatory control to the intended function of the NANP [2,3]. NANPs can be engineered to associate with specific targets which make them useful in diagnostics as well as the targeted delivery of therapeutic agents to detect and combat disease with fewer off-target impacts than experienced with traditional delivery systems [4]. For that, the sequences amenable to interactions with receptors on cellular surfaces to facilitate NANP uptake through receptor-mediated endocytosis [5]. A 2015 study by Narayan and colleagues established that class A scavenger receptors have enhanced affinity for spherical nucleic acid nanoparticles conjugates exhibiting a high guanine content [6]. High guanine content in the oligonucleotides of the conjugates facilitated adoption of a secondary structure that facilitated uptake of nanoparticles carrying camptothecin by A549 (human lung adenocarcinoma) cells which resulted in the significantly diminished viability of the cancerous cells seven days after treatment with a G-rich spherical conjugates [6].

NANPs are often used as a method of getting nucleic acids past "barriers" that exist in the body. Typically, carrier-free, naked, exogenous nucleic acids introduced without any chemical modifications would meet one of two fates within the body: rapid nuclease-mediated degradation and/or renal clearance. The engineered larger NANPs or NANPs mixed with the delivery agents assemble in such a way that the cargo they are delivering is protected from degradative nuclease activity which would otherwise stop the NANPs from achieving their therapeutic purposes. Additionally, the architectural parameters (size, shape, composition, functionalization with TNAs, etc.) of NANPs can be engineered in a way that is non-immunogenic and allows the payload to be shuttled to its intended target without triggering any immunological responses [7]. Once the problems of nuclease mediated degradation and immunogenicity are dealt with, NANPs can deliver gene regulatory RNA interference (RNAi) inducers to decrease the expression of overexpressed genes in pathogenic settings [4,8].

1.1. Nucleic Acid Nanoparticle Design and Functionalization

There is more than one way to design NANPs and the method used to design a particular NANP depends on the material delivered as well as its action upon delivery [9–11]. The physical properties of individual NANPs determine how well the NANP interacts with its intended target and consequently, impacts the efficacy of the NANP's payload delivery. Several computer-assisted approaches allowing for designing RNA and DNA NANPs have been introduced and explored [9,12–17]. Once a particular NANP is selected for further use, the addition of functional moieties to its structure can be achieved through a direct extension of 5′- or 3′-ends of individual strands that enter the NANP's composition. One way in which NANPs demonstrate specificity is through aptamers. Aptamers are single-stranded oligonucleotides that can adopt specific conformations [18]. Due to their specific conformational arrangements, aptamers can interact with cell components, namely receptors and ligands, in a level of specificity like that of antibodies to elicit a response [18,19]. Their ability to modulate pathways within cells makes aptamers an attractive candidate for therapeutics-particularly in an antagonistic indication [18].

In terms of NANP design, aptamers can be added to NANPs to enhance the specificity of NANP binding. This binding of NANPs can be to whole cells (on the cell surfaces), as well as proteins, and in some special cases, viruses [20]. NANPs provide therapeutic agents access to cells without leaving those therapeutic agents subject to degradation and less prone to triggering an immune response. The aptamers, which have undergone several rounds of Selective Evolution of Ligands by Exponential Enrichment (SELEX), are the key to getting the therapeutic agent to exactly where it is needed [19,20]. We can think of aptamers like a ZIP code on a letter such that the contents of the NANP could go to several different places, but when the aptamer is added, the potential delivery locations are narrowed down to a specific place within the cell.

1.2. Selection of Nucleic Acid for NANPs

In the previous section, it was stated that NANP design is influenced by the cargo that is delivered as well as the intended action. The same dependence on cargo and action must be considered when selecting the nucleic acid for a NANP. Both DNA and RNA NANPs have been developed for therapeutic and research applications, but the two have several distinct qualities that make them unique from one another. In general, DNA is typically regarded as the more stable of the nucleic acids, however, that does not mean that RNA is the "lesser" nucleic acid. In fact, RNA has some properties that make it favorable for use in NANPs.

A clear distinction between DNA and RNA nanoparticles is the base pairing capabilities of each nucleic acid. DNA is confined to Watson–Crick base pairing, meaning that A binds with T and C binds with G. While RNA often does form Watson–Crick base pairs, it also participates in non-canonical base pairing, such as G:U, G:A or C:A type as found in RNA structural folds [21]. These noncanonical base pairs play critical roles in RNA-folding to establish the three-dimensional structures required for diverse functions of RNA. Thus, RNA's ability to form non-canonical base pairs allows it to adopt motifs with discrete structure and function, setting it apart from DNA [21]. The motifs present in RNA provide it with enhanced thermal stability, which can be variable in relation to base pairs in DNA, creating differentials in thermal stability within the same DNA sequence [21,22]. This stability is further reflected in the free energy of RNA-RNA helices [21]. The free energy for RNA-RNA helices is the lower that of DNA-RNA hybrid helices and DNA-DNA helices [21,23,24].

For a NANP to carry out its intended function, it must be recognized by a receptor on the exterior of a target cell and endocytosed into the cell. Upon endocytosis, the NANP is inside of an endosome that functions to degrade molecules and reuse their components [21,25,26]. The endosome is acidic to aid in the degradation of endocytosed molecules [25–27]. In the case of DNA, this low pH environment leads to depurination-a process in which adenine and guanine (the purines) are protonated and subsequently lost from the sequence, leaving the DNA lacking purines and susceptible to degradation [21,28,29]. On the other hand, RNA is tolerant of lower pH levels than DNA, which will facilitate successful delivery through the endosome [21].

In summary, RNA nanoparticles are more stable thermodynamically and to pH ranges. Additionally, they have greater structural flexibility. When considered together, these features make RNAs preferred over DNAs as NANPs.

1.3. Small Interfering RNA (siRNA)

Small interfering RNA (or siRNA) is a short, non-coding, regulatory RNA [30]. Like other small RNAs, siRNAs are generally 20–30 nucleotides in length, and they can modulate the expression of genes [30,31]. siRNAs are usually exogenous to the organisms they regulate, and they can be designed by researchers to target a specific gene for downregulation [31]. siRNAs are commonly used in RNA interference (RNAi). RNAi has long existed in nature, however, more recently researchers have begun to harness its power to regulate gene expression [32]. siRNA is used to target specific sequences of RNA or DNA. siRNA originates from double-stranded RNA (dsRNA), which is not common in cells and triggers a cascade to eliminate the dsRNA [31]. The dsRNA is cleaved into small pieces, which become the siRNAs, by a protein called Dicer [31]. siRNAs are loaded into the RNA-Induced Silencing Complex, or RISC [31]. Once in the RISC, the strands of the siRNA are separated and only one strand is retained [31]. When the siRNA-bound RISC recognizes a mRNA sequence complementary to the siRNA strand, the RISC activates a protein called slicer to cleave the mRNA and render it untranslatable [31]. Sometimes, there is not perfect sequence complementarity, but it is still similar enough that the siRNA strand will bind the partially complementary RNA and block it from efficient translation (Figure 1A) [32]. In a therapeutic setting, researchers may introduce synthetically produced siRNAs for targeted knockdown of a gene of interest. In this case, the siRNA is not cleaved from a larger dsRNA inside the cell-it is already the appropriate size for association with the RISC [31].

NANPs are particularly useful in the delivery of siRNA due to their stability [33]. Typically, "naked" RNAs are quickly recognized as foreign and degraded by endogenous nucleases inside the cell or organism to which the RNA is delivered [33]. The stability provided to the siRNA payload delivered by the NANP increases the ability of the siRNA to reach its intended target without being destroyed by the recipient. Additionally, NANPs can be conjugated to aptamers which bind with high specificity to a desired target. The specificity conferred to the siRNA delivery by nanoparticles reduces the possibility of off-target binding/delivery and increases the concentration of siRNA at the desired target [19].

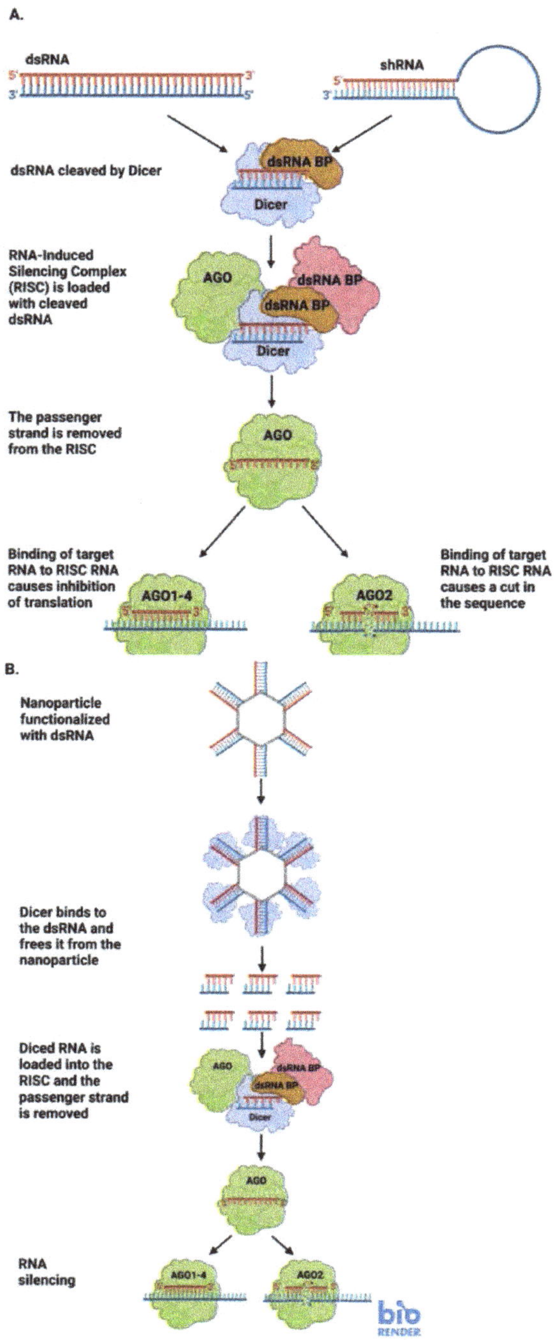

Figure 1. (**A**) Diagram detailing the steps involved in RNA interference (RNAi) through the production of small interfering RNAs (siRNAs). (**B**) A representative diagram of how siRNAs may be delivered to a cell via RNA-functionalized nanoparticles, and how the siRNAs can be used for RNAi upon delivery. Image created with BioRender.com.

1.4. NANP Applications in Medicine

In addition to the delivery of therapeutics to treat disease, NANPs can be used in disease prevention. RNA or DNA NANPs can be designed to deliver genetic information from pathogens in vaccines to prime the immune system for a natural exposure [34]. This is a new and exciting application of nanoparticles that we will undoubtedly see become more common as nucleic acid vaccines are further developed. NANPs can also be used as adjuvants that allow vaccine materials to be incorporated into the target tissue to stimulate an appropriate immune response [35]. On the flipside, nanoparticles can be altered to be immunologically inert and lack inflammatory activation, which can result in adverse effects to the patient. The future of NANP medicine will rely on careful optimization of these drugs and mitigation of their side effects and off-target effects.

2. Functionalized RNA Nanoparticles

2.1. RNAi

RNAi is the phenomenon of dsRNAs knocking down or silencing the expression of endogenous mRNAs initially described by Andrew Fire and Craig Mello nearly 25 years ago [36,37]. RNAi relies on small interfering RNAs (siRNAs) which silence encoded genes (Figure 1A). This occurs through a short double-stranded RNA binding with a target and preventing translation into functional protein [38]. While Fire and Mello characterized RNAi with an exogenous RNAi with dsRNA delivered from outside in C. elegans, the phenomenon is an evolutionarily conserved method of post-transcriptional gene silencing that has served as a means for organisms to protect themselves from exogenous threats as well as regulate gene expression [39]. RNAi has been characterized in plants as a sort of immune protection from plant viruses and insects [39]. A few years ago, Chejanovsky and colleagues used deep sequencing to detect the presence of perfect siRNA matches for three viruses that strongly contribute to colony collapse disorder in the genomes of honey bee colonies that had succumbed to colony collapse disorder [40]. In an experimental setting, RNAi has been used to tackle a wide range of biological problems. RNAi-mediated biological pest control has been used experimentally to protect valuable crops which will provide nutrients to countless people and livestock [41]. Building off of Chejanovsky's work in honey bees, RNAi has been experimentally employed to combat viruses that contribute to colony collapse disorder and threaten global food supplies [42]. NANPs can be used to mediate the delivery of exogenous RNAs for RNAi within cells. NANPs functionalized with the exogenous RNA deliver the RNA inside a cell to silence a target gene and prevent the formation of that gene's product [43]. Once the NANP has been internalized within the cell, the enzyme Dicer acts upon the attached RNAs, allowing them to participate in RNAi-mediated gene silencing (Figure 1B) [43].

Currently, RNAi is a subject of much research and development for clinical applications [44]. Despite a rocky start in the 2000s in which several RNAi therapeutic candidates were pulled from clinical trials due to unintended effects, Patisiran became the first RNAi drug to receive FDA approval in 2018 [37,45]. Patisiran treats hereditary transthyretin amyloidosis-a deadly genetic disorder which is characterized by the deposition of amyloid plaques of the protein transthyretin in key organs such as the heart and kidneys, leading to deterioration of quality of life for patients and eventual death [45–48]. Today, there are several RNAi drugs in clinical trials for treating cancers, inherited genetic disorders such as Sickle Cell Disease and familial hypercholesterolemia, and viruses such as HIV and hepatitis B [37].

Like with all new technologies, RNAi does not come without drawbacks. Naturally, we must consider immunogenicity. RNAi therapeutics work at their best when they are delivered to the appropriate tissue without degradation. When the RNAi drug triggers an immune response, the drug may never reach its intended target, or if it does, it could be in a less effective state [37]. In addition to unintended immunogenic responses, accumulation of RNAi therapeutics in unintended locations and subsequent toxicity are of major concern to RNAi drug developers [37]. While accumulation of RNAi therapeutics in off-target tissue is a very legitimate problem in and of itself, an additional unintended consequence of RNAi drug treatment is the drug acting as it is supposed to outside of its target tissue and causing interruption to normal tissue function [37].

To continue to develop the applications of RNAi therapeutics, researchers have gotten creative with how they have approached drawbacks. Researchers have incorporated base modifications into the RNA delivery system to get around the problem of immunogenic activation caused by RNAi therapeutics [5,37,49]. In 2007, Robbins and colleagues demonstrated that the addition of a 2′-O-Me modification to siRNA reduced the siRNA's immunogenic potential without diminishing its ability to decrease expression of a target gene [49]. The added benefit of the base modifications is increased RNA stability, which generally increases the amount of time it is able to remain in the body without becoming degraded or neutralized by the immune system [5,37]. When the circulation time is increased, the RNAi therapeutic has an increased probability of arriving at its target and carrying out its intended function.

To mitigate off-target RNAi activity, researchers propose some common-sense measures to enhance safety for RNAi therapeutic recipients. The first proposed measure occurs long before the RNAi therapeutic makes it to the patient-it is extensive quality control to ensure that the drug has as few targets in the human genome as possible [37,50]. In their 2019 review, Setten and colleagues uphold that some off-target binding is inevitable and patients receiving RNAi therapy should be administered the smallest dose capable of achieving the desired effect, while being closely monitored for signs of off-target activity that is threatening to the patient's wellbeing [37].

It becomes more challenging when on-target effects in off-target tissues are considered. The drug is doing what it is supposed to do, just in a suboptimal location, which can imperil the patient's health. Much like mitigating off-target effects in off-target tissue, researchers must design intentional, highly specific therapeutics to ensure that there are as few possible routes for the therapeutic to build up in a non-target location [37]. In addition to designing therapeutics with the awareness of these effects, drug developers have "reverse engineered" compounds to reverse the impacts of the siRNA therapeutics in the event that they excel at performing their intended functions outside of the optimal location [37,51].

2.2. NANP-Induced Immunogenicity

The use of NANPs in humans as well as other animals carries the potential for immune activation. The immunogenic response must be thoroughly evaluated by the investigators to ensure that it is not activated in an unintended manner (i.e., prior to the delivery of the NANP to the target tissue). Researchers do have some (though not complete) control in how immunostimulation proceeds through the design of their NANPs. A 2017 research article by Guo and colleagues examined the impact of NANP sequence as well as physical properties (size and shape) on their immunostimulatory effects [52]. In this study, the research team demonstrated that increases in the size of RNA squares resulted in enhancement of cytokine secretion, namely TNF-α and IL-6 [52]. Similarly, RNA squares with attached uniform RNA sequences showed the same effect-as the number of attached RNA sequences increased, so did the levels of cytokine secretion [52]. The same study indicated that three-dimensional RNA nanoparticles elicited higher levels of cytokine secretion than their planar counterparts [52]. Figure 2 provides a visual representation of the results that were described in Guo's 2017 article (Figure 2).

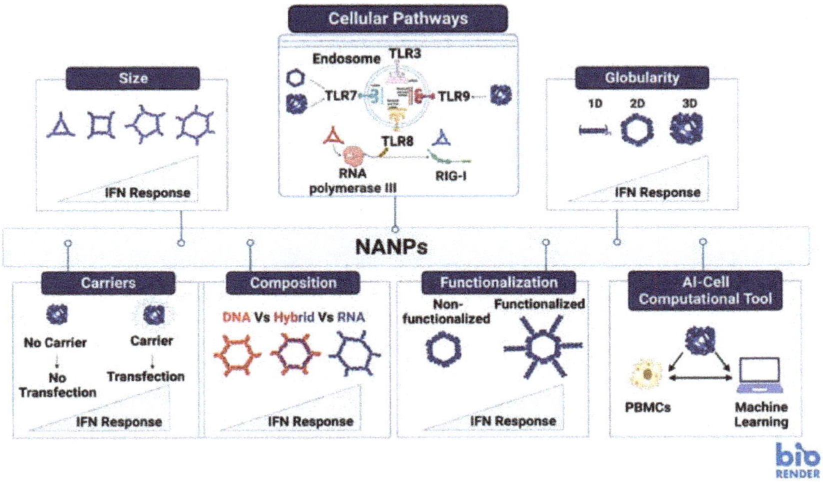

Figure 2. Structural parameters and other factors affecting immune stimulation by NANPs. Image created with BioRender.com.

2.3. Immuno-Adjuvant

The use of adjuvants offers a robust method where immunostimulatory compounds are employed to potentiate and modulate the immune response, when used together with vaccines. Among the recent advancements of the immunomodulatory NANPs, is their ability to regulate and modulate the immune responses when encountered. The rapid clinical development of this technology is impeded by several hurdles among which is the unknown immune stimulation by NANPs. However, recent studies have reported that mammalian cells use the same patterns of recognition established to defend themselves against viral and bacterial nucleic acids to process NANPs [2]. Further investigations showed that various interactions of NANPs with different immune cells elicit different immune responses. The regulation of the immune response by NANPs can be controlled by varying their design, specifically by altering numerous architectural parameters including the NANPs' size, shape, functionalization and composition (Figure 2) [2,8,53,54]. The correlation between those variations in NANPs' structures and their effect on the immune response is now being considered carefully during the design process aiding to develop NANPs that would either modulate the immune activation or stay immunoquiescent [54]. Other aspects considered to successfully translate NANP technologies into clinical settings would include the choice of delivery carriers and administration routes for NANP formulations. If NANPs complexed with a carrier are delivered via intravenous administration, they may induce undesired inflammation due to cytokine induction and complement activation. However, if the same system is administered locally, it would serve perfectly as immune-adjuvant as it will induce the same cytokine and interferon response along with complement activation, which would potentiate the vaccine efficacy and enhance immunotherapy efficacy [53].

2.4. NANPs with Regulated Immune Responses

Human cells have receptors for the recognition of foreign nucleic acids called pattern recognition receptors (PRRs), which can also distinguish and process NANPs. Therefore, understanding the underlying mechanisms of recognition and the NANPs structural parameters that affect their recognition process would allow the tunability of the immunostimulatory effects. This, in turn, will allow engineering the immunoquiescent NANPs intended for drug delivery, or NANPs with regulated immunological properties that could be used in immunotherapies [3]. In the case of immune-adjuvants, PRR agonists help to stimulate the innate cytokine and interferon production, which endorses the cellular antiviral defenses [3].

Extensive studies have been carried out to assess the immunostimulation of the representative library of NANPs introduced to freshly collected human peripheral blood mononuclear cells (PBMCs) [3,7,8,55–61]. PBMCs were chosen as a highly reliable model for prediction of cytokine storm toxicity in humans [57]. One of the parameters that affects the immunorecognition of NANPs is their chemical composition (e.g., DNAs vs. RNAs vs. DNA/RNA hybrids) (Figure 2-Composition) [55,62]. A study reported that altering the composition of NANPs can modulate the mechanisms and degree of elicited immune responses [63]. It also highlighted that NANPs made of RNA normally demonstrate significantly higher immune activations in comparison to their DNA counterpart, since RNA NANPs can trigger both TLR7 and RIG-I mediated cytokine and interferon response (Figure 2-Composition) [63–65]. Another parameter is NANP size; increases in the size of NANPs may lead to elevated immunostimulation (Figure 2-Size). One more parameter that plays a major role in immunostimulation is the dimensionality of NANPs. Several studies have confirmed that for RNA NANPs, fibrous structures (1D) demonstrate reduced immunostimulation when compared to planar NANPs (2D) and that the globular RNA NANPs (3D) produce the highest levels of immune responses amongst all of them (Figure 2-Globularity) [2,59,60]. It was reported that 2D and 3D RNA NANPs induced interferon production upon activation of TLR7, while 1D NANPs did not [62]. In addition, it was shown that the immunostimulation of NANPs functionalized with therapeutic nucleic acids (TNAs) induce higher production of type I and II IFNs when compared to non-functionalized NANPs and that the extent of activation can be regulated by relative orientation of the TNAs (Figure 2-Functionalization) [8,60].

2.5. *The Role of Carriers on NANPs Immunorecognition*

Another hurdle precluding broader clinical application of NANPs is their intracellular delivery [1,66,67]. One of the most essential factors for NANPs delivery and their efficacy is the use of carriers and complexation agents. Extensive studies had previously reported that carrier-free NANPs do not elicit any immune response as they are invisible to the cells, and the use of delivery platforms further tailors the immunorecognition of NANPs [2,53,68]. Hence, for efficient intracellular delivery of NANPs, various delivery agents such as lipid-based carriers [55], exosomes [69], polymeric agents [70], and inorganic materials [71,72] have been investigated. One study employed PBMCs to investigate the use of amine-terminated PAMAM dendrimers to deliver NANPs (e.g., RNA and DNA cubes) and compared to commercially available Lipofectamine 2000 (L2K), a well-established lipid-based delivery platform [55]. The results highlighted that the uptake of the NANPs by different human immune blood cells, and their cytokine responses varied based on the delivery system used [55]. NANPs complexed with dendrimers did not induce type I and type III IFNs as opposed to NANPs complexed with L2K. In addition, NANPs complexed with L2K did not induce cytokine production (IL-1α, IL-1 β, IL-6, TNFα), while NANPs complexed with dendrimers induced the production of these stress associated cytokines. The 3D RNA NANPs (RNA cubes) delivered by dendrimers also elicited a more potent profile of cytokine production when compared to their DNA counterparts, which aligned with previous findings highlighting the effect of NANP composition on immunorecognition [53]. Additionally, as was expected, the carrier-free NANPs did not elicit any immune responses [55]. To overcome the barrier of safe and efficient delivery of NANPs while avoiding the carrier-associated toxicity, naturally occurring nanovesicles involved in cellular communication (e.g., exosomes) can be utilized. The exosomes provide a stealth-coating for loaded NANPs, which prevents nuclease degradation of NANPs as well as exposure to PRRs. For example, exosome-mediated delivery of RNA cubes which are known to have high immunostimulatory effects on cells, showed negligible immune activation [69], as compared to other carriers.

As each set of NANPs holds a unique physicochemical and architectural profile, this creates a burden to predict the type of immune response and its magnitude. To overcome this challenge, a computational predictive tool called "artificial immune cell", or AI-cell, was developed to guide the design of NANPs to fit the desired immunological profiles. This unprecedented computational approach is fed by physicochemical and immunological profiles for an array of various NANPs and uses innovative transformer architectures to predict the immunological activity of NANPs based on the entered oligo and their sequence compositions [7]. This freely available web-based implementation is expected to advance the understanding of properties that contribute to immunomodulatory activity of NANPs and draw guidelines for their design principles. The AI-cell shall further promote the therapeutic nucleic acid nanotechnology even further by addressing the public health challenges related to the toxicities of nucleic acid therapies [7].

3. RNA Motifs and Domains, and Their Delivery via Nanoparticles

3.1. Motifs

In 1999, P.B. Moore defined an RNA motif as a "discrete sequence or combination of base juxtapositions found in naturally occurring RNAs in unexpectedly high abundance" [73]. Moore's definition of an RNA motif leaves room for RNA sequence motifs as well as structural motifs [74]. RNA motifs occur naturally, can exhibit three-dimensional structure, and can interact with other motifs in RNAs and protein domains to contribute to their overall functionality [73]. Researchers and pharmaceutical developers have taken advantage of naturally occurring RNA motifs and incorporated them into their nanoparticle designs for enhanced stability and increased capacity for payload delivery and tracking [5].

3.2. Domains

In a 2002 review of protein domains, Ponting and Russell provided three different perspectives from which protein domains could be defined: biochemical, structural, and sequential [75]. Structurally, they defined domains as "spatially distinct units" [75]. In the biochemical context, they were less concerned with structure and specified a domain as a region with a clear-cut function [75]. From a sequential standpoint, Ponting and Russell claim that a domain is characterized by homology to other sequences which achieve similar functions in different environments [75]. While each definition of the domain holds some truth, all three should be considered together to get the full understanding of the domain.

Ponting and Russell did provide a more modern definition of a domain as a structure that could adopt the necessary structural conformation for carrying out its function [75]. While this definition is applied to protein domains, it could be applied in the context of RNA as well. After all, RNA can adopt specific conformations that facilitate biological functions. RNA motifs, described earlier, have the capacity to build upon each other and interact with other motifs in ways that perform biological functions. Those interactions between motifs are key to the establishment of RNA domains. A 2011 review by Reiter, Chan and Mondragón described domains as complex, functional, three-dimensional structures in the RNA that are comprised of (and stabilized by) interacting RNA motifs [76]. Functional RNA motifs are the building blocks of larger functional RNA domains that have unique three-dimensional structures [76].

3.3. Motifs and Domains in Research

Since the late 1980s, RNA motif research has been rapidly growing. In a 1998 review article, Conn and Draper claim that there are only a few functional RNA motifs, but when these motifs are placed together in combination, the functions that can be carried out by the structured RNA and the specificity with which these functions can be executed is enormous [77]. Nanoparticles, which were in their early days at the time of Conn and Draper's review, take full advantage of combinatorial effects of RNA motifs. RNA motifs themselves are quite frequently incorporated into the designs of nanoparticles for their functionality [5].

RNA tectonics (or tectoRNAs) utilize naturally occurring RNA motifs to form hierarchically folded modular functional RNAs which can be used to construct RNA nanostructures [78,79] and illuminate the functions of already-existing RNA structures [80]. We can think of modular tectoRNAs as jigsaw puzzle pieces that when put together, provide us with the greater unified function of the specific nanostructure much like the pieces of an actual jigsaw puzzle show us an entire picture [79]. However, unlike a jigsaw puzzle, tectoRNAs (or pieces) can be reused in different nanostructures (puzzles) to make and execute entirely new functions [80].

3.4. RNA Functional Augmentation

Over the last nearly three years, the public has become increasingly aware of the use of RNA in medicine through the SARS-CoV2 vaccine. Two of the three mainstream COVID-19 vaccines contain the mRNA transcript encoding the spike protein, which facilitates viral entry into the cell [81]. The mRNA to be delivered is encased in lipid nanoparticles to prevent rapid degradation of the encoded instructions [81]. Once delivered, the mRNA becomes translated into the viral spike protein and takes on its unique conformation to elicit an immune response against the spike protein [82]. The nanoparticle delivery facilitates non-immunogenic delivery of the mRNA cargo which is selected for its ability to trigger immune activation.

Outside of the COVID-19 context, RNA nanoparticle delivery could be used for functional complementation studies. Previously, conjugates of gold nanoparticles were used as carriers for functional RNA structures in cells and these in-cell structures efficiently contribute to gene expression regulation [83]. X-ray crystallographic determination has provided further evidence that these nanostructures can fold into stable RNA motifs, such as kissing loops and T-junctions, that resemble natural RNA motifs [84]. This opens up the possibility of using RNA structural motifs as nanostructures for genetic complementation studies to restore a normal phenotype to mutants with some defect (Figure 2-Functionalization). In this arena, our team is exploring how crucial discrete structural domains of RNA can be used as nanostructures to compensate functional deficiency in parasitic disease caused by the protozoan parasite *Trypanosoma brucei*, the causative agent of African sleeping sickness. Telomerase RNA is a long noncoding RNA that is an integral functional subunit of a large RNA-protein complex responsible for synthesizing the G-rich strand of telomeres, the physical ends of linear chromosomes. Telomerase is critical for telomere length maintenance, thus preventing chromosome instability in eukaryotic cells [85,86]. Our previous and ongoing works with *T. brucei* telomerase RNA structural domain deletion mutants have demonstrated that certain domains of the telomerase RNA are vital to cell proliferation [87]. Once the functions of the *T. brucei* structural domains have been established, we aim to deliver the missing telomerase RNA structural domains to the domain-deletion mutants in an effort to restore their functions.

3.5. Domain Delivery and Associated Challenges

Several different types of RNA can be delivered to cells for purposes such as post-transcriptional regulation, enhancement of catalytic activity and augmentation of gene expression [88]. NANPs functionalized with aptamers can deliver cargos (including RNA domains) to highly specific locations within cells (Figure 3) [88]. Additionally, CRISPR (Clustered Regularly Interspaced Short Palindromic Repeats) technologies are showing great promise in targeting highly specific sequences for editing (Figure 4) [88]. CRISPR genome editing is inspired by a bacterial immune system to protect bacteria against viral invaders [89]. The CRISPR-associated (Cas9) restriction enzyme is directed by a guide RNA (gRNA) which binds to a protospacer adjacent motif (PAM) on a segment of DNA containing the target for editing [89,90]. Cas9 moves from the PAM to determine if there is base complementarity between the gRNA and the DNA target sequence [89,90]. When complementarity between the gRNA and target DNA is located, the Cas9 endonuclease creates a double strand break (DSB) in the target DNA sequence [89,91].

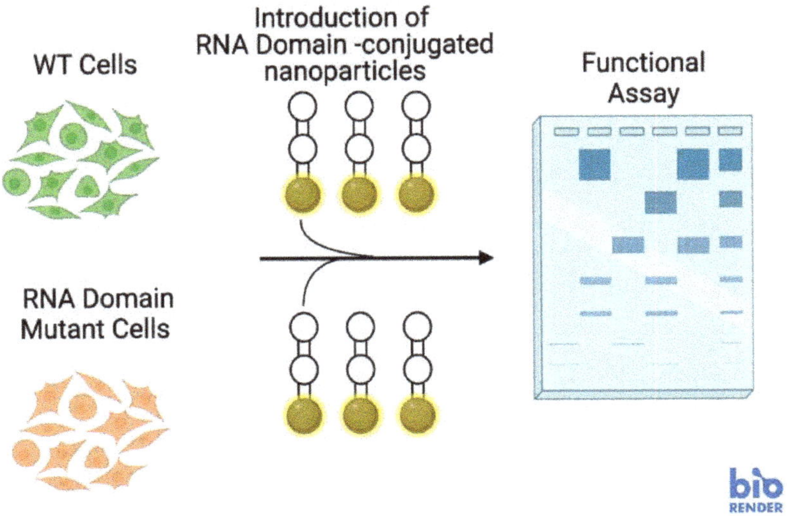

Figure 3. Hypothetical experimental schematic for delivering RNA structural domains to RNA domain-depleted cells for complementation studies with nucleic acid nanoparticles (NANPs). Image created with BioRender.com.

Once the DSB has been made, the host cell's DNA damage repair machinery is activated to prevent cell death [89,92]. Repair processes such as non-homologous end-joining (NHEJ) and homology directed repair (HDR) commonly activate and they cause alterations in the target sequence [89]. A single DSB repaired through NHEJ often incorporates extra base pairs which interrupt the coding sequence and result in the absence of the gene product (Figure 4) [89]. If two different gRNAs are used to make cuts at different places in the target sequence, a segment of the target DNA can be eliminated [89]. If a DNA template is also present in the CRISPR-Cas9 reaction involving two gRNAs, HDR will incorporate the DNA template into the target sequence [89].

Although NANPs are exceptionally promising as tools for modulating gene regulation, they come with considerable delivery and stability issues. NANPs exhibit diminished stability in mammalian serum [63]. In addition to the low stability of NANPs inside of the body, RNAs are rapidly degraded inside of cells by ribonucleases. When an unstable NANP is coupled with RNA domains that could be destroyed upon cellular entry, the prospects of the RNA domains reaching their intended target decrease dramatically. Furthermore, RNA carries a net negative charge, which makes it unlikely to achieve internalization within cells without modifications or incorporation within a carrier that is more amenable to traversing the plasma membrane [2,63]. NANPs carrying exogenous RNA are also quite effective activators of immune responses which can stimulate inflammation harmful to the patient's wellbeing, and to the successful delivery of the NANP payload.

The challenges of NANP stability and rapid RNase-mediated degradation of RNAs delivered by NANPs are not absolute. In fact, over the last decade, significant strides have been made in mitigating NANP serum instability and exogenous RNA degradation. In a 2020 article by Johnson and colleagues, they effectively demonstrated that the composition of the NANP itself carried a significant impact on how stable the NANP was in serum [63]. Replacement of the 2′-OH group in RNA by a 2′-F increased stability for triangular NANPs that contained either an RNA or DNA center [63]. In addition to enhancing the stability of NANPs, the substitution of the 2′ hydroxyl group for a fluorine assisted in mitigating RNase-mediated degradation of RNA NANPs [63,93].

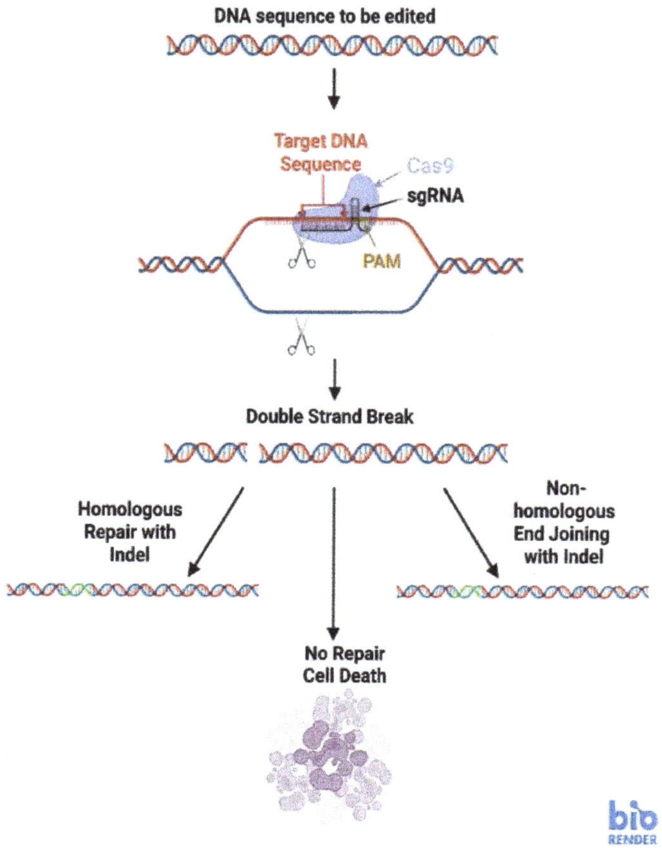

Figure 4. Basic diagram of the CRISPR-Cas9 genome editing system. Image created with BioRender.com.

4. Determination of RNA Structural Properties in Nanobiotechnology

4.1. Importance of RNA Structure Determination

It is a strong theme throughout biology that structure is crucial in determining function. The structure of RNA is dynamic and typically reflects the RNA's specific function [94]. Whether it is the delivery of RNA nanoparticles in cells or RNAs delivered via gold or other nanostructures, determining thermodynamically stable structures, such as three-way junction (3WJ) motifs or structural RNA domains are important for further investigations. With the advent of high-throughput sequencing based RNA probing and cell-penetrating chemical probes, it is now possible to determine structures of RNAs in vivo. In our research team's work, we have demonstrated the dynamic nature of RNA through different stages in the life cycle of *Trypanosoma brucei* [87]. The needs of cells can change throughout their life cycles and having an understanding of how the changing structure of RNA contributes to meeting those needs is of the utmost importance [95].

4.2. Methods for Structural Prediction

In some of our recent work, we employed a selective 2′-hydroxyl acylation analyzed by primer extension mutational profiling (SHAPE-MaP) technique to model telomerase RNA secondary structure at different stages of the life cycle in *Trypanosoma brucei* (Figure 5) [87,96]. We chose this technique for its ability to visualize RNA conformations within living cells, as well as its adaptability for immunoprecipitated RNA [87]. The SHAPE protocol works by applying a SHAPE chemical probe to a sample to facilitate the addition of bulky adducts to RNA bases that are not engaged in a binding arrangements with other bases (from the same RNA or a different RNA), protein, or DNA [96]. For the purpose of comparison, RNA extracts are also treated with a control that does not place bulky adducts on the unbound bases, which is usually the solvent for the SHAPE reagent [96]. Once the adducts are associated with the unbound bases, library preparation is started through the production of cDNA, which will induce mutations in the sequence at the location of SHAPE reagent-inflicted adducts. A second strand of DNA is made to stabilize the DNA libraries and prepare them for high-throughput DNA sequencing. The sequences from the SHAPE reagent treated samples is compared with the sequence data from the control treated samples, then aligned to identify where the SHAPE reagent induced mutations are located (Figure 5) [96]. The sequence data is processed by the SHAPE-MaP software program to calculate the flexibility (reactivity) at each base position [96]. Both the sequence data and the flexibility data are used in a structure prediction program (we used RNAstructure) to create minimum free energy models of the RNA secondary structure [87,96,97].

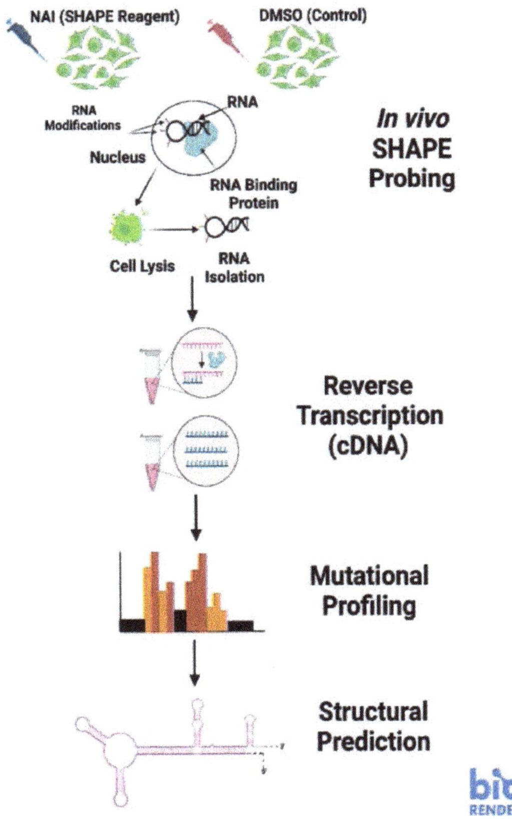

Figure 5. The SHAPE-MaP pipeline for RNA structural analysis. Image created with BioRender.com.

More recently, a new application of selective 2′-hydroxyl acylation analyzed by primer extension emerged as juxtaposed merged pairs protocol (SHAPE-JuMP) [98]. SHAPE-JuMP serves the purpose of supporting higher-order RNA structural prediction through crosslinking nearby structures in the same RNA transcript [98]. The crosslinker is a SHAPE reagent called trans-bis-isatoic anhydride (TBIA) which has two functional sites that interact with the 2′ hydroxyl groups of bases in RNA structural domains that are near to each other [98]. Once the RNA structures are crosslinked, RT-C8, a reverse transcriptase capable of "jumping" over the TBIA crosslinker, reverse transcribes the RNA into DNA while leaving out the RNA between the crosslinked bases [98]. The skipped region of RNA appears as a deletion when the sequence is aligned to a reference sequence [98].

4.3. Advantages and Limitations of Structural Modeling

A crucial advantage of using SHAPE-MaP techniques to predict the secondary structure of RNA is the ability to use it inside living cells in addition to outside of the cells and on deproteinized "naked" RNA [96]. In their protocol paper, SHAPE-MaP developers Smola and Weeks propose using both in-cell and cell-free SHAPE-MaP procedures to identify the locations of likely RNA-protein interactions [96]. Since the SHAPE-MaP software calculates base flexibility (proclivity for interaction with other nucleic acids or proteins) at the individual base level, RNA SHAPE-treated inside of cells can be compared with RNA SHAPE-treated samples after extraction and deproteinization to determine where there are differences in flexibility. The locations that exhibit base flexibility clue investigators into areas that warrant further evaluation to determine (1) if there is in fact some kind of interaction occurring between the RNA of interest and some other molecule(s), and (2) what those other molecules are if there is an interaction occurring [96].

SHAPE-JuMP builds upon the concept of SHAP-MaP. Its creators credit it with being better adapted at handling long range, RNA tertiary interactions that can be more prone to errors in traditional SHAPE-MaP protocols [98]. An additional benefit of the SHAPE-JuMP procedure of RNA modeling is the structural support provided to interacting structures within the transcript [98]. Essentially, TBIA freezes the interacting RNA structures in place for the purpose of reverse transcription and sequencing [98]. The immobilization of the interacting structure makes SHAPE-JuMP a particularly effective method of predicting the structures of large RNAs [98].

While quite powerful, SHAPE-MaP is not perfect. SHAPE-MaP software is unable to differentiate between multiple different isoforms of the same transcript. Different isoforms of RNAs can be predicted by several different RNA processing events which are the results of complex RNA interactions. These interactions have impacts on the RNA's ability to perform its intended function, whether that be translation into a functional protein or a regulatory role. In addition to the inability to differentiate between structural isoforms of an RNA transcript, SHAPE-MaP requires chemical probes that can cross the plasma membrane of the cells being studied [96]. There are several SHAPE probes commercially available, but they do not all have the same inclination to penetrate cellular plasma membranes [96]. SHAPE-MaP analysis relies on effective DNA library preparations and in RNA transcripts that are highly repetitive or structurally inaccessible, these sequencing results and predicted secondary structures are not as reliable as regions that are non-repetitive or structurally accessible [96]. SHAPE-JuMP is quite novel and not all limitations have been fully characterized. The SHAPE-JuMP creators did cite a less than optimal ability to identify tertiary contacts between interacting structures as well as a reliance on amplicon sequencing as major limitations of the protocol [98].

The applications of RNA structural modeling using SHAPE techniques are not limited to RNA in living systems. NANPs themselves take on discrete structures that are key to the effective delivery of their cargo to the appropriate location. The RNA-SHAPE techniques to produce structural models of RNA can be applied after NANP production in a quality control step to ensure that the nanoparticles accurately formed the intended structure. Additionally, SHAPE structural modeling techniques can be employed to evaluate how the structures of NANPs change when the cargo is delivered. Furthermore, we can use RNA-SHAPE to determine if the delivery process makes any changes to the structure of the nucleic acid cargo that may impact its ability to perform its intended function(s). These techniques can assist investigators in evaluating the structural stability and integrity of their delivery systems as well as cargo loads to enhance the efficacy of delivery and incorporation of the functional cargo.

5. Discussion

Nanotechnology has been a crucial area of research over the last decade. Nanoparticles are a growing area of academic and commercial investment, and the exploration of nucleic acid nanoparticles has opened up a wide array of research and therapeutic avenues. In the clinic, NANPs offer medical providers with potential preventative, diagnostic and therapeutic applications for patient care. Of course, as is the case with all novel technologies, there is room for nucleic acid nanoparticles to improve. Problems surrounding immunogenicity, off target activity, and unintended buildup all persist, and careful experimentation will lead researchers to optimizing specific NANPs for their unique purposes.

Outside of the clinic in a basic research capacity, NANPs are of great academic value. Investigators can use NANPs to deliver nucleic acids to knock down expression of genes to clarify the purpose of the gene. Additionally, in the arena of RNA domains, we propose the use of NANPs to deliver RNA structural domains to cells that have been depleted of the same structural domains. In this planned research, one could examine the impact of RNA structural domain loss followed by complementation of the cells with the missing domain to clarify the purpose of discrete RNA structural domains. Once optimized, this system will allow us to gain a better understanding of how RNA structure influences its function.

As research with nucleic acid nanoparticles progresses, investigators will need to address the challenges and disadvantages that come with working with them. Challenges such as unintended immune neutralization, NANP accumulation, and off-target effects are most common. The most powerful tool researchers have in mitigating these challenges is careful design of the NANPs so that they are non-immunogenic prior to target site delivery, do not build up in inappropriate locations, and do not act on inappropriate tissue. This can be done through robust screening of candidate NANPs to ensure that their specificity is as narrow as possible (i.e., it has only one complementary sequence). When designing NANPs for use in the clinic, specificity is absolutely crucial for proper delivery and appropriate immune activation. Additionally, investigators must consider the stability of the NANP they wish to design. While the NANP should be stable enough to travel to the intended target tissue, once there, it must be able to deliver its cargo and then be broken down and cleared to prevent accumulation of "spent" NANPs. The structure of the NANP is a key element in its activity and special attention must be paid to how the nanostructures will react upon arrival at the target site. Advances in NANP design technology (i.e., software like NanoTiler and SELEX) will undoubtedly assist researchers in creating NANPs that have highly specific sequences to limit target possibilities and the proper three-dimensional structures to act on a specific target, then undergo degradation and clearance [19,20,99].

Author Contributions: Conceptualization, K.K., Y.R. and K.C.; visualization, K.K., Y.R. and K.C.; writing—original draft preparation, K.K. and Y.R.; writing—review and editing, K.K., Y.R. and K.C.; supervision, K.C.; all authors critically revised, discussed, and edited the article until it reached its current form. All authors have read and agreed to the published version of the manuscript.

Funding: This work received no external funding.

Institutional Review Board Statement: Not Applicable.

Informed Consent Statement: Not Applicable.

Data Availability Statement: Not Applicable.

Acknowledgments: This review paper was written as a collaborative effort between the Chakrabarti Lab in the Department of Biological Sciences and the Afonin Lab in the Nanoscale Science Program within the Department of Chemistry at the University of North Carolina at Charlotte.

Conflicts of Interest: The authors declare no conflict of interest.

References

1. Afonin, K.A.; Dobrovolskaia, M.A.; Church, G.; Bathe, M. Opportunities, Barriers, and a Strategy for Overcoming Translational Challenges to Therapeutic Nucleic Acid Nanotechnology. *ACS Nano* **2020**, *14*, 9221–9227. [CrossRef] [PubMed]
2. Hong, E.; Halman, J.R.; Shah, A.B.; Khisamutdinov, E.F.; Dobrovolskaia, M.A.; Afonin, K.A. Structure and Composition Define Immunorecognition of Nucleic Acid Nanoparticles. *Nano Lett.* **2018**, *18*, 4309–4321. [CrossRef] [PubMed]
3. Johnson, M.B.; Chandler, M.; Afonin, K.A. Nucleic acid nanoparticles (NANPs) as molecular tools to direct desirable and avoid undesirable immunological effects. *Adv. Drug Deliv. Rev.* **2021**, *173*, 427–438. [CrossRef] [PubMed]
4. Afonin, K.A.; Viard, M.; Koyfman, A.Y.; Martins, A.N.; Kasprzak, W.K.; Panigaj, M.; Desai, R.; Santhanam, A.; Grabow, W.W.; Jaeger, L.; et al. Multifuncitonal RNA Nanoparticles. *Nano Lett.* **2014**, *14*, 5662–5671. [CrossRef]
5. Jasinski, D.; Haque, F.; Binzel, D.W.; Guo, P. Advancement of the Emerging Field of RNA Nanotechnology. *ACS Nano* **2017**, *11*, 1142–1164. [CrossRef]
6. Narayan, S.P.; Choi, C.H.J.; Hao, L.; Calabrese, C.M.; Auyeung, E.; Zhang, C.; Goor, O.J.G.M.; Mirkin, C.A. The Sequence-Specific Cellular Uptake of Spherical Nucleic Acid Nanoparticle Conjugates. *Small* **2015**, *11*, 4173–4182. [CrossRef]
7. Chandler, M.; Jain, S.; Halman, J.; Hong, E.; Dobrovolskaia, M.A.; Zakharov, A.V.; Afonin, K.A. Artificial Immune Cell, AI-cell, a New Tool to Predict Interferon Production by Peripheral Blood Monocytes in Response to Nucleic Acid Nanoparticles. *Small* **2022**, *18*, e2204941. [CrossRef]
8. Chandler, M.; Rolband, L.; Johnson, M.B.; Shi, D.; Avila, Y.I.; Cedrone, E.; Beasock, D.; Danai, L.; Stassenko, E.; Krueger, J.K.; et al. Expanding Structural Space for Immunomodulatory Nucleic Acid Nanoparticles (Nanps) via Spatial Arrangement of Their Therapeutic Moieties. *Adv. Funct. Mater.* **2022**, *32*, 2205581. [CrossRef]
9. Afonin, K.A.; Kasprzak, W.K.; Bindewald, E.; Kireeva, M.; Viard, M.; Kashlev, M.; Shapiro, B.A. In Silico Design and Enzymatic Synthesis of Functional RNA Nanoparticles. *Acc. Chem. Res.* **2014**, *47*, 1731–1741. [CrossRef]
10. Karlsson, J.; Luly, K.M.; Tzeng, S.Y.; Green, J.J. Nanoparticle designs for delivery of nucleic acid dtherapeutics as brain cancer therapies. *Adv. Drug Deliv. Rev.* **2021**, *179*, 113999. [CrossRef]
11. Lee, H.; Lytton-Jean, A.K.R.; Chen, Y.; Love, K.T.; Park, A.I.; Karagiannis, E.D.; Sehgal, A.; Querbes, W.; Zurenko, C.S.; Jayaraman, M.; et al. Molecularly self-assembled nucleic acid nanoparticles for targeted in vivo siRNA delivery. *Nat. Nanotechnol.* **2012**, *7*, 389–393. [CrossRef] [PubMed]
12. Afonin, K.A.; Bindewald, E.; Yaghoubian, A.J.; Voss, N.; Jacovetty, E.; Shapiro, B.A.; Jaeger, L. In vitro assembly of cubic RNA-based scaffolds designed in silico. *Nat. Nanotechnol.* **2010**, *5*, 676–682. [CrossRef] [PubMed]
13. Afonin, K.A.; Kasprzak, W.; Bindewald, E.; Puppala, P.S.; Diehl, A.R.; Hall, K.T.; Kim, T.J.; Zimmermann, M.T.; Jernigan, R.L.; Jaeger, L.; et al. Computational and experimental characterization of RNA cubic nanoscaffolds. *Methods* **2014**, *67*, 256–265. [CrossRef] [PubMed]
14. Afonin, K.A.; Grabow, W.W.; Walker, F.M.; Bindewald, E.; Dobrovolskaia, M.A.; Shapiro, B.A.; Jaeger, L. Design and self-assembly of siRNA-functionalized RNA nanoparticles for use in automated nanomedicine. *Nat. Protoc.* **2011**, *6*, 2022–2034. [CrossRef] [PubMed]
15. Grabow, W.W.; Zakrevsky, P.; Afonin, K.A.; Chworos, A.; Shapiro, B.A.; Jaeger, L. Self-assembling RNA nanorings based on RNAI/II inverse kissing complexes. *Nano Lett.* **2011**, *11*, 878–887. [CrossRef]
16. Veneziano, R.; Ratanalert, S.; Zhang, K.; Zhang, F.; Yan, H.; Chiu, W.; Bathe, M. Designer nanoscale DNA assemblies programmed from the top down. *Science* **2016**, *352*, 1534. [CrossRef]
17. Yingling, Y.G.; Shapiro, B.A. Computational design of an RNA hexagonal nanoring and an RNA nanotube. *Nano Lett.* **2007**, *7*, 2328–2334. [CrossRef] [PubMed]
18. Keefe, A.D.; Pai, S.; Ellington, A. Aptamers as Therapeutics. *Nat. Rev. Drug Discov.* **2010**, *9*, 537–550. [CrossRef] [PubMed]
19. Benedetto, G.; Hamp, T.J.; Wesselman, P.J.; Richardson, C. Identification of Epithelial Ovarian Tumor-Specific Aptamers. *Nucleic Acid Ther.* **2015**, *25*, 162–172. [CrossRef]
20. Panigaj, M.; Johnson, M.B.; Ke, W.; McMillan, J.; Goncharova, E.; Chandler, M.; Afonin, K.A. Aptamers as Modular Components of Therapeutic Nucleic Acid Nanotechnology. *ACS Nano* **2019**, *13*, 12301–12321. [CrossRef]
21. Guo, P. The emerging field of RNA nanotechnology. *Nat. Nanotechnol.* **2010**, *5*, 833–842. [CrossRef] [PubMed]
22. Guo, P.; Haque, F.; Hallahan, B.; Reif, R.; Li, H. Uniqueness, Advantages, Challenges, Solutions, and Perspectives in Therapeutics Applying RNA Nanotechnology. *Nucleic Acid Ther.* **2012**, *22*, 226–245. [CrossRef] [PubMed]
23. Searle, M.S.; Williams, D.H. On the stability of nucleic acid structures in solution: Enthalpy-entropy compensations, internal rotations and reversibility. *Nucleic Acids Res.* **1993**, *21*, 2051–2056. [CrossRef] [PubMed]
24. Sugimoto, N.; Nakano, S.-I.; Katoh, M.; Matsumura, A.; Nakamuta, H.; Ohmichi, T.; Yoneyama, M.; Sasaki, M. Thermodynamic Parameters To Predict Stability of RNA/DNA Hybrid Duplexes. *Biochemistry* **1995**, *34*, 11211–11216. [CrossRef] [PubMed]
25. Canton, I.; Battaglia, G. Endocytosis at the nanoscale. *R. Soc. Chem.* **2012**, *41*, 2718–2739. [CrossRef] [PubMed]
26. Gao, H.; Shi, W.; Freund, L.B. Mechanics of receptor-mediated endocytosis. *Proc. Natl. Acad. Sci. USA* **2005**, *102*, 9469–9474. [CrossRef]
27. Lee, R.J.; Wang, S.; Low, P.S. Measurement of endosome pH following folate receptor-mediated endocytosis. *Biochim. Biophys. Acta* **1996**, *1312*, 237–242. [CrossRef]
28. Pogocki, D.; Schöneich, C. Chemical Stability of Nucleic Acid-Derived Drugs. *J. Pharm. Sci.* **2000**, *89*, 443–456. [CrossRef]

29. An, R.; Jia, Y.; Wan, B.; Zhang, Y.; Dong, P.; Li, J.; Liang, X. Non-Enzymatic Depurination of Nucleic Acids: Factors and Mechanisms. *PLoS ONE* **2014**, *9*, e115950. [CrossRef]
30. Wilson, R.C.; Doudna, J.A. Molecular mechanisms of RNA interference. *Annu. Rev. Biophys.* **2013**, *42*, 217–239. [CrossRef]
31. Dana, H.; Chalbatani, G.M.; Mahmoodzadeh, H.; Rezvan, K.; Rezaiean, O.; Moradzadeh, A.; Mehmandoost, N.; Moazzen, F.; Mazraeh, A.; Marmari, V.; et al. Molecular Mechanisms and Biological Functions of siRNA. *Int. J. Biomed. Sci.* **2017**, *13*, 48–57. [PubMed]
32. Hu, B.; Zhong, L.; Weng, Y.; Peng, L.; Huang, Y.; Zhao, Y.; Liang, X.-J. Therapeutic siRNA: State of the art. *Signal Transduct. Target. Ther.* **2020**, *5*, 1–25. [CrossRef] [PubMed]
33. Bartlett, D.W.; Davis, M.E. Physicochemical and Biological Characterization of Targeted, Nucleic Acid Nanoparticles. *Bioconjug. Chem.* **2007**, *18*, 456–468. [CrossRef] [PubMed]
34. Durbin, J.K.; Miller, D.K.; Niekamp, J.; Khisamutdinov, E.F. Modulating Immune Response with Nucleic Acid Nanoparticles. *Molecules* **2019**, *24*, 3740. [CrossRef] [PubMed]
35. Dobrovolskaia, M.A. Nucleic Acid Nanoparticles at a Crossroads of Vaccines and Immunotherapies. *Molecules* **2019**, *24*, 4620. [CrossRef]
36. Fire, A.; Xu, S.; Montgomery, M.K.; Kostas, S.A.; Driver, S.E.; Mello, C.C. Potent and specific genetic interference by double-stranded RNA in Caenorhabditis elegans. *Nat. Commun.* **1998**, *391*, 806–811. [CrossRef]
37. Setten, R.L.; Rossi, J.J.; Han, S.-P. The current state and future directions of RNAi-based therapeutics. *Nat. Rev. Drug Discov.* **2019**, *18*, 421–446. [CrossRef]
38. Montgomery, M.K.; Fire, A. Double-stranded RNA as a mediator in sequence-specific genetic silencing and co-suppression. *Trends Genet.* **1998**, *14*, 255–258. [CrossRef]
39. Cheng, J.; Moore, T.B.; Sakamoto, K.M. RNA interference and human disease. *Mol. Genet. Metab.* **2003**, *80*, 121–128. [CrossRef]
40. Chejanovsky, N.; Ophir, R.; Schwager, M.S.; Slabezki, Y.; Grossman, S.; Cox-Foster, D. Characterization of viral siRNA populations in honey bee colony collapse disorder. *Virology* **2014**, *454–455*, 176–183. [CrossRef]
41. Yan, S.; Ren, B.; Zeng, B.; Shen, J. Improving RNAi efficiency for pest control in crop species. *BioTechniques* **2020**, *68*, 283–290. [CrossRef] [PubMed]
42. Hunter, W.; Ellis, J.; Vanengelsdorp, D.; Hayes, J.; Westervelt, D.; Glick, E.; Williams, M.; Sela, I.; Maori, E.; Pettis, J.; et al. Large-Scale Field Application of RNAi Technology Reducing Israeli Acute Paralysis Virus Disease in Honey Bees (Apis mellifera, Hymenoptera: Apidae). *PLoS Pathog.* **2010**, *6*, e1001160. [CrossRef] [PubMed]
43. Dao, B.N.; Viard, M.; Martins, A.N.; Kasprzak, W.K.; Shapiro, B.A.; Afonin, K.A. Triggering RNAi with multifunctional RNA nanoparticles and their delivery. *DNA RNA Nanotechnol.* **2015**, *2*, 1–12. [CrossRef] [PubMed]
44. Bumcrot, D.; Manoharan, M.; Koteliansky, V.; Sah, D.W.Y. RNAi therapeutics: A potential new class of therapeutic drugs. *Nat. Chem. Biol.* **2006**, *2*, 711–719. [CrossRef] [PubMed]
45. Kristen, A.V.; Ajroud-Driss, S.; Conceicao, I.; Gorevic, P.; Kyriakides, T.; Obici, L. Patisiran, an RNAi therapeutic for the treatment of hereditary transthyretin-mediated amyloidosis. *Neurodegener. Dis. Manag.* **2018**, *9*, 5–23. [CrossRef] [PubMed]
46. Adams, D.; Gonzalez-Duarte, A.; O'Riordan, W.D.; Yang, C.-C.; Udea, M.; Kristen, A.V.; Tournev, I.; Schmidt, H.H.; Coelho, T.; Berk, J.L.; et al. Patisiran, an RNAi Therapeutic, for Hereditary Transthyretin Amyloidosis. *N. Engl. J. Med.* **2018**, *379*, 11–21. [CrossRef]
47. Ando, Y.; Coelho, T.; Berk, J.L.; Waddington Cruz, M.; Ericzon, B.-G.; Ikeda, S.-I.; Lewis, W.D.; Obici, L.; Plante-Bordeneuve, V.; Rapezzi, C.; et al. Guideline of transthyretin-related hereditary amyloidosis for clinicians. *Orphanet J. Rare Dis.* **2013**, *8*, 31. [CrossRef]
48. Conceicao, I.; Gonzalez-duarte, A.; Obici, L.; Schmidt, H.H.-J.; Simoneau, D.; Ong, M.-L.; Amass, L. "Red-flag" symptom clusters in transthyretin familial amyloid polyneuropathy. *J. Peripher. Nerv. Syst.* **2016**, *21*, 5–9. [CrossRef]
49. Robbins, M.; Judge, A.; Liang, L.; McClintock, K.; Yaworski, E.; MacLachlan, I. 2′-O-methyl-modified RNAs act as TLR7 antagonists. *Mol. Ther.* **2007**, *15*, 1663–1669. [CrossRef]
50. Janas, M.M.; Schlegel, M.K.; Harbison, C.E.; Yilmaz, V.O.; Jiang, Y.; Parmar, R.; Zlatev, I.; Castoreno, A.; Xu, H.; Shulga-Morskaya, S.; et al. Selection of GalNAc-conjugated siRNAs with limited off-target-driven rat hepatotoxicity. *Nat. Commun.* **2018**, *9*, 1–10. [CrossRef]
51. Zlatev, I.; Castoreno, A.; Brown, C.R.; Qin, J.; Waldron, S.; Schlegel, M.K.; Degaonkar, R.; Shulga-Morskaya, S.; Xu, H.; Gupta, S.; et al. Reversal of siRNA-mediated gene silencing in vivo. *Nat. Biotechnol.* **2018**, *36*, 509–511. [CrossRef] [PubMed]
52. Guo, S.; Li, H.; Ma, M.; Fu, J.; Dong, Y.; Guo, P. Size, Shape, and Sequence-Dependent Immunogenicity of RNA Nanoparticles. *Mol. Ther. Nucleic Acids* **2017**, *9*, 399–408. [CrossRef] [PubMed]
53. Bila, D.; Radwan, Y.; Dobrovolskaia, M.A.; Panigaj, M.; Afonin, K.A. The Recognition of and Reactions to Nucleic Acid Nanoparticles by Human Immune Cells. *Molecules* **2021**, *26*, 4231. [CrossRef]
54. Chandler, M.; Johnson, M.B.; Panigaj, M.; Afonin, K.A. Innate immune responses triggered by nucleic acids inspire the design of immunomodulatory nucleic acid nanoparticles (NANPs). *Curr. Opin. Biotechnol.* **2020**, *63*, 8–15. [CrossRef] [PubMed]
55. Avila, Y.I.; Chandler, M.; Cedrone, E.; Newton, H.S.; Richardson, M.; Xu, J.; Clogston, J.D.; Liptrott, N.J.; Afonin, K.A.; Dobrovolskaia, M.A. Induction of Cytokines by Nucleic Acid Nanoparticles (NANPs) Depends on the Type of Delivery Carrier. *Molecules* **2021**, *26*, 652. [CrossRef]

56. Chandler, M.; Afonin, K.A. Smart-Responsive Nucleic Acid Nanoparticles (NANPs) with the Potential to Modulate Immune Behavior. *Nanomaterials* **2019**, *9*, 611. [CrossRef]
57. Dobrovolskaia, M.A.; Afonin, K.A. Use of human peripheral blood mononuclear cells to define immunological properties of nucleic acid nanoparticles. *Nat. Protoc.* **2020**, *15*, 3678–3698. [CrossRef]
58. Halman, J.R.; Satterwhite, E.; Roark, B.; Chandler, M.; Viard, M.; Ivanina, A.; Bindewald, E.; Kasprzak, W.K.; Panigaj, M.; Bui, M.N.; et al. Functionally-interdependent shape-switching nanoparticles with controllable properties. *Nucleic Acids Res.* **2017**, *45*, 2210–2220. [CrossRef]
59. Hong, E.; Halman, J.R.; Shah, A.; Cedrone, E.; Truong, N.; Afonin, K.A.; Dobrovolskaia, M.A. Toll-Like Receptor-Mediated Recognition of Nucleic Acid Nanoparticles (NANPs) in Human Primary Blood Cells. *Molecules* **2019**, *24*, 1094. [CrossRef]
60. Ke, W.; Hong, E.; Saito, R.F.; Rangel, M.C.; Wang, J.; Viard, M.; Richardson, M.; Khisamutdinov, E.F.; Panigaj, M.; Dokholyan, N.V.; et al. RNA-DNA fibers and polygons with controlled immunorecognition activate RNAi, FRET and transcriptional regulation of NF-kappaB in human cells. *Nucleic Acids Res.* **2019**, *47*, 1350–1361. [CrossRef]
61. Rackley, L.; Stewart, J.M.; Salotti, J.; Krokhotin, A.; Shah, A.; Halman, J.R.; Juneja, R.; Smollett, J.; Lee, L.; Roark, K.; et al. RNA Fibers as Optimized Nanoscaffolds for siRNA Coordination and Reduced Immunological Recognition. *Adv. Funct. Mater.* **2018**, *28*, 1805959. [CrossRef] [PubMed]
62. Tran, A.N.; Chandler, M.; Halman, J.; Beasock, D.; Fessler, A.; McKeough, R.Q.; Lam, P.A.; Furr, D.P.; Wang, J.; Cedrone, E.; et al. Anhydrous Nucleic Acid Nanoparticles for Storage and Handling at Broad Range of Temperatures. *Small* **2022**, *18*, e2104814. [CrossRef] [PubMed]
63. Johnson, M.B.; Halman, J.R.; Miller, D.K.; Cooper, J.S.; Khisamutdinov, E.F.; Marriott, I.; Afonin, K.A. The immunorecognition, subcellular compartmentalization, and physicochemical properties of nucleic acid nanoparticles can be controlled by composition modification. *Nucleic Acids Res.* **2020**, *48*, 11785–11798. [CrossRef]
64. Johnson, M.B.; Halman, J.R.; Burmeister, A.R.; Currin, S.; Khisamutdinov, E.F.; Afonin, K.A.; Marriott, I. Retinoic acid inducible gene-I mediated detection of bacterial nucleic acids in human microglial cells. *J. Neuroinflamm.* **2020**, *17*, 139. [CrossRef] [PubMed]
65. Johnson, M.B.; Halman, J.R.; Satterwhite, E.; Zakharov, A.V.; Bui, M.N.; Benkato, K.; Goldsworthy, V.; Kim, T.; Hong, E.; Dobrovolskaia, M.A.; et al. Programmable Nucleic Acid Based Polygons with Controlled Neuroimmunomodulatory Properties for Predictive QSAR Modeling. *Small* **2017**, *13*. [CrossRef] [PubMed]
66. Afonin, K.A.; Dobrovolskaia, M.A.; Ke, W.; Grodzinski, P.; Bathe, M. Critical review of nucleic acid nanotechnology to identify gaps and inform a strategy for accelerated clinical translation. *Adv. Drug Deliv. Rev.* **2022**, *181*, 114081. [CrossRef] [PubMed]
67. Rolband, L.; Beasock, D.; Wang, Y.; Shu, Y.G.; Dinman, J.D.; Schlick, T.; Zhou, Y.; Kieft, J.S.; Chen, S.J.; Bussi, G.; et al. Biomotors, viral assembly, and RNA nanobiotechnology: Current achievements and future directions. *Comput. Struct. Biotechnol. J.* **2022**, *20*, 6120–6137. [CrossRef]
68. Ke, W.; Chandler, M.; Cedrone, E.; Saito, R.F.; Rangel, M.C.; de Souza Junqueria, M.; Wang, J.; Shi, D.; Truong, N.; Richardson, M.; et al. Locking and Unlocking Thrombin Function Using Immunoquiescent Nucleic Acid Nanoparticles with Regulated Retention In Vivo. *Nano Lett.* **2022**, *22*, 5961–5972. [CrossRef] [PubMed]
69. Nordmeier, S.; Ke, W.; Afonin, K.A.; Portnoy, V. Exosome mediated delivery of functional nucleic acid nanoparticles (NANPs). *Nanomedicine* **2020**, *30*, 102285. [CrossRef]
70. Halman, J.R.; Kim, K.T.; Gwak, S.J.; Pace, R.; Johnson, M.B.; Chandler, M.R.; Rackley, L.; Viard, M.; Marriott, I.; Lee, J.S.; et al. A cationic amphiphilic co-polymer as a carrier of nucleic acid nanoparticles (Nanps) for controlled gene silencing, immunostimulation, and biodistribution. *Nanomed. Nanotechnol. Biol. Med.* **2020**, *23*, 102094. [CrossRef] [PubMed]
71. Cruz-Acuna, M.; Halman, J.R.; Afonin, K.A.; Dobson, J.; Rinaldi, C. Magnetic nanoparticles loaded with functional RNA nanoparticles. *Nanoscale* **2018**, *10*, 17761–17770. [CrossRef] [PubMed]
72. Juneja, R.; Lyles, Z.; Vadarevu, H.; Afonin, K.A.; Vivero-Escoto, J.L. Multimodal Polysilsesquioxane Nanoparticles for Combinatorial Therapy and Gene Delivery in Triple-Negative Breast Cancer. *ACS Appl. Mater. Interfaces* **2019**, *11*, 12308–12320. [CrossRef] [PubMed]
73. Moore, P.B. Structural Motifs in RNA. *Annu. Rev. Biochem.* **1999**, *68*, 287–300. [CrossRef] [PubMed]
74. Leontis, N.B.; Lescoute, A.; Westhof, E. The building blocks and motifs of RNA architecture. *Curr. Opin. Struct. Biol.* **2006**, *16*, 279–287. [CrossRef]
75. Ponting, C.P.; Russell, R.R. The Natural History of Protein Domains. *Annu. Rev. Biophys. Biomol. Struct.* **2002**, *31*, 45–71. [CrossRef]
76. Reiter, N.J.; Chan, C.W.; Mondragón, A. Emerging structural themes in large RNA molecules. *Curr. Opin. Struct. Biol.* **2011**, *21*, 319–326. [CrossRef] [PubMed]
77. Conn, G.L.; Draper, D.E. RNA structure. *Curr. Opin. Struct. Biol.* **1998**, *8*, 278–285. [CrossRef]
78. Nasalean, L.; Baudrey, S.; Leontis, N.B.; Jaeger, L. Controlling RNA self-assembly to form filaments. *Nucleic Acids Res.* **2006**, *34*, 1381–1392. [CrossRef]
79. Westhof, E.; Masquida, B.; Jaeger, L. RNA tectonics: Towards RNA design. *Fold. Des.* **1996**, *1*, R78–R88. [CrossRef]
80. Mitchell, C.; Polanco, J.A.; DeWald, L.; Kress, D.; Jaeger, L.; Grabow, W.W. Responsive self-assembly of tectoRNAs with loop-receptor interactions from tetrahydrofolate (THF) riboswitch. *Nucleic Acids Res.* **2019**, *47*, 6439–6451. [CrossRef]
81. Huang, Q.; Zeng, J.; Yan, J. COVID-19 mRNA vaccines. *J. Genet. Genom.* **2021**, *48*, 107–114. [CrossRef] [PubMed]
82. Cai, Y.; Zhang, J.; Peng, H.; Sterling, S.M.; Walsh, R.M., Jr.; Rawson, S.; Rits-Volloch, S.; Chen, B. Distinct conformational states of SARS-CoV-2 spike protein. *Science* **2020**, *369*, 1586–1592. [CrossRef] [PubMed]

83. Graczyk, A.; Pawlowska, R.; Chworos, A. Gold Nanoparticles as Carriers for Functional RNA Nanostructures. *Bioconjug. Chem.* **2021**, *32*, 1667–1674. [CrossRef] [PubMed]
84. Liu, D.; Shao, Y.; Piccirilli, J.A.; Weizmann, Y. Structures of artificially designed discrete RNA nanoarchitectures at near-atomic resolution. *Sci. Adv.* **2021**, *7*, eabf4459. [CrossRef]
85. Blackburn, E.H. Telomerases. *Annu. Rev. Biochem.* **1992**, *61*, 113–129. [CrossRef] [PubMed]
86. Förstemann, K.; Lingner, J. Telomerase limits the extent of base pairing between template RNA and telomeric DNA. *EMBO Rep.* **2005**, *6*, 361–366. [CrossRef]
87. Dey, A.; Monroy-Eklund, A.; Klotz, K.; Saha, A.; Davis, J.; Li, B.; Laederach, A.; Chakrabarti, K. In vivo architecture of the telomerasae RNA catalytic core in Trypanosoma brucei. *Nucleic Acids Res.* **2021**, *49*, 12445–12466. [CrossRef]
88. Parlea, L.; Puri, A.; Kasprzak, W.; Bindewald, E.; Zakrevsky, P.; Satterwhite, E.; Joseph, K.; Afonin, K.A.; Shapiro, B.A. Cellular Delivery of RNA Nanoparticles. *ACS Comb. Sci.* **2016**, *18*, 527–547. [CrossRef]
89. Barrangou, R.; Doudna, J.A. Applications of CRISPR technologies in research and beyond. *Nat. Biotechnol.* **2016**, *34*, 933–941. [CrossRef]
90. Anderson, E.M.; Haupt, A.; Schiel, J.A.; Chou, E.; Machado, H.B.; Strezoska, Z.; Lnger, S.; McClelland, S.; Birmingham, A.; Vermeulen, A.; et al. Systematic analysis of CRISPR-Cas9 mismatch tolerance reveals low levels of off-target activity. *J. Biotechnol.* **2015**, *211*, 56–65. [CrossRef]
91. Sternberg, S.H.; Haurwitz, R.E.; Doudna, J.A. Mechanism of substrate selection by a highly specific CRISPR endoribonuclease. *RNA* **2012**, *18*, 661–672. [CrossRef] [PubMed]
92. Gaj, T.; Gersbach, C.A.; Barbas, C.F., III. ZFN, TALEN, and CRISPR/Cas-based methods for genome engineering. *Trends Biotechnol.* **2013**, *31*, 397–405. [CrossRef] [PubMed]
93. Bramsen, J.B.; Kjems, J. *Chemical Modification of Small Interfering RNA*; van Rij, R.P., Ed.; Humana Press: Totowa, NJ, USA, 2011; Volume 721.
94. Al-Hashimi, H.M.; Walter, N.G. RNA dynamics: It is about time. *Curr. Opin. Struct. Biol.* **2008**, *18*, 321–329. [CrossRef] [PubMed]
95. Strobel, E.J.; Yu, A.M.; Lucks, J.B. High-throughput determination of RNA structures. *Nat. Rev. Genet.* **2018**, *19*, 615–634. [CrossRef] [PubMed]
96. Smola, M.J.; Weeks, K.M. In-cell RNA structure probing with SHAPE-MaP. *Nat. Protoc.* **2018**, *13*, 1181–1195. [CrossRef] [PubMed]
97. Reuter, J.S.; Mathews, D.H. RNAstructure: Software for RNA secondary structure prediction and analysis. *BMC Bioinform.* **2010**, *11*, 129. [CrossRef] [PubMed]
98. Christy, T.W.; Giannetti, C.A.; Houlihan, G.; Smola, M.J.; Rice, G.M.; Wang, J.; Dokholyan, N.V.; Laederach, A.; Holliger, P.; Weeks, K.M. Direct Mapping of High-Order RNA Interactions by SHAPE-JuMP. *Biochemistry* **2021**, *60*, 1971–1982. [CrossRef]
99. Kasprzak, W.; Bindewald, E.; Kim, T.-J.; Jaeger, L.; Shapiro, B.A. Use of RNA structure flexibility data in nanostructure modeling. *Methods* **2011**, *54*, 239–250. [CrossRef]

molecules

MDPI

Review

The Recognition of and Reactions to Nucleic Acid Nanoparticles by Human Immune Cells

Dominika Bila [1], Yasmine Radwan [2], Marina A. Dobrovolskaia [3], Martin Panigaj [1,*] and Kirill A. Afonin [2,*]

1 Faculty of Science, Institute of Biology and Ecology, Pavol Jozef Safarik University in Kosice, 04154 Kosice, Slovakia; dominika.bila@student.upjs.sk
2 Nanoscale Science Program, Department of Chemistry, University of North Carolina at Charlotte, Charlotte, NC 28223, USA; yradwan@uncc.edu
3 Nanotechnology Characterization Laboratory, Cancer Research Technology Program, Frederick National Laboratory for Cancer Research Sponsored by the National Cancer Institute, Frederick, MD 21702, USA; marina@mail.nih.gov
* Correspondence: martin.panigaj@upjs.sk (M.P.); kafonin@uncc.edu (K.A.A.); Tel.: +421-55-234-1205 (M.P.); +1-704-687-0685 (K.A.A.)

Abstract: The relatively straightforward methods of designing and assembling various functional nucleic acids into nanoparticles offer advantages for applications in diverse diagnostic and therapeutic approaches. However, due to the novelty of this approach, nucleic acid nanoparticles (NANPs) are not yet used in the clinic. The immune recognition of NANPs is among the areas of preclinical investigation aimed at enabling the translation of these novel materials into clinical settings. NANPs' interactions with the complement system, coagulation systems, and immune cells are essential components of their preclinical safety portfolio. It has been established that NANPs' physicochemical properties—composition, shape, and size—determine their interactions with immune cells (primarily blood plasmacytoid dendritic cells and monocytes), enable recognition by pattern recognition receptors (PRRs) such as Toll-like receptors (TLRs) and RIG-I-like receptors (RLRs), and mediate the subsequent cytokine response. However, unlike traditional therapeutic nucleic acids (e.g., CpG oligonucleotides), NANPs do not trigger a cytokine response unless they are delivered into the cells using a carrier. Recently, it was discovered that the type of carrier provides an additional tool for regulating both the spectrum and the magnitude of the cytokine response to NANPs. Herein, we review the current knowledge of NANPs' interactions with various components of the immune system to emphasize the unique properties of these nanomaterials and highlight opportunities for their use in vaccines and immunotherapy.

Keywords: nucleic acid nanoparticles (NANPs); immunorecognition; immunoreaction; Toll-like receptors; cytokine storm syndrome; complement activation-related pseudoallergy

Citation: Bila, D.; Radwan, Y.; Dobrovolskaia, M.A.; Panigaj, M.; Afonin, K.A. The Recognition of and Reactions to Nucleic Acid Nanoparticles by Human Immune Cells. *Molecules* **2021**, *26*, 4231. https://doi.org/10.3390/molecules26144231

Academic Editor: Ali Nazemi

Received: 1 June 2021
Accepted: 8 July 2021
Published: 12 July 2021

1. Introduction

Nucleic Acid Nanoparticles

Nanomedicine is an application of nanotechnology in medical settings for diagnosis, treatment, and prevention. It exploits unique chemical, physical, and biological properties of materials at the nanoscale. One of the perspective branches of nanomedicine is nucleic acid nanotechnology, which uses nucleic acids—DNA, RNA, and their various modifications—to design and formulate nanostructures for therapeutic applications [1].

Due to the programmability and the intrinsic functions of nucleic acids, single-stranded DNA or RNA molecules are rationally designed into modular nucleic acid nanoparticles (NANPs) that are easily customized into supramolecular three-dimensional structures exclusively made of nucleic acids. RNA and DNA form canonical and non-canonical base pairings to assemble into various higher-order structures that serve as a basis for the assembly of different nanostructures including rings, fibers, and polygons [1–6]. Advantageously, the choice of nucleic acid components provides tunability for the physicochemical properties, biological activities, and multifunctionality of NANPs. Many experiments in biotechnology and biomedicine propose applications of NANPs as carriers of bioactive compounds, molecular tools for imaging and biosensing, scaffolds for biochemical reactions, or multifunctional nanoparticles combining the previous functions into one complex [1,7–9]. The rapidly evolving field of nucleic acid nanotechnology had provided multiple synthesis methods for NANPs, established their characterization techniques in vitro and in vivo, and produced proof-of-concept data for using NANPs in various therapeutic applications [10–13].

NANPs can achieve biomedical functions by delivering therapeutic nucleic acids (TNAs) that are designed to perform key functions in gene regulation and expression and protein synthesis to serve in therapeutic applications. The modular functionalization of NANPs with aptamers, antibodies, or small molecules for their targeted delivery allows NANPs to integrate and deliver various TNAs into cells for synergistic therapeutic effects. However, despite these developments, NANPs have yet to advance to clinical translation due to concerns that need to be investigated and resolved including their specific delivery to target cells, their enzymatic degradation, and their ability to induce an immune response upon cellular uptake [4,12,14,15]. While targeting and stability are not immediate life-threatening issues, the excessive immune recognition of NANPs and overreaction by immune cells can have potentially deleterious effects. Thereby, the immunostimulatory properties of NANPs are being extensively investigated [14,16,17].

Several physicochemical properties of NANPs determine their recognition by the immune cells; the most notable properties are 3D structure, composition (RNA to DNA ratio), molecular size, and the NANP's sequence. In addition, the immune response could be modulated by the type of delivery carriers used [6,14,18–22]. The proper design of NANPs with respect to immunostimulatory properties has the potential to activate innate and adaptive immune responses by activating nucleic acid immune sensors, thus having high potential as vaccine adjuvants and pan-antivirals [2,14,20,21,23–25]. Our emerging knowledge about the individual immunostimulatory abilities of nucleic acids aids in the design of safe NANPs, but it must be stressed that because of the effects of structure, the immunological characteristics of NANPs are not the sum of their individual components. Therefore, each NANP assembly must be experimentally tested and safety validated.

2. Recognition and Reaction of Immune Cells to Nucleic Acids

Immune cells are equipped with an extensive portfolio of so-called pattern recognition receptors (PRRs) that detect pathogen-associated molecular patterns (PAMPs) and damage-associated molecular patterns (DAMPS). The first line of PRRs include Toll-like receptors (TLRs) located on cell membranes (TLR1, TLR2, TLR4, TLR5, TLR6, TLR10) and in the endosomal compartment (TLR3, TLR7, TLR8, and TLR9) followed by RIG-I-like receptors (RLRs) or DNA sensor cyclic GMP–AMP synthase (cGAS) situated in the cytosol [20,26].

TLR sensing of nucleic acids is specific for RNA or DNA recognition and resides in the endosomal compartment, where TLR3 is specific for double-stranded RNA, including small interfering RNA (siRNA), TLR7 functions as a single-stranded RNA receptor, TLR8 is specific for bacterial and viral RNA immune recognition, and finally, TLR9 responds to bacterial and viral DNA (Figure 1) [20,26]. Recognition of nucleic acids from non-cellular origins activates a complex network of signaling cascades that usually culminates in the expression of interferons (IFNs), including other cytokines and various chemokines. The general goal of the response is to alarm adjacent cells and recruit cells of adaptive immunity. The recognition of nucleic acids by TLRs causes signal transduction through Toll/interleukin-1 receptor (TIR)-containing signaling adaptors, TRIF, or MyD88 [27,28]. The downstream acceptor of these signals is NF-κB which, upon activation, translocates into the nucleus and induces the expression of pro-inflammatory genes [26,29]. NF-κB is functioning in both innate and adaptive immune cells. In addition to the mediation of macrophage inflammatory responses, NF-κB promotes the activation and differentiation of T cells and the maturation and differentiation of B cells [30,31]. Finally, the expression of

IFNs modulates further immune defense via paracrine and autocrine signaling through the transcription of IFN-stimulated genes (ISGs). The main effector functions of ISGs are to target pathways and functions required during the pathogens' life cycle as well as to enhance innate immune signaling. In addition, ISGs encode proapoptotic proteins that lead cells to apoptosis under specific conditions [32,33].

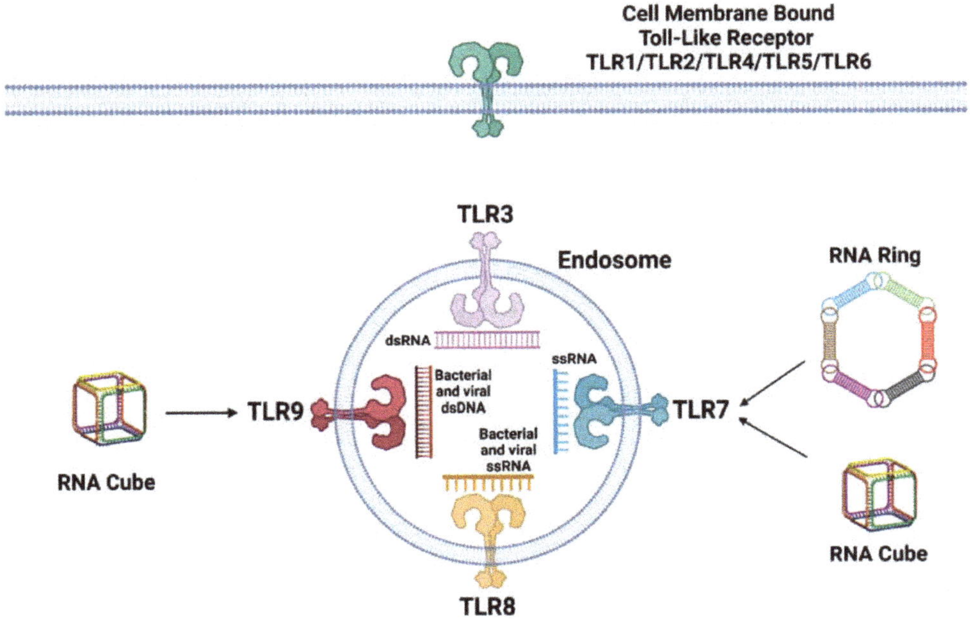

Figure 1. Toll-like receptors. Cell membrane-bound TLRs include TLR1, TLR2, TLR4, TLR5, and TLR6, while endosomal TLRs include TLR3, TLR7, TLR8, and TLR9. TLR3 recognizes double-stranded RNA (dsRNA). TLR8 recognizes bacterial and viral single-stranded RNA (ssRNA). TLR7 recognizes single-stranded RNA (ssRNA), as well as ring and cube RNA. TLR9 recognizes bacterial and viral double-stranded DNA (dsDNA), along with cube RNA. It is important to note that RNA cube triggers the activation of TLR9 and TLR7 only after its delivery inside the cell using a carrier such as L2K.

Intracellular surveillance of RNA is carried out by RLRs, mainly the retinoic acid-inducible gene-I protein (RIG-I) and melanoma differentiation-associated protein 5 (MDA5) located in the cytosol, although the presence of RIG-I has also been observed in the nucleus. RIG-I and MDA5 are activated by binding short double-stranded RNA (dsRNA) with a 5′-triphosphate and 5′-diphosphate or long dsRNA structures, respectively. Furthermore, for the most efficient activation of RIG-I, the blunt end is required as well as a short double-stranded sequence. Activated RIG-I interacts with the mitochondrial antiviral signaling protein (MAVS) residing on the mitochondrial membrane or peroxisomes. Finally, kinase complexes activated by MAVS induce transcription through IRF3, IRF7, and NF-κB. The main cytoplasmic sensors of dsDNA are cyclic GMP-AMP synthase (cGAS) and IFNγ-inducible protein 16 (IFI16), which is also located in the nucleus, where it probably detects naked viral DNA. After binding dsDNA, cGAS synthesizes the second messenger 2′3′-cyclic-GMP-AMP (cGAMP) that subsequently mobilizes the stimulator of IFN genes (STING) on the endoplasmic reticulum that again induces the transcription of antiviral genes through IRF3 and NF-κB [34–36].

3. Recognition and Reaction of Immune Cells to NANPs

NANPs demonstrate different interactions with various types of immune cells, that, unlike traditional nucleic acid therapeutics, are also determined by the type of carrier or complexation agent used for NANPs' intracellular delivery. Without such agents, plain NANPs are invisible to the immune cells and do not trigger cellular immunological responses. For example, flow cytometric analysis of freshly collected human peripheral blood mononuclear cells (PBMCs) treated with a carefully chosen panel of NANPs with various compositions (RNA, DNA) and connectivity (globular, planar, and fibrous) revealed that after complexation with Lipofectamine 2000 (L2K), most NANPs are associated with the monocyte fraction and less with lymphocytes. Subsequent confocal microscopy showed that in monocytes, L2K-complexed NANPs were located inside the cells. Using a dye labeling the endolysosomal compartment and an inhibitor of endosomal uptake, it was observed that unlike lymphocytes, monocytes transport L2K-complexed NANPs into their interiors via endosomes. Overall, phagocytosis and endosomal acidification are key processes for L2K-complexed NANPs' uptake by monocytes. A further functional study indicated that scavenger receptors (SRs) are the most probable receptors involved with binding and internalization of L2K-complexed NANPs. In addition, the inhibition of SRs also prevented the expression of IFN-α in response to L2K-complexed NANPs [21]. Scavenger receptors are a heterogenous group of cell surface receptors that recognize a broad range of ligands; therefore, we currently do not know the mechanism of how SRs recognize NANPs [37]. Without L2K, NANPs did not show any signs of internalization by immune cells present in PBMCs and did not trigger the activation of PRRs or interferon responses.

Plasmacytoid dendritic cells (pDCs) play a key role in linking the innate immune and adaptive response, and although they constitute less than 1% of the monocyte fraction, pDCs, in comparison with isolated monocytes and myeloid DCs, respond to L2K-complexed NANPs with the strongest expression of type I and III IFNs. While in all fractions, RNA cubes appear as the strongest inducer of IFN response, pDCs activated IFNs regardless of the composition (DNA vs. RNA) or 3D structure. The depletion of pDCs from PBMCs leads to a dramatic reduction of IFN production, which means that pDCs are the primary source of immune reaction to NANPs. Interestingly, the distinct expression profile of IFN-α, IFN-β, IFN-ω, and IFN-λ between whole PBMCs and isolated pDCs implies that most likely, there is cellular crosstalk among PBMC subpopulations, which determines the overall response to NANPs [21].

The next important question is, which PRRs are responsible for the recognition and triggering of signaling cascades? The application of a pan oligonucleotide inhibitor of endosomal TLR signaling completely prevented the induction of IFN response upon treatment of PBMCs with any L2K-complexed NANPs used in the study. Similar results were observed in purified pDCs. The model HEK293 cell lines overexpressing either TLR3, TLR7, TLR8, or TLR9 were used to rule out which TLR type recognizes respective NANPs. In this model, the globular NANPs (RNA cubes) were sensed by TLR7, and RNA fibers were sensed by the rest of the examined TLRs [21].

In a follow-up study, we downregulated TLR7 and TLR9 expression in PBMCs by a mix of siRNAs. TLR7 and TLR9 were chosen as TLRs expressed in pDCs that are the primary IFN producers in the PBMC pool. However, the interpretation of observed data is complicated by different levels of downregulation of TLRs among the cells isolated from different healthy donors. Even the extent of silencing between TLR7 and TLR9 in one donor varied. The possible explanation may lay in the inter-individual sequence heterogeneity or regulation of TLRs' expression. The significant reduction in IFN response for the L2K-complexed RNA cubes was observed in two out of three donors with silenced TLR7, while no decrease in IFN production was detected upon treatment with L2K-complexed RNA fibers or DNA cubes. The downregulation of TLR9 prevented IFN response only in culture from one donor treated with RNA cubes and from another donor treated with RNA rings [20]. Taken together, TLR7 is responsible for RNA rings' and cubes' immune recognition but not DNA cubes nor RNA fibers (Figure 1).

4. What Makes NANPs Immunostimulatory?

The recognition of NANPs by the cell defense system depends on several physicochemical characteristics, including composition, 3D structure, sequence, shape, size, and connectivity. One of the first observations that the composition of NANPs (number of RNA vs. DNA strands that enter the composition of a particular NANP) affects their immune recognition came from the earlier study of functionally interdependent shape-switching nanoparticles where we noted that all examined NANPs triggered an IFN-α response, but NANPs assembled from six RNA strands were the most immunostimulatory [38]. A similar trend was observed in a study implementing a new RNA tetra-U helix linking motif in triangles with different DNA vs. RNA composition. In a model of human microglia-like cells, the transfection of RNA triangles induced the highest level of IFN-β production, followed by hybrid DNA/RNA triangles. No expression of IFN-β was stimulated by DNA triangles [19].

Several structure–activity relationship models that link the physicochemical properties of NANPs to their immunostimulation have emerged from a larger analysis of 25 different NANPs [21]. First, globular RNA cubes proved to be the most immunostimulatory NANPs. In comparison to DNA cubes that have almost identical shape and size, RNA cubes induced not only IFN-α and IFN-ω as DNA NANPs did, but also IFN-β and type III IFNs (IFN-λ). In addition, RNA cubes were more immunostimulatory than any other RNA-based NANPs (planar rings or fibers), and planar DNA or RNA structures were more immunostimulatory than chemically corresponding fibrous nanoobjects (Figure 2). In all these examples, NANPs were delivered to the cells using L2K.

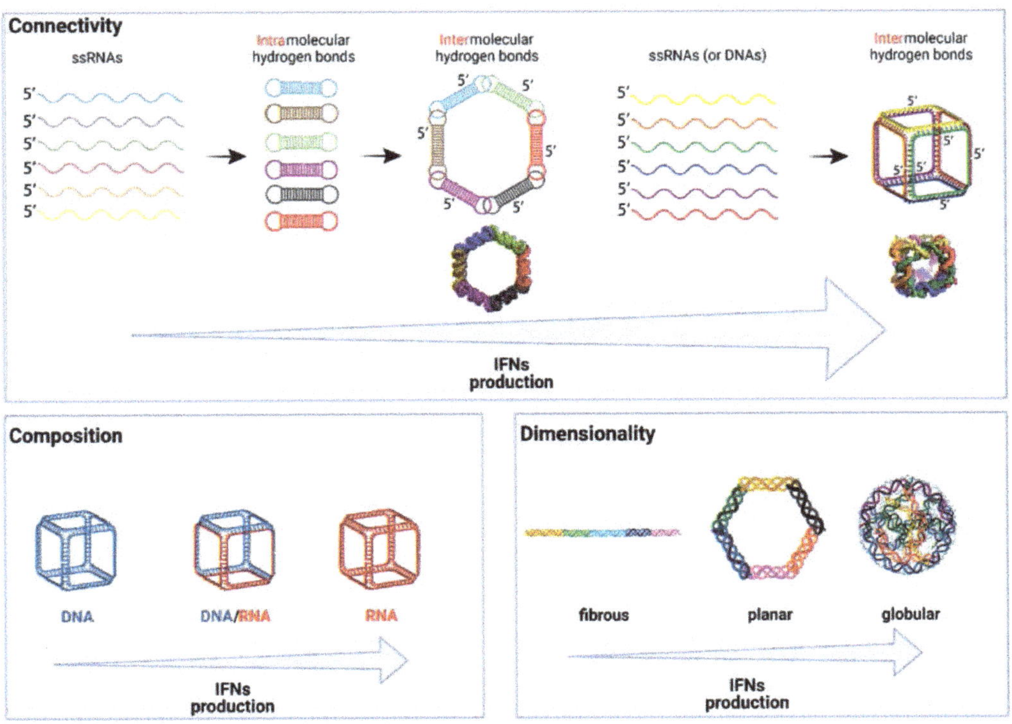

Figure 2. Influence of physicochemical properties on immune stimulation. The main characteristics of NANPs that affect their immunostimulation are connectivity (how individual NANP strands are assembled), composition (number of RNA strands vs. DNA), and dimensionality (3D shape).

The chemical complexity or diversity of assembled NANPs can be increased by the incorporation of modified bases in individual strands. Especially for RNA bases, the diverse modifications play significant roles in RNA stability and affect the immunostimulatory potential [39]. Various experiments have described that the modification of RNA (herein siRNA) helps to circumvent TLR signaling and renders modified RNA immunoquiescent [40]. Therefore, it is interesting that when used with a carrier (L2K or DOTAP), triangular NANPs that consisted of a DNA strand in their center and 2′fluoropyrimidine-modified RNA strands on their sides induced IFN-β and IL-6 production, unlike all DNA NANPs and NANPs composed of a DNA center and unmodified RNA sides. The results suggest that the presence of 2′fluoro-modification significantly enhances the immunoreactivity of DNA-containing NANPs. The NANPs with RNA in the center and 2′fluoropyrimidine-modified RNA sides stimulated IFN-β and IL-6 production similarly to all RNA NANPs and NANPs composed of an RNA center and DNA sides. This indicates that 2′fluoropyrmidine modification does not affect the immune mediator response. The fully 2′fluoropyrimide-modified RNA triangles stimulated significant IFN-β and IL-6 production similarly to NANPs with either an RNA center and 2′fluoropyrimidine-modified RNA sides or NANPs consisting of a DNA center and 2′fluoropyrimidine-modified RNA sides [13]. Surprisingly, the incorporation of 2′fluoro-modifications into RNA NANPs abrogated the activation of TLR7 in the HEK293 reporter cell line but failed to avoid RIG-I dependent immune responses [14].

The ability to design complementary NANPs (also called anti-NANPs) that are assembled from the reverse complementary strands of evaluated NANPs allows for examining the effects of the sequence of NANPs on the ability to activate an IFN response. NANPs and anti-NANPs had completely different sequences but nearly identical 3D shapes. The RNA rings and DNA cubes were able to stimulate similar levels of IFN to their anti-NANPs analogs and anti-RNA cubes maintained the high response, which indicates that the NANPs′ sequences are less important for immunostimulation than their 3D shape and composition (RNA vs. DNA). Except for the RNA rings and RNA fibers that are assembled from pre-formed monomers, all other studied NANPs (cubes, polygons, tetrahedrons, and DNA fibers) create intermolecular bonds (Figure 2). Indeed, free-unpaired nucleotides (ssUs) have enhancing effects on immunogenicity, but it appears only for globular NANPs such as RNA cubes. Interestingly, PBMCs from donors that demonstrated a higher IFN response to a TLR agonist (ODN 2216) reacted stronger to RNA cubes with nine ssUs in their corners than to cubes with a lower number of ssUS (three and six). On the other side, blood cells with lower reactions to the administered TLR agonist induced a similar IFN expression irrespective of the numbers of ssUs.

The size of the nanoparticles is one the main characteristics with potential impact on interactions with cells. Similar to the case of the number of free nucleotides in RNA cubes, the difference was observed only in donor cells with high reactions to ODN 2216, where hexagons activated the stronger response than three-, four-, or five-sided RNA polygons. Adjusting the mass of smaller polygons to be equal to or larger than that of the larger polygons had no effect on IFN production. In cells with low activation by ODN 2216, there was no observed difference between individual NANPs. In the case of DNA polygons, no significant differences were detected between different sizes of NANPs [21].

5. Delivery Method/Carrier: An Unexpected Immunomodulator

The immunostimulatory potential of NANPs is significantly influenced by the employed delivery method. The NANPs without a delivery agent are not efficiently internalized and thus do not induce IFN production. Even if naked NANPs are delivered to cells via electroporation, no production of IFNs was detected in response to any of the tested NANPs (Figure 3). Moreover, electroporated cells lose the ability to respond to other known inducers of IFN response, such as TLR9 agonist ODN 2216, although the addition of ODN2216 to the non-electroporated cells resulted in high levels of type I and III IFNs. The results suggest that electroporation negatively affects endosomal TLR signaling, thereby affecting the ability of cells to elicit an immune response [20].

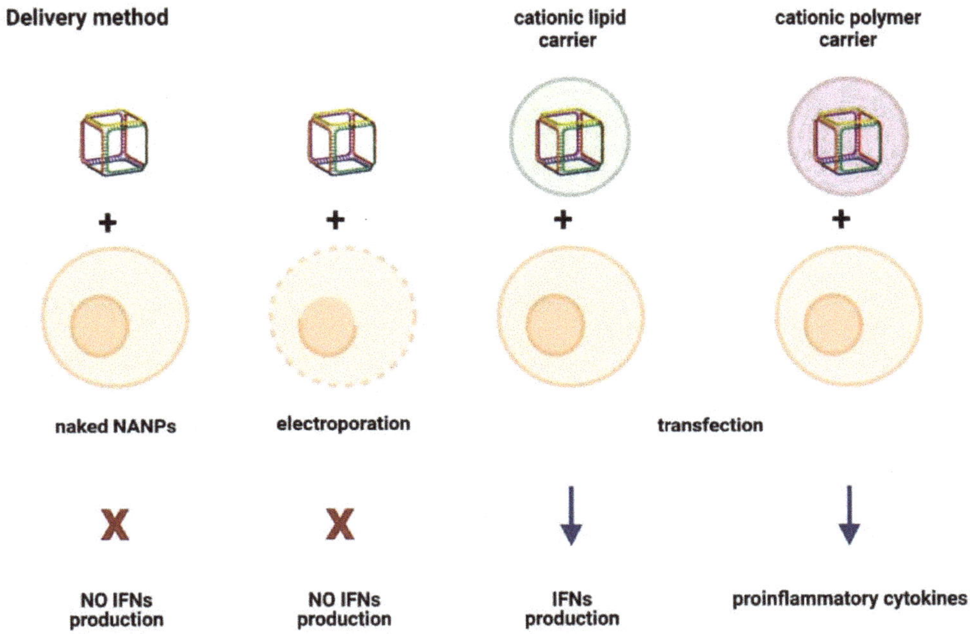

Figure 3. Delivery method and type of carrier affects cellular immune response. Naked NANPs do not trigger IFN response even upon transport to the cytosol via electroporation. Cellular defense is activated by NANPs only if they are in complex with carrier. The type of delivery carrier determines the spectrum of cytokines produced in response to NANPs.

The importance of complexing the NANPs with a carrier for immunorecognition was demonstrated in a study that tested the ability of RNA cubes to induce the type I IFN immune response. The NANPs added to the cell cultures without a delivery carrier were incapable of stimulating an IFN response, while the NANPs complexed with L2K showed the ability to induce the secretion of both type I and type III IFNs. On the other side, ODN 2216, which was used as a positive control, stimulated an IFN response regardless of its complexation with L2K. The application of carrier itself did not cause the induction of the IFN response [21]. L2K does not affect NANPs' structures. Not surprisingly, different carriers demonstrate distinct transfection efficiencies for the same NANP [14].

Although the delivery of NANPs remains a challenge, new carriers are constantly introduced and tested. For instance, the immunostimulatory ability of the lipid-based carrier versus a cationic amphiphilic copolymer was compared. The NANPs delivered via the lipid-based carrier stimulated the production of both IL-6 and IFN-β. In contrast, when the NANPs were delivered using an amphiphilic copolymer, no statistically significant presence of IL-6 or IFN-β was detected. The results suggest that the employment of a cationic amphiphilic copolymer as a delivery carrier can reduce the immunostimulation, therein decreasing off-target effects [41].

Another recent study compared a lipid-based carrier (L2K) and dendrimers (PAMAM) to determine whether the spectrum and the magnitude of the cytokine response to RNA and DNA cubes depend on the type of the utilized carrier. The results showed significant differences in the induction of type I and type III IFNs and pro-inflammatory cytokines between NANPs delivered utilizing a lipid-based carrier and those delivered via dendrimers. The NANPs complexed with L2K stimulated type I and type III IFNs, while the complexation of NANPs with dendrimers did not induce an IFN response. A remarkable difference was observed for cytokines associated with stress and danger (TNFα, IL-1 β, IL-6). The NANPs delivered via L2K did not stimulate a danger response, whereas those complexed with dendrimer induced the production of the stress- and danger-associated pro-inflammatory cytokines. The examination of chemokines (IL-8, MIP-1α, MIP-1β, MCP-1, MCP-2, and RANTES) showed that dendrimers alone did not stimulate any of the chosen chemokines, while the L2K carrier alone induced the production of all examined chemokines but MCP-2. The induction of MCP-2 was detected only when NANPs were complexed with the lipid-based carrier but not for dendrimer-complexed NANPs. Intriguingly, the induction of IL-8, MIP-1α, MCP-1, and RANTES was comparable between NANPs complexed with the lipid-based carrier and complexed with dendrimers. These results support the hypothesis that the type of carrier used for NANPs' delivery significantly alters their ability to stimulate the immune response, both quantitatively and qualitatively [4].

6. Complement Activation-Related Pseudoallergy (CARPA) and Cytokine Release Syndrome (CRS)

The systemic administration of pharmacologic or biologic agents can cause a strong and serious response in immune cells. Infusion-related reactions (IRs), a form of anaphylaxis or other hypersensitivity reactions occurring within minutes to hours of infusion, are immune-mediated adverse effects that occur after the administration of various products, including low-molecular-weight drugs, antibodies, and recombinant proteins, therapeutic nucleic acids, and nanotechnology-formulated products. Frequently observed symptoms in patients with IRs comprise flushing or rash, chest and back pain, dyspnea, wheezing, chills, or fever. These manifestations can lead to serious and potentially fatal consequences. Therefore, accurate assessments and early intervention are crucial when these symptoms occur. When IRs are triggered by the complement system, anaphylactoid reactions or CARPA occur. CARPA has the same symptoms and timeline of development as immediate type hypersensitivity (ITH) reactions. However, in contrast to the ITH, which are mediated by the antigen-specific IgE, CARPA is triggered by the complement. Both CARPA and CRS, also known as cytokine storm, are common, and the best understood mechanisms of IRs are associated with nanotechnology-formulated products [42].

The fundamental processes of CARPA include complement system activation, stimulation of blood cells and secretory cells, and the response of effector cells to mediator presence. The complement is activated via an initial trigger. The initial trigger can be radiocontrast agents, therapeutic antibodies, micellar and liposomal formulations, or nanoparticles. After the activation of the complement, anaphylatoxins are released. The anaphylatoxins are primary mediators that bind to target secretory cells (macrophages, mast cells, basophils, other phagocytic cells, and leukocytes), resulting in a release of secondary mediators that include cytokines, proteases, histamine, tryptase, prostaglandins, platelet-activating factor, thromboxane A2, and leukotrienes. The indications of CARPA are like those that occur with common allergies, with some unique exceptions. The most frequent symptoms are asthma, chest pain, chills, confusion, coughing, dermatitis, diaphoresis, dyspnea, edema, erythema, fever, headache, hypertension, hypotension, hypoxemia, nausea, rash, and wheezing [43]. The significant distinguishing feature is that the reaction arises after the first exposure to the drug and then decreases upon repeated exposure. In the case of NANPs, the lipid-based carrier is the most common cause of complement activation, which can subsequently lead to CARPA [44]. The large size and positive or negative surface charge of liposomes were shown to promote complement activation, whereas liposomes of a smaller size and neutral charge had reduced ability for activation [45]. In addition, the susceptibility of liposomes for complement activation was demonstrated to depend on dose and, in the case of PEGylated liposomes, on the presence of anti-PEG antibodies.

The CRS is a systemic inflammatory response caused by the excessive and rapid release of various pro-inflammatory molecules, including but not limited to INF-γ, TNF-α, IL-1, and IL-6. Macrophages, neutrophils, NK cells, and T cells are most often implicated in the pathogenesis of cytokine storm. The activation of primary T cells or immune cells' lysis initiates the production of IFN-γ and TNF-α, which stimulate macrophages, dendritic cells, other immune cells, and endothelial cells to release more pro-inflammatory cytokines (Figure 4). The production of IL-6 is essential for cytokine storm because IL-6 activates T cells and other immune cells, thereby creating a positive feedback loop. The trigger activating CRS can be traditional therapeutic proteins and nucleic acids as well as small molecular drug allergens, whereas nanocarriers can amplify their toxicity. The analysis comparing the ability of adenoviral vectors and lipid-based carriers to induce cytokine production showed that lipid-based carriers exhibit higher immunostimulatory potential than viral vectors. The clinical translation of numerous nanoformulations designed for nucleic acid delivery was terminated in part due to the immune-mediated adverse effects [46].

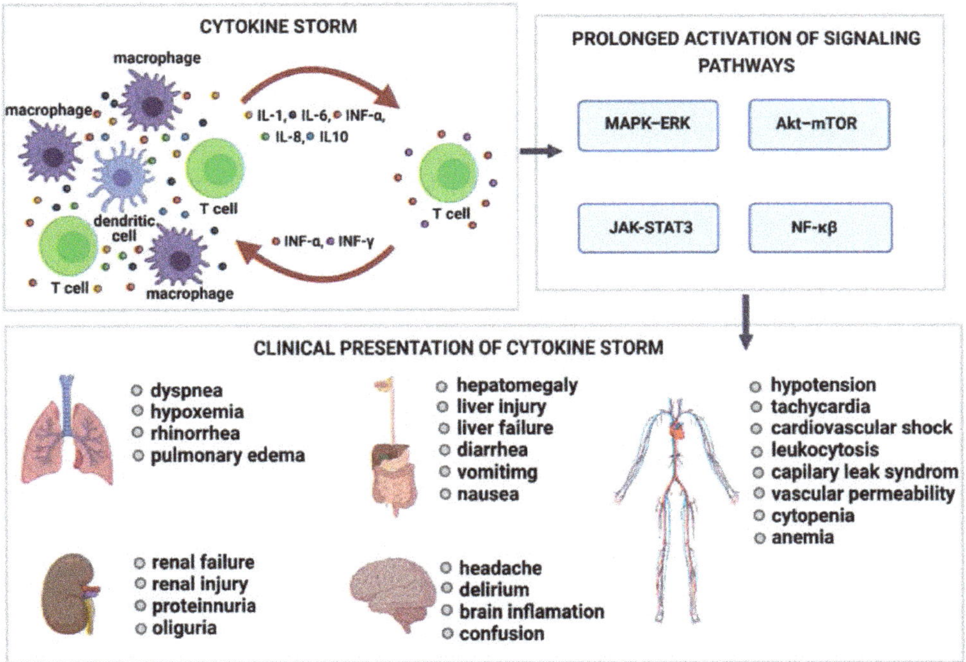

Figure 4. Cytokine storm. Cytokine storm is the result of the rapid release of numerous pro-inflammatory cytokines, including INF-γ, INF-α, IL-1, and IL-6. T cells, macrophages, neutrophils, and NK cells are most often involved in the cytokine storm pathogenesis. The activation of primary T cells or immune cells' lysis stimulates the production of IFN-γ and TNF-α, which activate other immune cells and endothelial cells to release more pro-inflammatory cytokines. The excessive production of IL-6 constantly activates the JAK–STAT3, Akt–mTOR, and MAPK–ERK signaling pathways. Their prolonged activation stimulates immune cells to produce more cytokines, which causes hyperinflammation and multiple organ failure. JAK–STAT3, Janus kinase-signal transducer and activator of transcription 3; MAPK, mitogen-activated protein kinase; mTOR, mammalian target of rapamycin; NF-κB, nuclear factor κB.

7. Conclusions

It is evident that the programmability, biological compatibility, and modularity of nucleic acids assembled into multifunctional NANPs promotes this class of biologically active molecules into an innovative class of personalized therapeutics. To successfully translate these materials to the clinic, one has to recognize the importance of the indication, route of administration, and complexation of NANPs with delivery carriers. If delivered with a carrier via intravenous administration, the induction of cytokines and/or interferons by NANPs may lead to undesirable inflammation. Moreover, some carriers such as liposomes may also trigger CARPA upon systemic administration. However, the same type of cytokine or interferon response and complement activation by the carrier upon local administration may contribute to vaccine efficacy and improve the efficacy of immunotherapy. Experimental data from our laboratory provide several ways for controlling NANPs' immunostimulatory properties. Among them are NANPs' physicochemical properties (e.g., size, shape, sequence, connectivity), complexation with a delivery agent (e.g., lipofectamine, dendrimers), and route of administration (e.g., i.c., vs. s.c. or i.d.). Since the relationship between NANPs' physicochemical/bioactive parameters and the immune system has just emerged, it is necessary to improve the current understanding of NANPs' immunostimulatory properties for their successful translation to the clinic. We believe that the recent onset of mRNA vaccines to fight the COVID-19 pandemic will boost the field of therapeutic nucleic acids, including NANPs.

Funding: Research reported in this publication was supported by the National Institute of General Medical Sciences of the National Institutes of Health under Award Numbers R01GM120487 and R35GM139587 (to K.A.A.). The content is solely the responsibility of the authors and does not necessarily represent the official views of the National Institutes of Health. M.P. involvement in this publication is the result of the project implementation: Open scientific community for modern interdisciplinary research in medicine (OPENMED), ITMS2014+: 313011V455 supported by the Operational Program Integrated Infrastructure, funded by the ERDF. The study was funded in part by federal funds from the National Cancer Institute, National Institutes of Health, under contract 75N91019D00024 (M.A.D.). The content of this publication does not necessarily reflect the views or policies of the Department of Health and Human Services, nor does the mention of trade names, commercial products, or organizations imply endorsement by the U.S. Government. All figures were created with BioRender.com (21 June 2021).

Conflicts of Interest: The authors declare no conflict of interest.

References

1. Afonin, K.A.; Viard, M.; Koyfman, A.Y.; Martins, A.N.; Kasprzak, W.K.; Panigaj, M.; Desai, R.; Santhanam, A.; Grabow, W.W.; Jaeger, L.; et al. Multifunctional RNA nanoparticles. *Nano Lett.* **2014**, *14*, 5662–5671. [CrossRef]
2. Khisamutdinov, E.F.; Li, H.; Jasinski, D.L.; Chen, J.; Fu, J.; Guo, P. Enhancing immunomodulation on innate immunity by shape transition among RNA triangle, square and pentagon nanovehicles. *Nucleic Acids Res.* **2014**, *42*, 9996–10004. [CrossRef]
3. Rackley, L.; Stewart, J.M.; Salotti, J.; Krokhotin, A.; Shah, A.; Halman, J.; Juneja, R.; Smollett, J.; Roark, B.; Viard, M.; et al. RNA Fibers as Optimized Nanoscaffolds for siRNA Coordination and Reduced Immunological Recognition. *Adv. Funct. Mater.* **2018**, *28*. [CrossRef]
4. Avila, Y.I.; Chandler, M.; Cedrone, E.; Newton, H.S.; Richardson, M.; Xu, J.; Clogston, J.D.; Liptrott, N.J.; Afonin, K.A.; Dobrovolskaia, M.A. Induction of cytokines by nucleic acid nanoparticles (NANPs) depends on the type of delivery carrier. *Molecules* **2021**, *26*, 652. [CrossRef]
5. Leontis, N.B.; Stombaugh, J.; Westhof, E. The non-Watson-Crick base pairs and their associated isostericity matrices. *Nucleic Acids Res.* **2002**, *30*, 3497–3531. [CrossRef] [PubMed]
6. Dobrovolskaia, M.A.; Afonin, K.A. Use of human peripheral blood mononuclear cells to define immunological properties of nucleic acid nanoparticles. *Nat. Protoc.* **2020**, *15*, 3678–3698. [CrossRef] [PubMed]
7. Delebecque, C.J.; Lindner, A.B.; Silver, P.A.; Aldaye, F.A. Organization of intracellular reactions with rationally designed RNA assemblies. *Science* **2011**, *333*, 470–474. [CrossRef] [PubMed]
8. Panigaj, M.; Johnson, M.B.; Ke, W.; McMillan, J.; Goncharova, E.A.; Chandler, M.; Afonin, K.A. Aptamers as modular components of therapeutic nucleic acid nanotechnology. *ACS Nano* **2019**, *13*, 12301–12321. [CrossRef] [PubMed]
9. Gong, L.; Zhao, Z.; Lv, Y.-F.; Huan, S.-Y.; Fu, T.; Zhang, X.-B.; Shen, G.-L.; Yu, R.-Q. DNAzyme-based biosensors and nanodevices. *Chem. Commun.* **2015**, *51*, 979–995. [CrossRef] [PubMed]
10. Kim, J.; Franco, E. RNA nanotechnology in synthetic biology. *Curr. Opin. Biotechnol.* **2020**, *63*, 135–141. [CrossRef] [PubMed]
11. Guo, P. The emerging field of RNA nanotechnology. *Nat. Nanotechnol.* **2010**, *5*, 833–842. [CrossRef]
12. Afonin, K.A.; Dobrovolskaia, M.A.; Church, G.; Bathe, M. Opportunities, barriers, and a strategy for overcoming translational challenges to therapeutic nucleic acid nanotechnology. *ACS Nano* **2020**, *14*, 9221–9227. [CrossRef]
13. Panigaj, M.; Dobrovolskaia, M.A.; Afonin, K.A. 2021: An immunotherapy odyssey and the rise of nucleic acid nanotechnology. *Nanomedicine* **2021**. [CrossRef]
14. Johnson, M.B.; Halman, J.R.; Miller, D.K.; Cooper, J.S.; Khisamutdinov, E.F.; Marriott, I.; Afonin, K.A. The immunorecognition, subcellular compartmentalization, and physicochemical properties of nucleic acid nanoparticles can be controlled by composition modification. *Nucleic Acids Res.* **2020**, *48*, 11785–11798. [CrossRef]

15. Juneja, R.; Vadarevu, H.; Halman, J.; Tarannum, M.; Rackley, L.; Dobbs, J.; Marquez, J.; Chandler, M.; Afonin, K.; Vivero-Escoto, J.L. Combination of Nucleic Acid and Mesoporous Silica Nanoparticles: Optimization and Therapeutic Performance In Vitro. *ACS Appl. Mater. Interfaces* **2020**, *12*, 38873–38886. [CrossRef]
16. Jasinski, D.; Haque, F.; Binzel, D.W.; Guo, P. Advancement of the Emerging Field of RNA Nanotechnology. *ACS Nano* **2017**, *11*, 1142–1164. [CrossRef] [PubMed]
17. Johnson, M.B.; Chandler, M.; Afonin, K.A. Nucleic acid nanoparticles (NANPs) as molecular tools to direct desirable and avoid undesirable immunological effects. *Adv. Drug. Deliv. Rev.* **2021**, *173*, 427–438. [CrossRef] [PubMed]
18. Johnson, M.B.; Halman, J.R.; Satterwhite, E.; Zakharov, A.V.; Bui, M.N.; Benkato, K.; Goldsworthy, V.; Kim, T.; Hong, E.; Dobrovolskaia, M.A.; et al. Programmable Nucleic Acid Based Polygons with Controlled Neuroimmunomodulatory Properties for Predictive QSAR Modeling. *Small* **2017**, *13*. [CrossRef] [PubMed]
19. Bui, M.; Johnson, M.; Viard, M.; Satterwhite, E.; Martins, A.; Li, Z.; Marriott, I.; Afonin, K.; Khisamutdinov, E. Versatile RNA tetra-U helix linking motif as a toolkit for nucleic acid nanotechnology. *Nanomed. Nanotechnol. Biol. Med.* **2017**, *13*, 1137–1146. [CrossRef]
20. Hong, E.; Halman, J.R.; Shah, A.; Cedrone, E.; Truong, N.; Afonin, K.A.; Dobrovolskaia, M.A. Toll-like receptor-mediated recognition of nucleic acid nanoparticles (NANPs) in human primary blood cells. *Molecules* **2019**, *24*, 1094. [CrossRef]
21. Hong, E.; Halman, J.R.; Shah, A.B.; Khisamutdinov, E.F.; Dobrovolskaia, M.A.; Afonin, K.A. Structure and Composition Define Immunorecognition of Nucleic Acid Nanoparticles. *Nano Lett.* **2018**, *18*, 4309–4321. [CrossRef]
22. Ke, W.; Afonin, K.A. Exosomes as natural delivery carriers for programmable therapeutic nucleic acid nanoparticles (NANPs). *Adv. Drug Deliv. Rev.* **2021**. [CrossRef] [PubMed]
23. Chandler, M.; Panigaj, M.; Rolband, L.; Afonin, K. Challenges in optimizing RNA nanostructures for large-scale production and controlled therapeutic properties. *Nanomedicine* **2020**, *15*, 1331–1340. [CrossRef] [PubMed]
24. Chandler, M.; Johnson, M.; Panigaj, M.; Afonin, K. Innate immune responses triggered by nucleic acids inspire the design of immunomodulatory nucleic acid nanoparticles (NANPs). *Curr. Opin. Biotechnol.* **2020**, *63*, 8–15. [CrossRef] [PubMed]
25. Guo, S.; Li, H.; Ma, M.; Fu, J.; Dong, Y.; Guo, P. Size, Shape, and Sequence-Dependent Immunogenicity of RNA Nanoparticles. *Mol. Ther. Nucleic Acids* **2017**, *9*, 399–408. [CrossRef]
26. Kawasaki, T.; Kawai, T. Toll-like receptor signaling pathways. *Front. Immunol.* **2014**, *5*. [CrossRef] [PubMed]
27. Kawai, T.; Akira, S. TLR signaling. *Cell Death Differ.* **2006**, *13*, 816–825. [CrossRef]
28. Kawai, T.; Akira, S. The role of pattern-recognition receptors in innate immunity: Update on Toll-like receptors. *Nat. Immunol.* **2010**, *11*, 373–384. [CrossRef]
29. Deguine, J.; Barton, G.M. MyD88: A central player in innate immune signaling. *F1000Prime Rep.* **2014**, *6*, 97. [CrossRef] [PubMed]
30. Liu, T.; Zhang, L.; Joo, D.; Sun, S.C. NF-kappaB signaling in inflammation. *Signal Transduct. Target Ther.* **2017**, *2*. [CrossRef]
31. Dorrington, M.G.; Fraser, I.D.C. NF-kappaB Signaling in Macrophages: Dynamics, Crosstalk, and Signal Integration. *Front. Immunol.* **2019**, *10*, 705. [CrossRef] [PubMed]
32. Schneider, W.M.; Chevillotte, M.D.; Rice, C.M. Interferon-stimulated genes: A complex web of host defenses. *Annu. Rev. Immunol.* **2014**, *32*, 513–545. [CrossRef]
33. Yang, E.; Li, M.M.H. All About the RNA: Interferon-Stimulated Genes That Interfere With Viral RNA Processes. *Front. Immunol.* **2020**, *11*, 605024. [CrossRef] [PubMed]
34. Rehwinkel, J.; Gack, M.U. RIG-I-like receptors: Their regulation and roles in RNA sensing. *Nat. Rev. Immunol.* **2020**, *20*, 537–551. [CrossRef] [PubMed]
35. Chan, Y.K.; Gack, M.U. Viral evasion of intracellular DNA and RNA sensing. *Nat. Rev. Microbiol.* **2016**, *14*, 360–373. [CrossRef]
36. Unterholzner, L.; Almine, J.F. Camouflage and interception: How pathogens evade detection by intracellular nucleic acid sensors. *Immunology* **2019**, *156*, 217–227. [CrossRef]
37. Canton, J.; Neculai, D.; Grinstein, S. Scavenger receptors in homeostasis and immunity. *Nat. Rev. Immunol.* **2013**, *13*, 621–634. [CrossRef]
38. Halman, J.R.; Satterwhite, E.; Roark, B.; Chandler, M.; Viard, M.; Ivanina, A.; Bindewald, E.; Kasprzak, W.K.; Panigaj, M.; Bui, M.N.; et al. Functionally-interdependent shape-switching nanoparticles with controllable properties. *Nucleic Acids Res.* **2017**, *45*, 2210–2220. [CrossRef]
39. Kariko, K.; Buckstein, M.; Ni, H.; Weissman, D. Suppression of RNA recognition by Toll-like receptors: The impact of nucleoside modification and the evolutionary origin of RNA. *Immunity* **2005**, *23*, 165–175. [CrossRef]
40. Morrissey, D.V.; Lockridge, J.A.; Shaw, L.; Blanchard, K.; Jensen, K.; Breen, W.; Hartsough, K.; Machemer, L.; Radka, S.; Jadhav, V.; et al. Potent and persistent in vivo anti-HBV activity of chemically modified siRNAs. *Nat. Biotechnol.* **2005**, *23*, 1002–1007. [CrossRef]
41. Halman, J.R.; Kim, K.T.; Gwak, S.J.; Pace, R.; Johnson, M.B.; Chandler, M.R.; Rackley, L.; Viard, M.; Marriott, I.; Lee, J.S.; et al. A cationic amphiphilic co-polymer as a carrier of nucleic acid nanoparticles (Nanps) for controlled gene silencing, immunostimulation, and biodistribution. *Nanomedicine* **2020**, *23*, 102094. [CrossRef] [PubMed]
42. Lenz, H.J. Management and preparedness for infusion and hypersensitivity reactions. *Oncologist* **2007**, *12*, 601–609. [CrossRef] [PubMed]
43. Szebeni, J. Complement activation-related pseudoallergy: A new class of drug-induced acute immune toxicity. *Toxicology* **2005**, *216*, 106–121. [CrossRef]

44. Szebeni, J. Complement activation-related pseudoallergy caused by liposomes, micellar carriers of intravenous drugs, and radiocontrast agents. *Crit. Rev. Ther. Drug Carrier Syst.* **2001**, *18*, 567–606. [CrossRef] [PubMed]
45. Mohamed, M.; Abu Lila, A.S.; Shimizu, T.; Alaaeldin, E.; Hussein, A.; Sarhan, H.A.; Szebeni, J.; Ishida, T. PEGylated liposomes: Immunological responses. *Sci. Technol. Adv. Mater.* **2019**, *20*, 710–724. [CrossRef] [PubMed]
46. Sakurai, H.; Kawabata, K.; Sakurai, F.; Nakagawa, S.; Mizuguchi, H. Innate immune response induced by gene delivery vectors. *Int. J. Pharm.* **2008**, *354*, 9–15. [CrossRef] [PubMed]

molecules

Review

Innate Immunity Modulating Impurities and the Immunotoxicity of Nanobiotechnology-Based Drug Products

Claire K. Holley and Marina A. Dobrovolskaia *

Nanotechnology Characterization Lab, Cancer Research Technology Program, Frederick National Laboratory for Cancer Research Sponsored by the National Cancer Institute, Frederick, MD 21702, USA; claire.holley@nih.gov
* Correspondence: marina@mail.nih.gov

Abstract: Innate immunity can be triggered by the presence of microbial antigens and other contaminants inadvertently introduced during the manufacture and purification of bionanopharmaceutical products. Activation of these innate immune responses, including cytokine secretion, complement, and immune cell activation, can result in unexpected and undesirable host immune responses. These innate modulators can also potentially stimulate the activation of adaptive immune responses, including the formation of anti-drug antibodies which can impact drug effectiveness. To prevent induction of these adverse responses, it is important to detect and quantify levels of these innate immunity modulating impurities (IIMIs) that may be present in drug products. However, while it is universally agreed that removal of IIMIs from drug products is crucial for patient safety and to prevent long-term immunogenicity, there is no single assay capable of directly detecting all potential IIMIs or indirectly quantifying downstream biomarkers. Additionally, there is a lack of agreement as to which of the many analytical assays currently employed should be standardized for general IIMI screening. Herein, we review the available literature to highlight cellular and molecular mechanisms underlying IIMI-mediated inflammation and its relevance to the safety and efficacy of pharmaceutical products. We further discuss methodologies used for direct and indirect IIMI identification and quantification.

Keywords: immunity; bionanopharmaceuticals; impurities; immunotoxicity; immunogenicity; bioassays; nanomedicine

Citation: Holley, C.K.; Dobrovolskaia, M.A. Innate Immunity Modulating Impurities and the Immunotoxicity of Nanobiotechnology-Based Drug Products. *Molecules* **2021**, *26*, 7308. https://doi.org/10.3390/molecules26237308

Academic Editor: Alejandro Baeza

Received: 2 November 2021
Accepted: 24 November 2021
Published: 1 December 2021

1. Introduction

The body's primary "innate" defense against foreign invaders is triggered by an immediate but relatively non-specific localized immune response including both cellular and biochemical components. The cells contain pathogen recognition receptors (PRRs) capable of tightly binding pathogen-associated molecular patterns (PAMPs) common to several classes of infectious agents [1]. PAMP binding by cognate PRRs triggers immune cell activation, chemokine/cytokine secretion, and biochemical mediators, including the complement system (both systemically produced by the liver and cellularly produced by the activated immune cells), ficolins, pentraxins, and the coagulation system. The coordinated function of these components leads to the hallmark signs of acute inflammation: redness due to increased blood flow and tissue permeability, swelling caused by increased leukocyte (neutrophil, basophil, monocyte) recruitment and subsequent fluid retention in affected tissues, heat (local), and fever (systemic) to decrease pathogen replication and activate production of complement proteins for pathogen opsonization, and pain from the previous effects which act as a warning to the host of tissue damage and infection [2,3]. Together, these processes work to destroy invaders as well as prevent and repair any further tissue damage.

Lastly, innate immune effectors promote the secondary "education" of the immune system against similar future attacks. For this, microbial antigens generated via pathogen phagocytosis are displayed on the surface of antigen-presenting cells (APCs), specifically macrophages and dendritic cells (DCs). Through co-stimulation by pro-inflammatory cytokines and APC-antigen presentation, T-cells differentiate into specialized subsets responsible for promoting enhanced B-cell activation ($CD4^+$ helper T-cells), direct pathogen degradation ($CD8^+$ cytotoxic T-cells), and immune modulation (regulatory T-cells (T_{regs})) [3,4]. Upon B-cell activation, gene rearrangement produces large quantities of highly variable and specific antibodies. While this "adaptive" immune response is slow compared to the innate immune response, these antibody-producing plasma cells are maintained long-term, the "memory" of which allows for more rapid recognition and a stronger, more specific immune response upon secondary antigen exposure [1].

Unlike the epigenetic recombination required by the adaptive immune response, trained immunity is a form of non-specific, T-cell independent innate immunity, which relies mainly upon macrophage activation and pro-inflammatory cytokine production for long-term functional reprogramming of innate immune cell responses. Therefore, secondary antigen exposure can lead to temporarily altered cellular responses, either enhanced or reduced, compared to the primary response [5]. Depending on the degree of "training," protection can be conferred against reinfection by a specific microorganism and some additional non-specific protection against other unrelated pathogens [5].

To prevent inadvertent activation of these immune responses, new pharmaceutical compounds must go through several phases of investigation and regulatory review, consisting of discovery/development, preclinical testing, clinical testing, and approval, before being introduced to the market. Drug discovery/development encompasses the isolation (or fabrication) and subsequent characterization of a new compound, whether a molecule, nucleic acid sequence, or peptide/protein, for therapeutic use. This new compound is then subjected to preclinical (laboratory) testing, during which chemical or genetic analysis, pharmacological tools, and animal models are used to determine the safety and effectiveness of this drug towards a specific disease/condition. Due to the need for new drug compounds, half of all drug-related research and development expenditures occur during this stage, even though only one out of every thousand compounds progress to the next stage [6,7]. After successful testing in animal models, a new drug candidate is then deemed ready for clinical testing in humans. The clinical trial phases determine (I) the drug's metabolic and pharmacological actions, side effects, and effective dosage in healthy patients; and then (II) the drug's effectiveness in "diseased" patients as an improvement upon available treatments, if any. Of the compounds entering clinical trials, approximately 90% fail to pass the clinical phase I/II safety and efficacy requirements [7]. Those few compounds that do advance to clinical trial phase III are tested on a larger cohort of diseased patients to find the best balance between drug safety and effectiveness (dosage regimen, duration, etc.). Finally, once a therapeutic candidate has successfully passed these experimental hurdles, it must undergo final approval by a regulatory health agency (e.g., Food and Drug Administration (FDA) in the US) before being registered and sold as an available treatment [6]. Overall, from start to finish, the process of bringing a drug from the bench to the patient's bedside can cost over USD 800 million and take 8–10 years of effort with no guarantee of final approval [6,8].

Due to the financial and societal costs of the extensive process required for drug development, testing, and approval, it is essential that any potential product "failure" not be the result of the inadvertent inclusion of innate immunity modulating impurities (IIMIs, a.k.a innate immune response modulating impurities, IIRMIs [9]), components of a biotherapeutic treatment other than the target product that can potentially trigger the development of an immune response in the recipient [9,10]. Herein, we review the available literature to highlight cellular and molecular mechanisms underlying IIMI-mediated inflammation and its relevance to the safety and efficacy of pharmaceutical products, and to discuss methodologies used for IIMI identification. Challenges with the detection and understanding of the immunotoxic effects of drug products arising from intrinsic immunological properties (e.g., immunosuppression, immunostimulation, immunomodulation, immunogenicity) of activating pharmaceutical ingredients (APIs) or intended formulation components (e.g., carriers and excipients) are not covered in this review as they have been extensively discussed elsewhere [11–18].

2. Innate Immunity Modulating Impurities

IIMIs encompass everything from live microbial contamination and pathogen-derived antigens (proteins, sugars, nucleic acids) to compounds introduced during the nanobiotherapeutic manufacturing and purification processes (Figure 1) [19,20]. The first source of IIMIs is adventitiously introduced microbial contaminants including live bacteria, mycoplasma, fungi, viruses, or their by-products. While the most common source of these impurities is contaminated raw materials [10], other sources include non-sterile equipment, improper handling practices, or contaminated facilities, though these sources are less likely in a highly controlled facility that employs appropriate sterilization procedures [10]. The second source of IIMI contamination is from host-cell proteins (HCPs), proteins produced by modified host organisms that are unrelated to the intended recombinant product. The population of HCPs produced during biopharmaceutical manufacture depends on host cell type and strain, location of expressed product (cytoplasm, periplasm, external culture medium), physiochemical properties and modalities expressed by-product (charge, hydrophobicity, structure, post-translational modifications, etc.), and the techniques employed during recovery and purification [21]. Due to the limited subset of physicochemical properties optimized for purification, a sub-population of HCPs with similar attributes to the target product will normally co-purify regardless of the process employed [21]. In addition, the use of chemical additives needed to maintain these modified host cells (e.g., growth medium, transferrin, albumin, insulin), as well as chemical additives and selective pressure agents applied for increased product production and modification (e.g., methotrexate, antibiotics, guanidine HCl) can result in adverse patient reactions and can potentially lead to the formation of antibiotic resistant bacterial strains [10]. Lastly, even processes employed for product filtration and purification can inadvertently introduce impurities that leach into the final product. Common "leachates" include bacterial protein A which is normally used for isolating antibodies, as well as hydroxyapatite, tungsten and stainless-steel fragments, glass and cellulose fibers, surfactants, and silicones which can be introduced by filters or containers used during the manufacture and purification processes [10,22]. Ideally, detection of such "leachates" in a biopharmaceutic product will result in modification and/or augmentation of purification processes, such as the use of high-quality resins, to prevent introducing these impurities [10]. Overall, at each stage biopharmaceutical production, there is the potential to introduce IIMIs which may have little/no impact on the function of the resulting drug product but are potent immune activators that have the potential to trigger an undesirable host immune response [23].

When in the presence of these IIMIs (Figure 2), immune cells (e.g., DCs, macrophages, monocytes, neutrophils, and some epithelial cells) recognize these antigens via a variety of pattern recognition receptors (PRRs) containing leucine-rich repeats (LRR) [24,25], including toll-like receptors (TLRs), nod-like receptors (NLRs), retinoic acid-inducible gene-I (RIG)-I-like receptors (RLRs), and C-type lectin receptors (CLRs). Each of these receptor families binds highly conserved microbial structures containing pathogen-associated molecular patterns (PAMPs), or endogenous structures containing damage-associated molecular patterns (DAMPs) released via cell rupture which are important for augmenting the elimination of pathogens and pathogen-damaged cells [9,26,27].

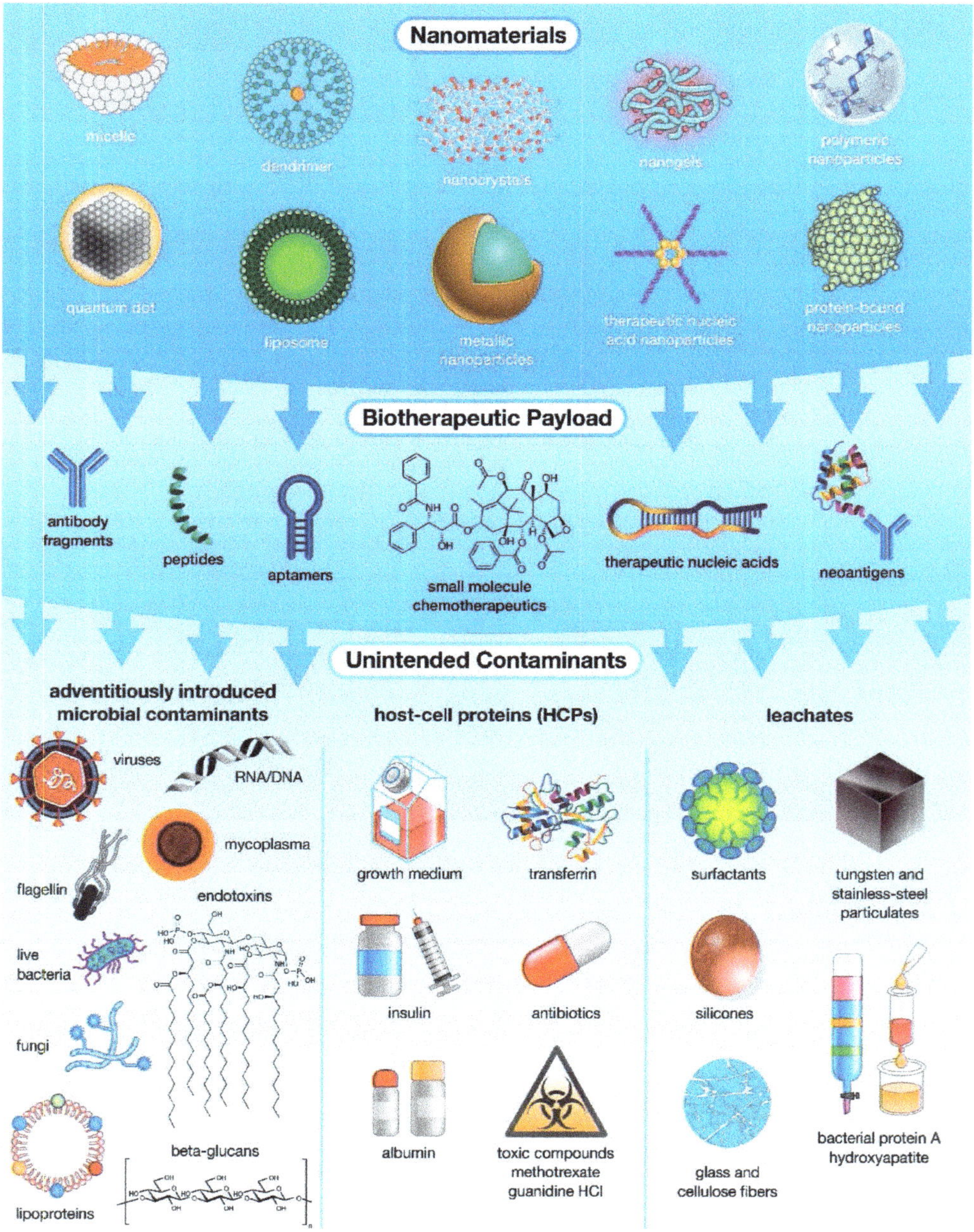

Figure 1. The many levels of possible unintended contamination in drug products. While most often associated with microbial contamination, unintended impurities can actually be introduced into pharmaceutical products from multiple sources, including raw materials and specialized host-cell reagents, and at various stages of production, ranging from fabrication and payload encapsulation in nanocarriers to purification of the final formulation.

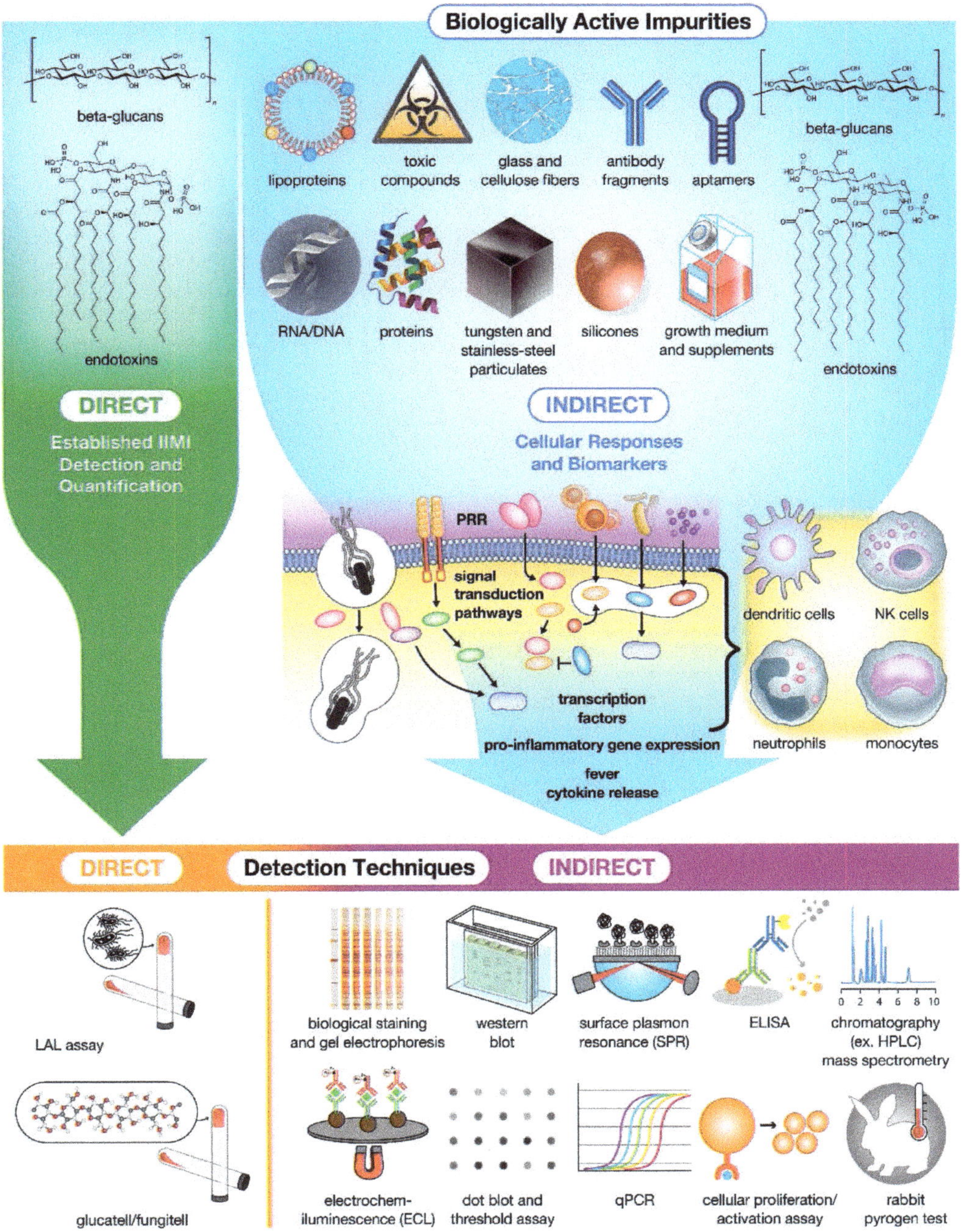

Figure 2. Impurities in drug products trigger innate cellular responses and produce biomarkers for bioassay detection and Quantification. Currently, only β-glucans and endotoxins can be detected and quantified directly using specialized assays. The remaining population of impurities must instead be detected and quantified indirectly using downstream biomarkers (e.g., proteins, peptides, and nucleic acids) and immune cell activation as hallmarks of contamination.

The most studied and diverse family of PRRs, TLRs are a family of highly varied signaling receptors, each of which binds to a different set of microbial structures to trigger intracellular signaling resulting in cytokine secretion and lymphocyte activation [24,26]. Membrane-tethered TLRs, which often require dimerization for appropriate antigen binding and subsequent intracellular signaling, bind to molecules found on bacterial surfaces, including triacyl lipopeptides/proteins, glycolipids, and peptidoglycans, all of which bind to either the TLR1/2 heterodimer or the Dectin1/TLR2 heterodimer; diacyl lipopeptides, lipoteichoic acid, or zymosan which bind to TLR2/6; lipopolysaccharides (LPS) or endotoxins, which bind to MD2, an extracellular adaptor protein for TLR4; and flagellin, which binds TLR5 [24,25,28]. Several DAMPs can also bind membrane-tethered TLRs, including but not limited to hyaluronic acid and other fatty acids, high-mobility group protein B1 (HMGB1), heat shock proteins, S100 proteins, fibrinogen, and tenascin-C which bind to TLR4 [29,30] and serum amyloid A protein, which binds the TLR2/6 heterodimer [27,31]. On the other hand, intracellular TLRs bind to microbial components released after pathogen endocytosis and phagocytosis, including viral double-stranded (ds) RNA containing poly(I:C) motifs which binds TLR3; unmethylated CpG-rich DNA which binds to TLR9; and Guanosine/Uridine-rich single-stranded (ss) RNA and anti-viral imidazoquinoline compounds that mainly bind to TLR8 but can also bind TLR7 [24,25,27,28] Many intracellular TLRs also recognize DAMPs. For example, TLR7 and TLR9 distinguish between snRNP immunocomplexes vs. immunocomplexes of self-DNA or histones respectively [27].

With the assistance of a variety of signaling adaptor proteins (TIRAP, TRAM) and TRIF/TRAF transcription factors [24,25], all antigen-bound TLRs, except TLR3, activate intracellular signaling through a myeloid differentiation primary response protein (MyD88)-dependent NFκB pathway resulting in the secretion of pro-inflammatory cytokines, including type II interferons (IFNs) (e.g., IFNγ), interleukins (ILs) (e.g., IL-1β, IL-6, CXCL8/IL-8, IL-12, and IL-18) and tumor necrosis factor α (TNFα); priming of caspase-1; and the activation of local lymphocytes and vascular endothelium, eventually resulting in antibody production [24,26]. Meanwhile, MyD88-independent activation of IRF3/7 leads to the type I IFNs (IFNα) response critical for antiviral defense [25,30]. However, the continuous stimulation of these PRRs, especially the "bipolar" PRRs involved in DAMP recognition, can lead to inflammatory dysregulation leading to the development of autoimmune and chronic inflammatory diseases [27], as well as blunted responses, also known as tolerance [32]. As such, these pathways are tightly controlled, with some TLRs (TLR2 and TLR4) even having decoy receptors designed to dampen innate responses during severe infection by blocking the interactions between the bacterial ligands and the active TLRs [26].

TLR function also overlaps and integrates with other PRR signaling pathways, including NLRs, RLRs, and CLRs. NLRs, such as NOD1 and NOD2, act as intracellular bacterial sensors by recognizing peptidoglycans (e.g., mDAP and MDP respectively) resulting in inflammasome-mediated NFκB activation leading to the production of IL-1β [26]. The TLR and NLR pathways are clearly integrated for producing IL-1β, as effective NLR activation requires both PAMP activation of the inflammasome and TLR priming, to initiate an inflammatory response [26]. Other NLRs are responsible for triggering the activation and regulation of pro-inflammatory caspase-1 and caspase-5. RLRs, on the other hand, are intracellular viral sensors, binding specifically to dsRNA. Like NLRs, these receptors contain caspase-recruitment domains (CARD) responsible for recruiting adaptor proteins resulting in IRF3 and NFκB activation, leading to the production of type I IFNs (IFNα/β) and pro-inflammatory cytokines (e.g., TNFα, IL-1β, IL-6). Due to these similarities with viral-sensing TLRs (i.e., TLR3, 7, 8, and 9), it is likely that TLRs and RLRs also function together to provide ubiquitous anti-viral protection [26]. Lastly, CLRs are carbohydrate-binding receptors located mainly on the surface of DCs [33,34]. Group I CLRs, which bind mannose and fucose, aid in pathogen phagocytosis, degradation, and antigen presentation to T-cells [33]. Group II CLRs, which bind glucan and dectin, appear to be more immunomodulatory; they induce upregulation of IL-10 and the secretion of cytokines (specifically IL-1β, IL-6, IL-12, and IL-13) required for T-cell polarization into the T_H1 or T_H17 subsets [33,34]. CLRs also act in collaboration with other TLRs (TLR2, 4, 5, 7, and 9) to amplify preceding TLR-mediated NFκB activation and cytokine induction, in addition to triggering the complement cascade through β-1,3-glucan binding complement receptor-3 (CR3, CD11b/CD18), located in the membrane of many phagocytic cells [24,33,34].

Overall, while the binding domains and adaptor proteins vary, there is a significant overlap between the downstream signaling domains employed by each of these pathways. However, these pathways are far from redundant. While TLR7 and TLR9 are expressed on the endosomes of many cells including DCs, eosinophils, basophils, and B-cells, TLR3 and TLR8 are only expressed by natural killer (NK) cells [24]. In the same way, where TLRs are located mainly on leukocytes (macrophages, DCs, neutrophils, etc.), NLRs and RLRs can be found on all cells except DCs [26]. The complex signaling interplay between these pathways, in response to bacterial and viral antigens, highlights the importance of pro-inflammatory cytokines and PAMP-PRR detection in providing a tailored front-line defense against a wide variety of invading pathogens [26]. Further, the interplay between these PRR signaling pathways also drives the induction of effective adaptive immune responses, in that IL-1R and caspase-1 play a crucial role in development of both $CD4^+$ and $CD8^+$ T-cells, as well as antibody responses [25]. As such, IIMI-induced immune responses in the presence of biological therapeutics can lead to immunogenicity toward the administered biologic and potentially to other similar endogenous proteins [19], which can result in loss of treatment efficacy as well as severe and potentially lethal clinical consequences including anaphylaxis, serum sickness, and the formation of autoimmunity [19].

3. Impact of IIMIs on the Immunotoxicity of Drug Products

In the presence of IIMIs, activated phagocytes secrete both stimulatory and inhibitory cytokines to drive and regulate the immune response (Figure 2). These small proteins, which include interferons (IFNs), interleukins (ILs), tissue necrosis factors (TNFs), and chemokines, create a multilevel signaling network that elicits inflammatory responses, angiogenesis, as well as cellular activation, proliferation, and differentiation. IFNs play a central role in innate immunity to viruses and other microbial pathogens [2,29]. ILs function mainly as immune system regulators, responsible for immune cell differentiation and activation [2,29]. Multifunctional TNFs activate vascular endothelium permeability to allow entry of complement proteins and effector cells; increase fluid drainage to lymph nodes to clear pathogens and educate T/B-cells; and stimulate the production of IL-6 responsible for systemic fever, metabolite mobilization, and shock [2,29]. As the largest family of cytokines, chemokines have many diverse functions, ranging from controlling cell migration (e.g., recruitment and activation of local neutrophils and basophils to the site of infection), to such diverse processes as embryogenesis, innate and adaptive immune system development and function, and cancer metastasis [2,3].

Under normal circumstances, cytokine-driven immunostimulation is protective, such as when it is triggered by adjuvants to increase vaccine potency. However, when immune stimulation is unexpected or uncontrolled, especially in the presence of therapeutic compounds, it leads to unintended cellular immune responses and/or antibody production in response to that drug product. Such immunotoxicity encompasses 'any adverse effect on the structure or function of the immune system, or other systems affected by the same biological mediators (e.g., nervous and endocrine systems), as a result of immune system dysfunction' [35]. Immunotoxicity is further classified by the level of response, including (1) non-specific immunostimulation, (2) uncontrolled hypersensitivity (allergy, autoimmunity, and chronic inflammation) leading to tissue damage, and (3) immunosuppression [35].

In the most general terms, immunostimulation is the normal, controlled activation of an immune response ("sensitivity") to an antigen, an important prerequisite for immunogenicity [36,37]. Weak antigen sensitivity responses due to the simple presence of an antigen often fail to elicit sufficient immune activation required to trigger humoral or cellular immunity and subsequent clinical effects [36]; whereas moderate immunostimulatory responses, which might require the assistance of an adjuvant for additional phagocyte activation and cytokine secretion, can result in the eventual downstream production of neutralizing antibodies leading to therapeutic immunogenicity [22,38]. The most common symptoms of immunostimulatory reactions are fever, chills, malaise, hypotension, and localized tissue inflammation (redness, heat, swelling, and pain) around application [2,3,39]. These symptoms are often quickly resolved or can be controlled through the application of immunosuppressive agents such as recombinant chemokines or monoclonal antibodies [36].

Inappropriate or inadequately controlled immunostimulation may lead to hypersensitivity reactions (HSRs) [37]. While no universal classification of HSRs exists, the system proposed by Gell and Coombs, which classifies HSR reactions based on underlying mechanisms, time of symptom occurrence, mediators, and clinical manifestations, is frequently used [40]. Type I HSRs, or classic "acute allergic" reactions such as asthma or food allergies, result from antigen binding to immunoglobulin E (IgE) antibodies on the surface of granulocytes (basophils, mast cells), triggering cellular degranulation and an immediate release of histamine, leukotrienes, and other mediators [40–42]. While also antibody-driven, type II HSRs lead to the production of IgM and IgG antibodies as well as the activation of complement, natural killer (NK) cells, neutrophils, and macrophages [41], all of which result in cellular cytotoxicity and tissue damage. These types are reactions are commonly seen in response to medications such as penicillin, thiazides, or cephalosporins. Type III HSRs are driven by uncontrolled systemic complement activation, resulting in large deposits of IgM immuno-complexes and anaphylatoxins C3a and C5a in tissues which can trigger cell death and compromise organ function [40,43]. Examples of this type of HSR include serum sickness and autoimmune diseases such as rheumatoid arthritis and lupus erythematosus [42]. As both type I and type III HSRs result in the degranulation of basophils and mast cells, true IgE-mediated type I allergy reactions, which are referred to as "anaphylaxis" even though they lack complement involvement, are often difficult to distinguish from IgE-independent complement-activation related pseudoallergy (CARPA) reactions, also known as anaphylactoid or pseudoallergy, which do rely on complement anaphylatoxins C3a and C5a [42,44–46]. Lastly, type IV HSRs such as contact dermatitis or drug sensitivities, are delayed T-cell and macrophage-mediated reactions characterized by increased cytokine release and lymphocyte stimulation [40,42].

Anaphylatoxins and activation of immune cells by PAMPs and DAMPs, also trigger cytokine responses. Since cytokines are pleiotropic and have overlapping functions, they are normally very effective for small-scale localized responses [3]; however, whatever the antigenic trigger, the unregulated overproduction of cytokines due to strong/hyper-immunostimulation (a.k.a cytokine storm or cytokine-response syndrome) can quickly spread unchecked throughout the body via the circulation, resulting in overwhelming systemic inflammation, catastrophic tissue damage, disseminated intravascular coagulation (DIC), and death [2,22,24,38]. Due to their systemic nature, cytokine storms are most often associated with severe, widespread infections, high levels of IIMI contamination (e.g., endotoxins at doses above 5 EU/kg), or massive tissue damage (e.g., shock/trauma) [2,47–49].

Cases of delayed unregulated cytokine secretion coupled with prolonged tissue infiltration by activated macrophages and lymphocytes can also lead to other serious immunological consequences, such as the formation of chronic inflammatory or autoimmune diseases [35,39]. While differentiated by the source of the inflammatory trigger, either endogenous (autoimmune) or exogenous (chronic inflammatory), the general result is the same. Excess TNF production is associated with a number of chronic inflammatory and autoimmune diseases [2,29] while an over-activation of the complement system has been implicated in the pathophysiology of asthma and acute respiratory distress syndrome [43]. Similarly, prolonged exposure to over-activated immune cells, cytokines, and antibody/immune complexes can trigger the formation of granulomas, a common defense mechanism in which harmful components are isolated away from healthy tissue. These chronic HSRs are debilitating as well as life-threatening, since the cells of the immune system are continuously attacking healthy tissues resulting in chronic pain, injury, and eventually organ failure [35].

Lastly, effective immune responses are normally a delicate and tightly controlled balance between stimulation and suppression. The systemic production of IL-10 is associated with the downregulation of neutrophil and monocyte function, working as an anti-inflammatory response following systemic inflammation [2,29]. While this natural counterbalance is conceptually beneficial in controlling systemic responses to local infections, immunotoxicity can occur when immunosuppression or dysregulation leads to an inappropriately reduced immune response resulting in frequent and serious adverse effects [35]. Since the majority of destructive immune responses are associated with HSRs, as previously discussed, many immunosuppressive therapeutics attempts to dampen overactive pro-inflammatory responses but instead have been reported to exacerbate asthma, eczema, and psoriatic lesions [2,39]. Dampening/deficiency of normal immune functions, such as the inhibition of T-cell function and adaptive immune responses, has also been associated with more frequent opportunistic concomitant infections (e.g., pneumonia, Candida, Kaposi's sarcoma, etc.) [35].

After activation by pro-inflammatory cytokines and PRR binding, local antigen-presenting cells (APCs), such as macrophages and dendritic cells (DCs), endocytose and degrade invading pathogens. APCs present fragments of these degraded pathogens on their membrane-bound major histocompatibility complex (MHC) receptors, which bind to and activate T-cells, initiating their downstream activation of B-cells [3,4,24,33]. The fate of activated T-cells is determined by the levels and types of cytokines induced during the inflammatory response, as well as the type and dose of antigen, type and affinity of MHC binding, route of administration, presence of other adjuvants, and patient genetic predisposition [4]. Major classes of T-cells include $CD4^+$ helper (T_H) T-cells activated by MHC class II antigen presentation, $CD8^+$ cytotoxic T-cells activated by MHC class I presentation, and regulatory T-cells (T_{regs}) [3,4,33]. In the presence of either IFNγ or a combination of IL-4, IL-6, and PGE-2, naive $CD4^+$ helper T-cells are further differentiated into specialized subsets of $CD4^+$ helper T-cells which are responsible for cell-mediated (T_H1) or humoral (T_H2) responses respectively [3,4]. T_H1 T-cells secrete large quantities of IFNγ, in addition to IL-2, IL-3, IL-12, IL-18, GM-CS, and TNFβ, to regulate the inflammatory response and fight intracellular pathogens and viruses [3,4,33]. These cytokines promote macrophage activation and the production of opsonizing and complement-fixing antibodies. However, if not properly regulated, T_H1-dependent immune reactions can also lead to antibody-dependent cellular toxicity and delayed HSRs, the most predominant of which can include autoimmune disorders, acute allograft rejection, and chronic inflammatory disorders [2,4,39]. On the other hand, T_H2 T-cells secrete large quantities of IL-4, IL-5, IL-13, in addition to IL-3, IL-6, IL-9, IL-10, GM-CSF, and TNF, to induce humoral responses and mucosal immunity, as well as fight helminths and extracellular pathogens [3,33]. These cytokines promote the proliferation of mast cells and eosinophils, favor the differentiation of IgE and IgG-producing B-cells, and facilitate the synthesis of mucosal IgA [3,4]. While T_H2 cells predominate in transplantation tolerance, they can also lead to chronic graft vs. host disease, systemic sclerosis, and allergen-reactive atopic disorders [4,39,43].

While it has been observed that cytokines from specific T_H cell subsets (e.g. IFNγ from T_H1 cells and IL-10 from T_H2 cells) usually inhibit the action of the other types of T-cells and their companion phagocytes [3,4], this classic binary model does not account for instances where an immunological response is triggered without any significant shift in T_H1/T_H2 balance, such as is the case with omega-3 fatty acids, or alternatively where there is T_H1/T_H2 activation with minimal immunological pathogenesis, such as with melanin, probiotics and zinc [3]. In addition, other sub-classes of T-cells have been identified which were not previously represented by this model, including but not limited to: T_H17 cells, which secrete IL-17 to mobilize phagocytes against extracellular fungi and bacteria; and T_{regs}, which produce FoxP3 to control the activity of the other effector T_H cells and maintain immunological tolerance to self-antigens [3,19,23,33]. However, increased levels of regulatory (T_H17, T_{reg}) cytokines such as IL-10 or IL-17 can also be an indication of adverse patient effects such as autoimmune diseases or advantageous concomitant infections [3].

4. Sources of Immunotoxicity in Nanotechnology-Based Products

The use of nanoscale platforms (e.g., dendrimers, liposomes, nanoparticles, nanotubes, nanogels, etc.) has become a popular technique to reduce drug immunotoxicity while improving therapeutic solubility, biodistribution, and cell-specific delivery compared to the traditionally formulated versions of these drugs. However, it has been noted that some nanocarriers can themselves be immunomodulatory (Figure 1), such as RNA nanoparticles which have been shown to induce pro-inflammatory cytokine secretion and enhance inflammation [11,50]. The raw materials used for nano-platform fabrication can have various immunological effects, either due to previously discussed contamination or due to the chemical properties of the material itself. Some nanomaterials are immunostimulatory, such as lipid-based nanocarriers and carbon nanotubes which have been shown to induce cytokine production and inflammation [50–52], while other nanomaterials are immunosuppressive including PEGylated NPs which lead to TLR9 inhibition and immune cell avoidance [50,51,53]. Similarly, the processes employed during nanocarrier synthesis and purification often use immunotoxic reagents, such as surfactants such as cetyltrimethylammonium bromide (CTAB); peptizing agents such as polystyrene sulfonate (PSS); or complexing agents such as nickel, to improve drug loading or enable molecule crosslinking [54]. While these chemicals are not generally intended to be in the final product, trace elements ("leachates") that remain after washing and filtration can induce cytokine production and inflammation, compounding the other immunomodulatory aspects of the nanocarrier [54].

Once fabricated, the physical properties of the nano-formulation, including size, shape, and surface charge, can also alter immunotoxicity. Nanoparticle interactions with the immune system have been extensively discussed elsewhere [11–13,50,55–58]. Here, we will use some examples to demonstrate structure–activity relationship between nanoparticle physicochemical characteristics and their immunological properties. First, several studies have shown that smaller particles (<500 nm) promote humoral T_H2 responses, compared to very large particles (>1 μm) which have been found to stimulate cell-mediated T_H1 responses. In addition, very small particles (<100 nm) are associated with increased $CD8^+$ and $CD4^+$ T-cell activation compared to their larger (>500 nm) counterparts, who induce good antibody responses [59]. Thus, small particles may invoke virus-like responses and larger particles induce bacteria-like responses [59]. Second, compared to spherical nanocarriers, oval-shaped liposomes and carbon nanotubes have been shown to activate complement and platelet aggregation with membrane rupture, respectively [50,60]. Finally, cationic carriers are more immunostimulatory than anionic or neutral carriers, triggering cytokine secretion (TNF, IL-12, IFNγ); activation of DCs, T-cells, and neutrophils; and procoagulant leukocyte and platelet activation which can potentially lead to DIC [12,50,61]. Taken together, while a nanocarrier is often designed to reduce the immunotoxicity of a therapeutic payload, the chemical and physical properties of that nanocarrier along with it being a source of undesirable IIMIs contamination may lead to an exaggeration of the immunotoxicity of the final drug product. For example, cationic polyamidoamine (PAMAM) dendrimers in the presence of low amounts of endotoxin have a variety of immunotoxic effects that neither dendrimers nor low levels of endotoxin alone have [11,50]. Therefore, the use of nanomaterial platforms should be considered as yet another source of IIMIs. Translational and regulatory challenges arising from immunomodulatory properties of nanocarriers and their ability to exaggerate immunotoxicity of low levels of IIMIs (e.g., endotoxin) have been extensively discussed elsewhere [12,13,50,56,61]. Immunogenicity of nanoparticles alone and in the context of IIMIs along with nanoparticle contribution to the immunogenicity of protein-based therapeutics have also been reviewed earlier [22].

5. IIMIs Commonly Found in Pharmaceutical Products

5.1. Microbial Components

When it comes to assessing biotherapeutic purity, the only current consensus is that it is important for manufacturers to minimize the potential for their formulation to trigger adverse patient reactions and future immunogenicity by removal of microbial or host cell-related impurities, as summarized by testing standards (Table 1) [29,35,62–68]. Currently only a fraction of the potential IIMIs, specifically lipopolysaccharide (LPS), β−glucan, flagellin, HMGB1, and nucleic acids, are routinely measured during immunotoxicity screening of biotherapeutics [69] to confirm that the levels of these IIMIs fall within the FDA-approved 1–100 ppm range [21]. In addition, due to the breadth and complexity of potential IIMIs, there is currently no single assay that can provide a profile of all IIMIs present within a biotherapeutic [70]. Other than the fact that any assays used to detect IIMIs and evaluate possible immunotoxicity should be tailored to the specific contaminant [62], there is currently very little agreement as to which analytical assays should be standardized for IIMI screening [21]. Therefore, most studies use a series of assays to broadly cover the detection of all possible IIMIs present in biopharmaceuticals [21,71], including single analyte mechanistic assays, basic staining/gel-based assays, immunoassays, and cellular-based assays (Figure 2).

Table 1. Examples of guidance documents and international standards for the measurement of impurities in therapeutic products. International standards (IS) and Guidance for Industry (GI) provided through the U.S. FDA, the U.S. Pharmacopeia (USP), and the International Organization for Standardization (ISO) describe the risks of endotoxin and pyrogen contamination in therapeutic products and outline the assays, protocols, and detection limits which have been standardized and approved for universal application in therapeutic safety and purity measurements.

Document	Type	Purpose	Reference
USP 85 Bacterial Endotoxins Test	GI	Describes method validation and sample preparation requirements for turbidity, chromogenic and gel-clot LAL assay	[63]
USP 151 Pyrogen Test	GI	Describes method validation and sample preparation requirements for the rabbit pyrogen test	[64]
FDA Immunotoxicity Testing Guidance (FDA-modified version of ISO-10993)	GI	Summarizes general types of toxicity and subsequent testing that should be considered for medical devices or constituent materials	[35]
FDA Guidance for Industry: Pyrogen and Endotoxin Testing: Questions and Answers	GI	Provides bacterial endotoxin and pyrogen testing recommendations (gel-blot, photometric, and kinetic tests) and acceptance criteria	[65]
FDA Endotoxin Testing Recommendations for Single-Use Intraocular Ophthalmic Devices	GI	Provides recommended endotoxin limits for the release of intraocular devices and single-use intraocular ophthalmic surgical instruments/accessories	[66]
FDA Questions and Answers on Quality Related Controlled Correspondence	GI	Provides answers to common scientific and regulatory questions around the manufacture and quality control of generic drug production including endotoxin testing	[67]
FDA Immunogenicity Assessment for Therapeutic Protein Products	GI	Outlines approaches to evaluate and mitigate adverse immune responses/immunogenicity associated with therapeutic protein products; discusses the importance of IIMI detection	[62]
ISO-10993-1 Biological Evaluation and Testing Standards for Medical Devices (prepared by ISO/TC 149)	IS	Outlines the potential biological risks arising from the use of medical devices and provides a framework to plan biological evaluation, testing methods, and acceptance criteria	[68]
ISO-29701 Endotoxin Standard (prepared by ISO/TC 229)	IS	Describes application of LAL assay for evaluation of endotoxin levels in nanomaterials intended for use in vitro	[29]
ISO-21582 Pyrogenicity Standard (prepared by ISO/TC 149)	IS	Specifies the principles and methods for pyrogen testing of medical devices and their materials	[72]

5.2. Whole Microbes

After biopharmaceutical manufacture and microbial inactivation via low pH adjustment, heat, and solvent/detergent treatments [10], filtration is used for the removal of bulk impurities such as neutralized pathogens (bacteria, viruses), destabilized protein aggregates, or other bulk contaminants [10,22]. Due to the comparatively large size of these impurities, microscopy techniques such as transmission electron microscopy (TEM) [10], have been used to assess the effectiveness of these initial filtration steps. These high-resolution microscopy techniques employ lasers or electrons beams and extensive sample preparation to achieve a 0.1–1 mm visualization limit [73,74], which makes them time- and cost-prohibitive. Further, given their inability to provide accurate IIMI quantification, microscopy techniques such as TEM can only provide an indication as to what additional filtration and purification steps may be required; as these filtration techniques may not be sufficient to completely remove all traces of IIMIs, more accurate IIMI detection and quantification must then employ antigen-specific assays [31,73].

5.3. Leachates

After filtration, a range of chromatographic techniques are used for drug concentration and purification, to remove impurities such as drug by-products, unprocessed raw materials, and other leachates that may have been introduced into the formulation during the manufacturing process [10,22]. For complete sample separation, chromatography exploits the physical characteristics of the target protein/peptide in solution, including size, mass, ionic charge, binding affinity, pH, and electrokinetics, to partition it away from other components that may be present in the solution after fabrication [75]. Some of the chromatography techniques previously used for assessing biotherapeutic purity include ion exchange, size exclusion, capillary electrophoresis (CE), micellar electrokinetic chromatography (MEKC), and reverse-phase high-performance liquid chromatography (HPLC) [75]. Often referred to as "high pressure" liquid chromatography due to how the sample in the mobile phase is pressurized before injection into the absorbent stationary phase column, HPLC has become one of the most popular chromatography methods due to its high-performance detection, separation, and quantification of very small volumes (5–50 μL) of samples including degradation by-products, IIMIs, and unprocessed raw materials. HPLC is often used to separate molecules that are not large enough or charged enough for adequate separation by traditional size-exclusion chromatography or ion exchange-chromatography respectively [75]. While separation efficiency and quantification analysis are highly accurate, this technique requires extensive protocol optimization for the best results [75] in addition to specialized equipment and a trained operator. Additionally, chromatography can typically only separate one IIMI at a time, though multidimensional chromatographic separations paired with fluorescence detection are currently being pursued [71].

Sub-visible particles, which can include anything from small molecules to the components of protein aggregates, can also be identified using mass spectrometry (MS) techniques [21,76]. MS separates charged molecules or fragments by accelerating them through an electric or magnetic field, which separates the molecules based on their mass-to-charge ratio and then identifies them by correlation with known molecule masses and fragmentation patterns. This technique is especially important in identifying the relative concentrations of impurities and degradation products relative to target drug products during pharmaceutical development [77]. As a pivotal technique in the process of molecule structure elucidation [77], high-resolution MS/MS is now also being used to identify and quantify larger, more complex impurities and proteins that can be isolated from the bands of an electrophoresis gel or sampled directly from solution using liquid chromatography-tandem mass spectrometry (LC-MS/MS) [20,21]. Due to improvements in high-throughput capabilities combined with improved sample preparation (e.g., chromatography fractionation and 2D gel electrophoresis), LC-MS/MS is now also being used for complete proteomic characterization and identification of complex therapeutic samples [21]. MS analysis is more precise than immunoassays but requires specialized equipment and analysis software, as well as trained personnel [76].

5.4. Host Cell Proteins

The most difficult IIMIs to isolate and quantify are host-cell proteins (HCPs) due to the diversity and complexity of the potential protein repertoire, as well as HCP similarities to the target drug product [69]. As there is currently no single assay that can detect and quantify all possible HCP-based IIMIs within a biotherapeutic formulation [70] nor any absolute control limits required by pharmaceutical regulators [21], most quality assurance uses a combination of methodologies to confirm drug product purity. A typical strategy often includes generic IIMI clearance studies such as the *Limulus* amebocyte lysate (LAL) test or mass spectrometry; sensitive silver staining (and immunoblotting) of electrophoretic gels; and quantitative HCP-specific immunoassays such as ELISAs [71], all of which will be discussed below.

6. Immune-Mediated Adverse Effects to Pharmaceutical Products

The combination of a strong immunostimulatory response [3,35,43] and the activation of specialized subsets of T-cells leads to target-specific destruction of pathogens and cancer cells, either by direct interaction with $CD8^+$ T-cells and natural killer (NK) cells or by $CD4^+$ T-cell activation and proliferation of B-cells to produce antigen-specific antibodies [19,23,24,78]. This IIMI-driven immunogenicity can lead to the formation of antibodies of different isotypes (e.g., IgM vs. IgG vs. IgE), allotypes (e.g., reflecting genetic differences between IgG of biologically unrelated individuals), and idiotypes (e.g., reflecting binding to specific epitopes within antibody variable sites) [19,23,79–81], resulting in anti-drug antibodies (ADAs) with varying impacts on drug effectiveness. Binding antibodies attach to a non-active portion of the therapeutic and therefore have little/no effect on therapeutic function, whereas cross-reactive neutralizing antibodies bind to therapeutic active sites, thereby neutralizing therapeutic function while also binding similar endogenous proteins and breaking immunological tolerance [19,23,82–84]. The presence of these ADAs can also have different functional consequences to the host including the HSR/anaphylaxis and autoimmune responses previously discussed [19,23,35,79–81]. The relationship between the occurrence of a specific antibody type and the impact on the patient are inversely related; binding antibodies are the most common but have the lowest clinical impact, while cross-reacting neutralizing antibodies are rare but have the highest clinical impact [23,79–81,85]. Therefore, it is important to understand, measure, and prevent this response from being induced.

During the fabrication and production of drug compounds, there are many potential sources for the introduction of IIMIs into the final biotherapeutic formulation (Figure 1) [19,20]. In addition to the impurities/contaminants previously discussed, there are also several product-related and host-related factors that may have little/no impact on the function of the resulting drug product but have been shown to impact the immunotoxicity and immunogenicity of biotherapeutics [19,23,78]. Product-related factors include structural properties of the drug (sequence, epitopes, post-translational modifications), exposure to antigenic sites, solubility, formulation stability and storage, downstream processing, presence of impurities/contaminants that might be introduced during processing [19,78]. These factors can be mostly controlled through careful optimization and modification of the fabrication/purification processes. Further compounding the risk of immunogenicity are host-related factors, including host genetic predisposition, endogenous protein genetic variants, concomitant illnesses (e.g., kidney or liver diseases), host immune status (e.g., autoimmunity, prior exposure) as well as the treatment dose, duration, and route of administration [19,23,78].

7. Methods for IIMI Detection

7.1. Direct Detection Methods

The first bioassay used to measure the presence of bacterial contamination was the rabbit pyrogen test (RPT) which detected pyrogens, any contaminant that induces a histamine response, fever, chills, and other unwanted inflammatory side effects. The rabbit pyrogen test detects all pyrogens, so it is subject to high variability and low selectivity, in addition to being expensive and requiring extensive use of animals [10,31]. As an improvement, the *Limulus* amebocyte lysate (LAL) test detects the hemolymph coagulation of the American horseshoe crab *Limulus polyphemus* when in the presence of bacterial endotoxin/LPS and is used as a standard for bacterial contamination [86,87]. However, this assay is specific for endotoxin, not general pyrogens [31], and has reduced specificity in the presence of fungal β-glucans because the horseshoe crab lysate used for this assay contains two proteins that trigger activation of the proteolytic cascade: factor C is specific to the presence of endotoxin while factor G is specific to β-glucans [88,89]. Knowing this, a modified version of the LAL assay containing glucan-blocking reagents or recombinant factor C overcomes β-glucan interference during endotoxin detection [90].

While β-(1,3)-d-glucans are not as immunologically potent as bacterial endotoxins, requiring μg/mL concentrations as compared to the endotoxin pg/mL concentrations to elicit an immunomodulatory response, they are a common IIMI present in many pharmaceutical products and solutions [89]. Moreover, while there is currently no compendial standard for β-glucan detection or acceptable levels, a modified version of the LAL assay is growing in popularity [90]. Since LAL factor G is specific to β-glucans, factor C depletion from the LAL lysate enhances the assay's sensitivity solely to β-glucan detection [89]. It is important to note that β-glucans are naturally introduced in a person's diet, so data generated from β-glucan quantification assays need to be from clinically relevant doses of the drug formulation [89].

Challenges with endotoxin and beta-glucan detection in nanoformulations stemming from carrier-, excipient-, or drug-mediated interferences, mechanisms of interferences, and ways for overcoming them have been identified and extensively discussed earlier [11,89,91–95].

7.2. Indirect Detection Methods

For the development of effective assays, an appropriate biomarker can consist of any compound (e.g., metals, solvents, pathogens, etc.) or useful characteristic, such as a mechanistic by-product, which can be measured or evaluated, either directly or indirectly, and used as an indicator of normal biological, pathogenic, or pharmacologic processes [83,84]. Therefore, any of the product- or host-related impurities previously discussed, as well as raw materials used during the product's manufacture and purification, can technically be considered a potential biomarker [85]. During method development, quantitative assays must be validated using appropriate controls and quantification must employ a standard curve of known analyte concentrations to determine the range of conditions under which appropriate levels of confidence can be attributed to the reproducibility and accuracy of the data [84,96]. Further, the validated assay must then demonstrate both sensitivity and specificity for the biomarker [84], such that the biomarker is correctly identified (i.e., true positive, sensitivity) at clinically relevant (ng/mL to pg/mL) concentrations [96] without also reacting to residual therapeutics or other impurities likely to be present within the therapeutic formulation (i.e., true negative, specificity). Reduced sensitivity can result in mistakenly missing the presence of IIMIs in a formulation (i.e., false negative) resulting in possible dangerous clinical manifestations and immunogenicity, while reduced specificity can result in misidentification of inert compounds as IIMI (i.e., false positive) leading to incorrect quantification and product disposal rather than administration to patients. Overall, when balancing these two parameters, increased sensitivity is often preferred to increased specificity.

7.3. Biological Staining and Gel-Based Methods

Biological staining is a common technique to detect and visualize the presence of HCPs and other impurities. This technique utilizes Coomassie Blue or silver staining to highlight the presence of protein analyzed by multidimensional (2D or 3D) gel electrophoresis [71] or fixed in histological samples, respectively [21]. While the sensitivity of these staining techniques is quite high, selectivity is not; this technique cannot discriminate between types or sources of proteins so other techniques need to be employed to further identify and quantify the protein contaminants [10]. Newer versions of this method employ fluorescent stains, such as SyproRuby, for 10–100 times increased sensitivity compared to previous stains since these stains are not dependent upon the protein composition [21]. Other stains also have improved specificity by binding to specific cellular elements (i.e., nucleic acids, carbohydrates, chromatin, etc.) though this method is still largely qualitative [21]. Gel electrophoresis and protein staining have progressed to the use of the more quantitative Western blot, a common antibody-dependent detection method [21] that has merit for identifying low (pg/mL) concentrations of protein impurities. Contaminating HCPs and product-related impurities are separated from the target biologic by gel electrophoresis [10], and then transferred to a PVDF or nitrocellulose membrane. Primary antibodies raised against HCPs are incubated with the membrane to allow for the formation of antigen–antibody complexes, which are then detected through secondary enzymatic or fluorescent labeling [21]. While this technique is both sensitive and specific, it requires the use of separate polyclonal antibodies against each impurity for optimal detection, which can be time and cost prohibitive in the long run [10]. In addition, this technique needs to be supplemented with additional immunoassays to help distinguish between process- or product-related impurities and impurities that might comigrate with the product [10].

7.4. Antibody-Based Enzymatic Methods

Surface plasmon resonance (SPR) uses antigen-ligand binding on a sensor chip to generate a signal due to a change in the refractive index caused by a difference in mass as the analyte binds to the ligand. Most often used to detect the presence of antibodies rather than antigens, this assay is capable of continuous measurements of binding interactions in 'real-time' [84]. For the detection of immunotoxic antigens, SPR assays tend to be less sensitive, less tolerant to therapeutics, and have lower throughput compared to enzyme-linked immunosorbent assays (ELISAs). In fact, SPR is capable of characterizing early immune responses by detecting and isotyping low-affinity antibodies, which other assays might miss, which makes it much more suitable for immunogenicity assays [70,97]. Furthermore, unlike other immunoassays where the reagents are cost-prohibitive, here the detection equipment is expensive and vendor specific [70,97].

Electrochemiluminescence (ECL) also uses antibodies to bind target impurities. However, unlike the commonly used enzyme-labeled secondary antibodies previously discussed, this technique employs a ruthenium-conjugated protein and tripropylamine (TPA) to produce a detectable, quantifiable luminescent signal. Ruthenium labels are stable, non-radioactive, and offer a choice of convenient coupling chemistries [70]. This is a highly sensitive and selective technique; however, this method requires the production and use of specific antibodies for analyte immobilize and detection, indicating that each impurity must be detected separately [98]. In addition, this technique requires the use of specialized, costly equipment containing carbon electrode plates for detection, which are not necessarily standard in most labs [10,70].

Enzyme-based (EIA) or fluorescence-based (FIA) microtiter plate assays were developed to circumvent the need for method-specific instrumentation and resources experienced with ECL and SPR [10]. This assay involves incubating the sample with a couple of biotinylated antigen-specific antibodies which, after binding and forming immunocomplexes, are removed from solution by association with streptavidin-coated paramagnetic beads. Thereafter, the beads are incubated with enzyme-labeled or fluorescence-labeled antibodies for colorimetric development. By substituting the paramagnetic beads for a solid-substrate surface, the traditional EIA/FIA was transformed into the enzyme-linked immunosorbent assay (ELISA), the most practically useful and commonly employed immunoassay [21]. As previously described, this type of assay employs a series of antibodies to capture specific target antigens. The bound antigen is then complexed with a secondary antibody modified to undergo an enzymatic reaction (colorimetric, fluorescent, or luminescent) for detection via spectrophotometer [70]. However, unlike the EIA/FIA, the use of a solid-substrate surface enables the assay to be set up in various configurations (e.g., sandwich, indirect, bridging, competitive, etc.) for optimal IIMI detection and quantification. ELISAs are relatively sensitive with a detection range of 12–200 ng/mL [10,99]; modern ELISAs have been optimized to improve their sensitivity and allow the detection of analytes at lower (e.g., pg/mL) levels. ELISAs also have high specificity due to their use of analyte-specific antibodies and can be performed relatively quickly (completed in one day) [10,21]. However, the dependency on highly specific antibodies also means that each analyte must be known and analyzed individually, which can be cost-prohibitive [70]. Common HCPs detected via ELISA include anaphylatoxins such as complement C3a [100]; inflammatory cytokines such as IL-1β, IL-6, IL-8, and TNFα [2,101]; and other IIMIs including HMGB1 and flagellin.

Due to antibody specificity combined with the progression of fluorophore technology, a large number of biomolecules can now be captured from the same small (μL to mL) sample and then detected simultaneously [3,71]. These "multiplex" assays are usually modified ELISA assays, though the EIA/FIA assay format can similarly be multiplexed, as is often used in flow cytometry [3]. Each analyte is then tagged with either a different fluorescent label or organized in a known array for detection via spectrophotometer. As the basic principles of the assay are unchanged, the sensitivity and specificity are still high, though fluorescence bleed-through increases as the number of analytes and fluorophores with similar excitation/emission spectrums increases. In addition, multiplexed assays are less time consuming and labor intensive, while providing higher throughput analysis, compared to an individual ELISA [3].

7.5. Nucleic Acid Hybridization Methods

For the detection of nucleic acids in pharmaceutical samples, hybridization techniques such as the dot blot or immunoligand assay (ILA) are often used. The ILA (a.k.a "Threshold Assay") reliably detects very small amounts of DNA and impurities in liquid solution [102]. This assay employs a biotinylated single-stranded binding (SSB) protein and general anti-ssDNA antibody to complex with any host ssDNA available in the sample. Streptavidin filtration then captures any biotinylated complexes on a specialized matrix-embedded silicon chip, after which the DNA is quantified via enzymatic hydrolysis and subsequent light-addressable potentiometric sensor (LAPS) detection [99]. This method has been shown to be 10–100 times more sensitive than traditional colorimetric or ELISA assays, with a detection range of 5–40 ng/mL [99], requires only small amounts of sample, removes steric binding or stability issues inherent in solid-phase systems, and comes in two formats (sandwich or competitive) depending on the size of the analyte being detected [102] though optimal ssDNA fragments tend to be larger than 600 base pairs [99]. However, this method has reduced specificity due to its sequence-independent binding by general ssDNA antibodies. Furthermore, this technique can be expensive as it requires the use of proprietary equipment, software, and consumables (e.g., silicon chips, specialized buffers, etc.) for quantification [10]. On the other hand, the dot blot employs a substrate covered with immobilized "randomly primed" DNA probes from a known microbial source tagged with radio or fluorescent labels. The probes are exposed to the drug sample allowing for binding between host-cell DNA present in the sample and the specific DNA probes. This binding is then detected and quantified to 3–800 pg/mL against a calibration curve by phosphor- or fluorescence-imaging systems [99].

The more popular method of detecting and identifying bacterial and viral nucleic acids is through reverse transcriptase (RT) and quantitative polymerase chain reaction (qPCR) assays [10]. For these assays, trace amounts of DNA or RNA are collected and then amplified through the PCR or RT-PCR method respectively, resulting in many identical copies of the target DNA. The levels of target DNA are then quantified and nucleic acid concentration in the original sample is derived from target copy numbers [99]. Innate immune activation can similarly be assessed by quantifying the levels of pro-inflammatory cytokines, such as IFNγ, IL-1β, IL-6, or other downstream biomarkers by quantifying the levels of target mRNA, amplified as cDNA, which are compared to standard housekeeper genes such as GAPDH or 18S [9,69,103] to determine the fold increase or decrease of the target genes [9,69,103]. Since this process uses specific DNA primers for PCR amplification, the resultant quantification is highly sensitive and specific for the target sequence [99]. However, this also means that species-specific primers must be known. Additionally, as amplification of each nucleic acid fragment requires its own primers, these reactions need to be carried out separately; though, like the previously discussed, multiplexing and proteomics analyses coupled with improvements in high-throughput capabilities have produced arrays of many immobilized primers used to amplify, identify, and quantify many different DNA sequences at the same time [9,21]. This standardization increases the amount of data produced while reducing the required time and labor of these assays [76].

7.6. Cell-Based Methods

Since the long-term goal of these studies is the prevention of patient immunotoxicity and possible immunogenicity, more recent assays focus on the in vitro and in vivo impact of IIMIs. These cellular assays detect immune cell activation and proliferation or quantify levels of secreted innate immunity biomarkers (e.g., cytokines, prostaglandins, complement), which may contribute to the process of immunogenicity by priming the immune cells.

Cellular proliferation assays examine the activation and proliferation of specific immune cell subsets, usually, macrophages, neutrophils, or lymphocytes, when treated with the biotherapeutic, compared to control cells and the potential adjuvant effect of known IIMIs [31]. For example, T-cells are activated by concanavalin A or phytohemagglutinin, while B-cells proliferate in response to LPS. While it has long been established that immune cell proliferation in vitro is correlated with cell-mediated immunity, these assays have not been extensively standardized and validated [36]. In addition to needing a skilled technician and the appropriate facilities to support these studies, this assay is time prohibitive as culturing these cells takes at least 48–72 h [36].

For a more specific way to determine the type of IIMIs present in a drug formulation, a model of HEK-BLUE cells containing a secreted embryonic alkaline phosphatase (SEAP) reporter inducible by NFκB, transfected with individual TLR receptors, can be used. When bound with their specific agonists alone or in mixtures of IIMIs, the observed NFκB activation for each TLR can be quantified through a colorimetric change. This reporter system has high sensitivity and specificity, similar to what was observed in normal human PBMCs [69,88]. Since therapeutic biologics could mask or interfere with the response of these cell lines, this model necessitates the use of additional inhibition controls. In addition, while this model is effective for detecting TLR-specific IIMIs, it does not yet cover innate immune responses that can be triggered solely through alternate pathways such as CLRs, NLRs, and RLRs. As such, the reporter cells were modified to contain different reporter systems (SEAP, THP-1, and MM6) that would be expressed in the presence of NFκB, TNFα, and mRNA from IL-6 or IL-8 respectively, thereby covering the activation of multiple innate immune responses [69].

Other in vitro models instead directly quantify the levels of cell-secreted immune modulators, such as cytokines and complement proteins (e.g., C3a, C5a), or antibodies [3]. While all of these soluble mediators play an integral role in host defense against microbial invasion, the network of cytokine interactions is responsible for maintaining cellular homeostasis, making them a popular biomarker for gauging the potential immunotoxicity and immunogenicity of new biotherapeutic compounds, especially when compared to normal (untreated) controls [3,36]. Increased levels of cytokines after application of a new drug product can therefore be associated with a product's immunotoxic effects (either stimulatory or inhibitory), which can lead to adverse patient reactions and reduced therapeutic efficacy due to the formation of ADAs [3]. As such, pharmaceutical immunogenicity is often determined through the use of commercially available multiplexed ELISA assays, chosen based on convenience, affordability, and availability [3], which typically quantify a limited panel of pro-inflammatory cytokines (IL-1, IL-6, or TNFα) [2,87] or subsequent T-effector (T_H1/T_H2) cytokines, including IL-2, IL-12, IFNγ or IL-4, IL-5, and IL-6 respectively, even though this may bias analysis towards specific immune pathways [2,3]. Despite the sensitivity and specificity of the multiplex ELISAs used for these analyses, the pleiotropic nature of cytokines and their overlapping activation pathways on numerous target cells [36] often make the results difficult to interpret. Hence, there is a lack of consensus as to which cytokines should be measured to accurately characterize the immunological effects of a new drug.

7.7. In Vivo Methods

A more recent study performed by Haile *et al.* employed an in vivo macaque skin model to better characterize the relationship between type and dose of IIMIs, patterns of innate immune receptors, and pathways triggered by these impurities, and immunogenicity. This model was developed due to the similarity of macaque PBMCs to human PBMCs, and to increase sensitivity compared to traditional murine models which are known to have less sensitive immune cells than those of humans [31]. These studies used mRNA collected after application of known IIMIs, as a basis for comparison to Rasburicase, as a model therapeutic, and measured by qRT-PCR to track the expression of 48 genes involved in the innate immune response, including ILs, TNFs, CD40, GAPDH, etc. [31]. This study demonstrated that, while an increased innate immune response is dependent upon the dose of IIMI administered, the presence of these impurities acted as an adjuvant during co-administration with a protein therapeutic, thereby increasing its immunogenicity. However, it was noted that even trace amounts of IIMIs triggered the transcription of multiple innate immunity genes, emphasizing the need to assess biotherapeutics for a wide variety of possible contaminants and related downstream biomarkers in a more thorough, relevant model, rather than just quantifying levels of specific IIMIs [31]. While the use of animal models is cost-, labor-, and time-prohibitive, these models can provide more applicable data as to the immunotoxicity and immunogenicity of biotherapeutics in humans.

Overall, methods of cell growth and stimulation are more or less optimized and standardized, and cell-based assays (both in vitro and in vivo) provide the most relevant data on IIMI and drug interactions with the immune system [3,76]. However, they are labor intensive, and the evaluation of cell-secreted biomarkers is subjective due to the cross-reactivity of most immunological pathways and the potential confounding influence of other substances that may modulate the activity of the target substance [3,76].

8. Conclusions and Future Directions

It is well documented that the presence of IIMIs in a biotherapeutic formulation can trigger immunotoxicity and, with repeated exposure, immunogenicity against the therapeutic [31,80–82]. To prevent these adverse patient reactions, the FDA currently requires quantification of five key IIMIs: LPS, HMGB1, β-glucan, flagellin, and nucleic acids [62,69], to demonstrate biotherapeutic safety, quality, and clinical performance. These guidelines aim to mitigate the formation of future ADAs through commonly activated innate immune receptors, specifically TLRs, CLRs, and complement. However, these guidelines do not necessarily account for potential immunotoxic responses to other IIMIs that may be present in the formulation. As such, the FDA panel of IIMIs required for quantification should be expanded to cover a much broader repertoire of impurities, including microbial antigens that can potentially trigger other innate immunity pathways, common manufacturing leachates, and solvents, and toxic additives required for maintaining host cells. The list of possible leachates, solvents, and host cell additives will be extensive, requiring tailoring to the specific processes employed during manufacturing and purification [62]. As for the microbial IIMIs, most innate immunity receptors and pathways can be covered using ten common IIMIs, some of which are already required and discussed, including flagellin, FSL-1, zymosan, ODN2006, and ODN2216, both high- and low-molecular-weight poly(I:C), MDP, CLO75, and LPS. While this ten IIMI panel necessitates more laboratory testing before new drugs can gain approval, adhering to the ppm levels for these required IIMIs will demonstrate that little/no immunotoxicity will result from trace levels of substances present in the drug formulation, therefore reducing the potential for immunogenicity.

Second, to measure and quantify IIMIs present in biotherapeutic formulations, a variety of available assays have been discussed. As genomic and proteomic technology advances, these assays have become more sensitive and specific, enabling improved detection and quantification of IIMIs. In addition, many of these assays are now being coupled into high-throughput formats which can produce more data with reduced sample and reagent volumes, as well as cost and labor expenditures. However, due to the variety of potential IIMIs, there is currently no single assay that can provide a profile of all IIMIs present within a biotherapeutic [70]. Moreover, there is a lack of agreement as to which analytical assays should be standardized [21] so most studies use a series of assays to broadly cover the detection of all possible IIMIs present in biopharmaceuticals [21,71]. To better standardize results across experiments and laboratories, the use of a single high-throughput platform capable of detecting a wide panel of biomarkers of the same class (small molecules, proteins, or nucleic acids) in parallel, such as multiplexed ELISAs, MS, or genomic arrays, should be employed.

Finally, given that immunostimulation is the overall concern, the use of newer cell-based assays which track levels of biomarkers (e.g., cytokines, transcription factors, mRNA [62]) affected by the presence of IIMIs, rather than the individual IIMIs themselves, can provide a stronger connection between the applied biotherapeutic and its impact on immunotoxicity and immunogenicity [31]. Past cellular studies focusing on a limited selection of cytokines and chemokines, usually, a combination of pro-inflammatory IL-1, IL-8, IL-6, TNFs, and IFNs, have failed to adequately interrogate the entire immune cascade [2,36]. Since immunotoxicity can cover a range of patient responses from immunostimulation and HSR to immunosuppression, measuring a wider assortment of cytokines, including but not limited to IFNs (α, γ, λ); ILs ($1\alpha/\beta$, 2, 6, 8, 10, 12, 17); interferon-gamma inducible protein (IP-10); TNFα, prostaglandin-E_2 (PGE-2), macrophage inflammatory protein (MIP-1α), and monocyte chemoattractant protein (MCP-1), can provide a more complete picture as to the type and degree of immunotoxic response that can potentially be triggered by a new biotherapeutic formulation.

Author Contributions: All authors (C.K.H. and M.A.D.) researched literature and wrote manuscript. All authors have read and agreed to the published version of the manuscript.

Funding: This study was supported in whole by federal funds from the National Cancer Institute, National Institutes of Health, under contract 75N91019D00024. The content of this publication does not necessarily reflect the views or policies of the Department of Health and Human Services, nor does mention of trade names, commercial products, or organizations imply endorsement by the U.S. Government.

Conflicts of Interest: The authors declare no conflict of interest.

References

1. Janeway, C.A., Jr.; Medzhitov, R. Innate immune recognition. *Annu. Rev. Immunol.* **2002**, *20*, 197–216. [CrossRef]
2. Tisoncik, J.R.; Korth, M.J.; Simmons, C.P.; Farrar, J.; Martin, T.R.; Katze, M.G. Into the eye of the cytokine storm. *Microbiol. Mol. Biol. Rev.* **2012**, *76*, 16–32. [CrossRef]
3. Elsabahy, M.; Wooley, K.L. Cytokines as biomarkers of nanoparticle immunotoxicity. *Chem. Soc. Rev.* **2013**, *42*, 5552–5576. [CrossRef]
4. Singh, V.; Mehrotra, S.; Agarwal, S. The paradigm of Th1 and Th2 cytokines. *Immunol. Res.* **1999**, *20*, 147–161. [CrossRef]

5. Netea, M.G.; Domínguez-Andrés, J.; Barreiro, L.B.; Chavakis, T.; Divangahi, M.; Fuchs, E.; Joosten, L.A.; van der Meer, J.W.; Mhlanga, M.M.; Mulder, W.J. Defining trained immunity and its role in health and disease. *Nat. Rev. Immunol.* **2020**, *20*, 375–388. [CrossRef] [PubMed]
6. Abrantes-Metz, R.M.; Adams, C.; Metz, A. Pharmaceutical development phases: A duration analysis. *FTC Bur. Econ. Work. Pap.* **2004**, *274*, 30. [CrossRef]
7. Hay, M.; Thomas, D.W.; Craighead, J.L.; Economides, C.; Rosenthal, J. Clinical development success rates for investigational drugs. *Nat. Biotechnol.* **2014**, *32*, 40–51. [CrossRef]
8. Kaitin, K.I.; DiMasi, J.A. Pharmaceutical innovation in the 21st century: New drug approvals in the first decade, 2000–2009. *Clin. Pharmacol. Ther.* **2011**, *89*, 183–188. [CrossRef] [PubMed]
9. Verthelyi, D.; Wang, V. Trace levels of innate immune response modulating impurities (IIRMIs) synergize to break tolerance to therapeutic proteins. *PLoS ONE* **2010**, *5*, e15252. [CrossRef] [PubMed]
10. DiPaolo, B.; Pennetti, A.; Nugent, L.; Venkat, K. Monitoring impurities in biopharmaceuticals produced by recombinant technology. *Pharm. Sci. Technol. Today* **1999**, *2*, 70–82. [CrossRef]
11. Dobrovolskaia, M.A. Pre-clinical immunotoxicity studies of nanotechnology-formulated drugs: Challenges, considerations and strategy. *J. Control. Release* **2015**, *220*, 571–583. [CrossRef]
12. Dobrovolskaia, M.A.; McNeil, S.E. Immunological properties of engineered nanomaterials. *Nat. Nanotechnol.* **2007**, *2*, 469. [CrossRef] [PubMed]
13. Dobrovolskaia, M.A.; Shurin, M.; Shvedova, A.A. Current understanding of interactions between nanoparticles and the immune system. *Toxicol. Appl. Pharm.* **2016**, *299*, 78–89. [CrossRef] [PubMed]
14. Mohammadpour, R.; Ghandehari, H. Mechanisms of Immune Response to Inorganic Nanoparticles and Their Degradation Products. *Adv. Drug Deliv. Rev.* **2021**, *180*, 114022. [CrossRef] [PubMed]
15. Reijers, J.A.A.; Malone, K.E.; Bajramovic, J.J.; Verbeek, R.; Burggraaf, J.; Moerland, M. Adverse immunostimulation caused by impurities: The dark side of biopharmaceuticals. *Br. J. Clin. Pharm.* **2019**, *85*, 1418–1426. [CrossRef] [PubMed]
16. Rosenberg, A.S.; Sauna, Z.E. Immunogenicity assessment during the development of protein therapeutics. *J. Pharm Pharmcol.* **2018**, *70*, 584–594. [CrossRef]
17. Sauna, Z.E.; Lagassé, D.; Pedras-Vasconcelos, J.; Golding, B.; Rosenberg, A.S. Evaluating and Mitigating the Immunogenicity of Therapeutic Proteins. *Trends Biotechnol.* **2018**, *36*, 1068–1084. [CrossRef] [PubMed]
18. Sauna, Z.E.; Richards, S.M.; Maillere, B.; Jury, E.C.; Rosenberg, A.S. Editorial: Immunogenicity of Proteins Used as Therapeutics. *Front. Immunol.* **2020**, *11*, 614856. [CrossRef]
19. Kessler, M.; Goldsmith, D.; Schellekens, H. Immunogenicity of biopharmaceuticals. *Nephrol. Dial. Transplant.* **2006**, *21*, v9–v12. [CrossRef] [PubMed]
20. Jawa, V.; Joubert, M.K.; Zhang, Q.; Deshpande, M.; Hapuarachchi, S.; Hall, M.P.; Flynn, G.C. Evaluating immunogenicity risk due to host cell protein impurities in antibody-based biotherapeutics. *AAPS J.* **2016**, *18*, 1439–1452. [CrossRef]
21. Wang, X.; Hunter, A.K.; Mozier, N.M. Host cell proteins in biologics development: Identification, quantitation and risk assessment. *Biotechnol. Bioeng.* **2009**, *103*, 446–458. [CrossRef] [PubMed]
22. Ilinskaya, A.N.; Dobrovolskaia, M.A. Understanding the immunogenicity and antigenicity of nanomaterials: Past, present and future. *Toxicol. Appl. Pharmacol.* **2016**, *299*, 70–77. [CrossRef]
23. Jawa, V.; Cousens, L.P.; Awwad, M.; Wakshull, E.; Kropshofer, H.; De Groot, A.S. T-cell dependent immunogenicity of protein therapeutics: Preclinical assessment and mitigation. *Clin. Immunol.* **2013**, *149*, 534–555. [CrossRef] [PubMed]
24. Parham, P. *The Immune System*, 3rd ed.; Garland Science, Taylor & Francis Group, LLC: New York, NY, USA, 2009.
25. Kumar, H.; Kawai, T.; Akira, S. Toll-like receptors and innate immunity. *Biochem. Biophys. Res. Commun.* **2009**, *388*, 621–625. [CrossRef]
26. Creagh, E.M.; O'Neill, L.A. TLRs, NLRs and RLRs: A trinity of pathogen sensors that co-operate in innate immunity. *Trends Immunol.* **2006**, *27*, 352–357. [CrossRef] [PubMed]
27. Jounai, N.; Kobiyama, K.; Takeshita, F.; Ishii, K.J. Recognition of damage-associated molecular patterns related to nucleic acids during inflammation and vaccination. *Front. Cell. Infect. Microbiol.* **2013**, *2*, 168. [CrossRef]
28. Gorden, K.B.; Gorski, K.S.; Gibson, S.J.; Kedl, R.M.; Kieper, W.C.; Qiu, X.; Tomai, M.A.; Alkan, S.S.; Vasilakos, J.P. Synthetic TLR agonists reveal functional differences between human TLR7 and TLR8. *J. Immunol.* **2005**, *174*, 1259–1268. [CrossRef]
29. Technical Committee ISO/TC 229, N. ISO 29701:2010(en) Nanotechnologies—Endotoxin Test on Nanomaterial Samples for In Vitro systems—Limulus Amebocyte Lysate (LAL) Test. Available online: https://www.iso.org/obp/ui/es/#iso:std:iso:29701:ed-1:v1:en (accessed on 17 November 2021).
30. Yu, L.; Wang, L.; Chen, S. Endogenous toll-like receptor ligands and their biological significance. *J. Cell. Mol. Med.* **2010**, *14*, 2592–2603. [CrossRef]
31. Haile, L.A.; Puig, M.; Polumuri, S.K.; Ascher, J.; Verthelyi, D. In vivo effect of innate immune response modulating impurities on the skin milieu using a macaque model: Impact on product immunogenicity. *J. Pharm. Sci.* **2017**, *106*, 751–760. [CrossRef]
32. Medvedev, A.E.; Sabroe, I.; Hasday, J.D.; Vogel, S.N. Invited review: Tolerance to microbial TLR ligands: Molecular mechanisms and relevance to disease. *J. Endotoxin Res.* **2006**, *12*, 133–150. [CrossRef]
33. Geijtenbeek, T.B.; Gringhuis, S.I. Signalling through C-type lectin receptors: Shaping immune responses. *Nat. Rev. Immunol.* **2009**, *9*, 465–479. [CrossRef] [PubMed]

34. Goodridge, H.S.; Wolf, A.J.; Underhill, D.M. β-glucan recognition by the innate immune system. *Immunol. Rev.* **2009**, *230*, 38–50. [CrossRef] [PubMed]
35. Center for Devices and Radiological Health, U.S. FOOD & DRUG. Immunotoxicity Testing Guidance. 1999; pp. 1–16. Available online: https://www.fda.gov/regulatory-information/search-fda-guidance-documents/immunotoxicity-testing-guidance (accessed on 17 August 2021).
36. Descotes, J. Methods of evaluating immunotoxicity. *Expert Opin. Drug Metab. Toxicol.* **2006**, *2*, 249–259. [CrossRef] [PubMed]
37. Abbas, A.K.; Lichtman, A.H.; Pillai, S. Hypersensitivity Disorders. In *Cellular and Molecular Immunology E-book*, 8th ed.; Elsevier Health Sciences: Philadelphia, PA, USA, 2014.
38. Rosenberg, A. Immunogenicity of biological therapeutics: A hierarchy of concerns. *Dev. Biol.* **2003**, *112*, 15–22.
39. Vial, T.; Choquet-Kastylevsky, G.; Descotes, J. Adverse effects of immunotherapeutics involving the immune system. *Toxicology* **2002**, *174*, 3–11. [CrossRef]
40. Gell, P.G.H.; Coombs, R.R.A. The Classification of Allergic Reactions Underlying Disease. In *Clinical Aspects of Immunology*; Blackwell: Oxford, UK, 1963; pp. 317–337.
41. Krishna, M.; Nadler, S.G. Immunogenicity to biotherapeutics–the role of anti-drug immune complexes. *Front. Immunol.* **2016**, *7*, 21. [CrossRef] [PubMed]
42. Szebeni, J.; Simberg, D.; González-Fernández, Á.; Barenholz, Y.; Dobrovolskaia, M.A. Roadmap and strategy for overcoming infusion reactions to nanomedicines. *Nat. Nanotechnol.* **2018**, *13*, 1100–1108. [CrossRef] [PubMed]
43. Sarma, J.V.; Ward, P.A. The complement system. *Cell Tissue Res.* **2011**, *343*, 227–235. [CrossRef]
44. Szebeni, J. Complement activation-related pseudoallergy: An innate response to nanomedicines acting as pseudo-viruses. *Eur. J. Nanomed.* **2015**, *7*, 203–205. [CrossRef]
45. Finkelman, F.D.; Khodoun, M.V.; Strait, R. Human IgE-independent systemic anaphylaxis. *J. Allergy Clin. Immunol.* **2016**, *137*, 1674–1680. [CrossRef]
46. Ali, H. Regulation of human mast cell and basophil function by anaphylatoxins C3a and C5a. *Immunol. Lett.* **2010**, *128*, 36–45. [CrossRef]
47. Fennrich, S.; Hennig, U.; Toliashvili, L.; Schlensak, C.; Wendel, H.P.; Stoppelkamp, S. More than 70 years of pyrogen detection: Current state and future perspectives. *Altern. Lab. Anim.* **2016**, *44*, 239–253. [CrossRef]
48. Binkowska, A.M.; Michalak, G.; Pilip, S.; Kopacz, M.; Słotwiński, R. The diagnostic value of early cytokine response in patients after major trauma-preliminary report. *Cent. Eur. J. Immunol.* **2018**, *43*, 33–41. [CrossRef] [PubMed]
49. Administration, U.S.F. a. D. Inspection Technical Guides: Bacterial Endotoxins/Pyrogens. Health and Human Services, D. a. D. P. Ed. 1985; pp. 1–9. Available online: https://www.fda.gov/inspections-compliance-enforcement-and-criminal-investigations/inspection-technical-guides/bacterial-endotoxinspyrogens (accessed on 4 October 2021).
50. Dobrovolskaia, M.A.; McNeil, S.E. Strategy for selecting nanotechnology carriers to overcome immunological and hematological toxicities challenging clinical translation of nucleic acid-based therapeutics. *Expert Opin. Drug Deliv.* **2015**, *12*, 1163–1175. [CrossRef] [PubMed]
51. Sakurai, H.; Kawabata, K.; Sakurai, F.; Nakagawa, S.; Mizuguchi, H. Innate immune response induced by gene delivery vectors. *Int. J. Pharm.* **2008**, *354*, 9–15. [CrossRef] [PubMed]
52. Shvedova, A.A.; Kisin, E.R.; Mercer, R.; Murray, A.R.; Johnson, V.J.; Potapovich, A.I.; Tyurina, Y.Y.; Gorelik, O.; Arepalli, S.; Schwegler-Berry, D. Unusual inflammatory and fibrogenic pulmonary responses to single-walled carbon nanotubes in mice. *Am. J. Physiol.-Lung Cell. Mol. Physiol.* **2005**, *289*, L698–L708. [CrossRef]
53. Tsai, C.-Y.; Lu, S.-L.; Hu, C.-W.; Yeh, C.-S.; Lee, G.-B.; Lei, H.-Y. Size-dependent attenuation of TLR9 signaling by gold nanoparticles in macrophages. *J. Immunol.* **2012**, *188*, 68–76. [CrossRef]
54. Crist, R.M.; Grossman, J.H.; Patri, A.K.; Stern, S.T.; Dobrovolskaia, M.A.; Adiseshaiah, P.P.; Clogston, J.D.; McNeil, S.E. Common pitfalls in nanotechnology: Lessons learned from NCI's Nanotechnology Characterization Laboratory. *Integr. Biol. (Camb.)* **2013**, *5*, 66–73. [CrossRef]
55. Dobrovolskaia, M.A. Nucleic Acid Nanoparticles at a Crossroads of Vaccines and Immunotherapies. *Molecules* **2019**, *24*, 4620. [CrossRef]
56. Dobrovolskaia, M.A.; Germolec, D.R.; Weaver, J.L. Evaluation of nanoparticle immunotoxicity. *Nat. Nanotechnol.* **2009**, *4*, 411–414. [CrossRef]
57. Shah, A.; Dobrovolskaia, M.A. Immunological effects of iron oxide nanoparticles and iron-based complex drug formulations: Therapeutic benefits, toxicity, mechanistic insights, and translational considerations. *Nanomedicine* **2018**, *14*, 977–990. [CrossRef]
58. Zolnik, B.S.; González-Fernández, A.; Sadrieh, N.; Dobrovolskaia, M.A. Nanoparticles and the immune system. *Endocrinology* **2010**, *151*, 458–465. [CrossRef]
59. Xiang, S.D.; Scholzen, A.; Minigo, G.; David, C.; Apostolopoulos, V.; Mottram, P.L.; Plebanski, M. Pathogen recognition and development of particulate vaccines: Does size matter? *Methods* **2006**, *40*, 1–9. [CrossRef]
60. Simak, J.; De Paoli, S. The effects of nanomaterials on blood coagulation in hemostasis and thrombosis. *Wiley Interdiscip. Rev. Nanomed. Nanobiotechnol.* **2017**, *9*, e1448. [CrossRef] [PubMed]
61. Dobrovolskaia, M.A.; Patri, A.K.; Simak, J.; Hall, J.B.; Semberova, J.; De Paoli Lacerda, S.H.; McNeil, S.E. Nanoparticle size and surface charge determine effects of PAMAM dendrimers on human platelets in vitro. *Mol. Pharm.* **2012**, *9*, 382–393. [CrossRef] [PubMed]

62. Center for Drug Evaluation and Research, U.S. FOOD & DRUG. Guidance for Industry: Immunogenicity Assessment for Therapeutic Protein Products. 2014; Volume FDA-2013-D-0092, pp. 1–36. Available online: https://www.fda.gov/regulatory-information/search-fda-guidance-documents/immunogenicity-assessment-therapeutic-protein-products (accessed on 4 October 2021).
63. ⟨85⟩ Bacterial Endotoxins Test. In USP 35-NF 30, (USP-NF). Available online: http://www.houshiji.com/Uploads/Download/5a27396519552.pdf (accessed on 17 November 2021).
64. ⟨151⟩ Pyrogen Test. In USP 40-NF 35, (USP-NF). Available online: http://www.uspbpep.com/usp31/v31261/usp31nf26s1_c151.asp (accessed on 17 November 2021).
65. Guidance for Industry: Pyrogen and Endotoxins Testing: Questions and Answers. Available online: https://www.fda.gov/regulatory-information/search-fda-guidance-documents/guidance-industry-pyrogen-and-endotoxins-testing-questions-and-answers (accessed on 17 November 2021).
66. Center for Devices and Radiological Health, U.S. FOOD & DRUG. Endotoxin Testing Recommendations for Single-Use Intraocular Ophthalmic Devices. 2015. Available online: https://www.fda.gov/media/88615/download (accessed on 17 November 2021).
67. Questions and Answers on Quality Related Controlled Correspondence. 2021. Available online: https://www.fda.gov/media/152281/download (accessed on 17 November 2021).
68. Biological Evaluation of Medical Devices–Part 1: Evaluation and Testing within a Risk Management Process. Available online: https://www.iso.org/obp/ui/#iso:std:iso:10993:-1:ed-5:v2:en (accessed on 17 November 2021).
69. Haile, L.A.; Puig, M.; Kelley-Baker, L.; Verthelyi, D. Detection of innate immune response modulating impurities in therapeutic proteins. *PLoS ONE* **2015**, *10*, e0125078.
70. Wadhwa, M.; Knezevic, I.; Kang, H.-N.; Thorpe, R. Immunogenicity assessment of biotherapeutic products: An overview of assays and their utility. *Biologicals* **2015**, *43*, 298–306. [CrossRef]
71. Eaton, L.C. Host cell contaminant protein assay development for recombinant biopharmaceuticals. *J. Chromatogr. A* **1995**, *705*, 105–114. [CrossRef]
72. Pyrogenicit–Principles and Methods for Pyrogen Testing of Medical Devices. Available online: https://www.iso.org/obp/ui/#iso:std:iso:tr:21582:ed-1:v1:en (accessed on 17 November 2021).
73. Singh, S.K.; Afonina, N.; Awwad, M.; Bechtold-Peters, K.; Blue, J.T.; Chou, D.; Cromwell, M.; Krause, H.-J.; Mahler, H.-C.; Meyer, B.K.; et al. An Industry Perspective on the Monitoring of Subvisible Particles as a Quality Attribute for Protein Therapeutics. *J. Pharm. Sci.* **2010**, *99*, 3302–3321. [CrossRef] [PubMed]
74. Carpenter, J.; Cherney, B.; Lubinecki, A.; Ma, S.; Marszal, E.; Mire-Sluis, A.; Nikolai, T.; Novak, J.; Ragheb, J.; Simak, J. Meeting report on protein particles and immunogenicity of therapeutic proteins: Filling in the gaps in risk evaluation and mitigation. *Biologicals* **2010**, *38*, 602–611. [CrossRef]
75. Pacáková, V.; Suchánková, J.; Štulík, K. Separation of biologically active peptides by capillary electrophoresis and high-performance liquid chromatography. *J. Chromatogr. B Biomed. Sci. Appl.* **1996**, *681*, 69–76. [CrossRef]
76. Baumann, A. Early development of therapeutic biologics-pharmacokinetics. *Curr. Drug Metab.* **2006**, *7*, 15–21. [CrossRef]
77. Liu, Y.; Romijn, E.P.; Verniest, G.; Laukens, K.; De Vijlder, T. Mass spectrometry-based structure elucidation of small molecule impurities and degradation products in pharmaceutical development. *TrAC Trends Anal. Chem.* **2019**, *121*, 115686. [CrossRef]
78. Barbosa, M.D.; Kumar, S.; Loughrey, H.; Singh, S.K. Biosimilars and biobetters as tools for understanding and mitigating the immunogenicity of biotherapeutics. *Drug Discov. Today* **2012**, *17*, 1282–1288. [CrossRef]
79. Jawa, V.; Terry, F.; Gokemeijer, J.; Mitra-Kaushik, S.; Roberts, B.J.; Tourdot, S.; De Groot, A.S. T-cell dependent immunogenicity of protein therapeutics pre-clinical assessment and mitigation–updated consensus and review 2020. *Front. Immunol.* **2020**, *11*, 1301. [CrossRef] [PubMed]
80. Lee, J.W.; Salimi-Moosavi, H. Bioanalysis of target biomarker and PK/PD relevancy during the development of biotherapeutics. *Bioanalysis* **2012**, *4*, 2513–2523. [CrossRef]
81. Wang, J.; Lozier, J.; Johnson, G.; Kirshner, S.; Verthelyi, D.; Pariser, A.; Shores, E.; Rosenberg, A. Neutralizing antibodies to therapeutic enzymes: Considerations for testing, prevention and treatment. *Nat. Biotechnol.* **2008**, *26*, 901–908. [CrossRef] [PubMed]
82. Van Regenmortel, M.H. Antigenicity and immunogenicity of synthetic peptides. *Biologicals* **2001**, *29*, 209–213. [CrossRef]
83. Group, B.D.W.; Atkinson Jr, A.J.; Colburn, W.A.; DeGruttola, V.G.; DeMets, D.L.; Downing, G.J.; Hoth, D.F.; Oates, J.A.; Peck, C.C.; Schooley, R.T. Biomarkers and surrogate endpoints: Preferred definitions and conceptual framework. *Clin. Pharmacol. Ther.* **2001**, *69*, 89–95.
84. Wagner, J.A. Strategic approach to fit-for-purpose biomarkers in drug development. *Annu. Rev. Pharmacol. Toxicol.* **2008**, *48*, 631–651. [CrossRef]
85. Fisher, A.C.; Lee, S.L.; Harris, D.P.; Buhse, L.; Kozlowski, S.; Yu, L.; Kopcha, M.; Woodcock, J. Advancing pharmaceutical quality: An overview of science and research in the US FDA's Office of Pharmaceutical Quality. *Int. J. Pharm.* **2016**, *515*, 390–402. [CrossRef] [PubMed]
86. Neun, B.W.; Dobrovolskaia, M.A. NCL Method STE-1.2: Detection and Quantification of Gram Negative Bacterial Endotoxin Contamination in Nanoparticle Formulations by Kinetic Turbidity LAL Assay. Available online: https://ncl.cancer.gov/sites/default/files/protocols/NCL_Method_STE-1.2.pdf (accessed on 24 August 2021).

87. Nakagawa, Y.; Maeda, H.; Murai, T. Evaluation of the in vitro pyrogen test system based on proinflammatory cytokine release from human monocytes: Comparison with a human whole blood culture test system and with the rabbit pyrogen test. *Clin. Diagn. Lab. Immunol.* **2002**, *9*, 588–597. [CrossRef] [PubMed]
88. Kikkert, R.; de Groot, E.R.; Aarden, L.A. Cytokine induction by pyrogens: Comparison of whole blood, mononuclear cells, and TLR-transfectants. *J. Immunol. Methods* **2008**, *336*, 45–55. [CrossRef] [PubMed]
89. Neun, B.W.; Cedrone, E.; Potter, T.M.; Crist, R.M.; Dobrovolskaia, M.A. Detection of Beta-Glucan Contamination in Nanotechnology-Based Formulations. *Molecules* **2020**, *25*, 3367. [CrossRef]
90. Neun, B.W.; Dobrovolskaia, M.A. NCL Method STE-4: Detection and Quantification of β-(1,3)-D-Glucan Contamination in Nanoparticle Formulations by Factor C Depleted LAL (Glucatell®) Assay. Available online: https://ncl.cancer.gov/sites/default/files/NCL_Method_STE-4.pdf (accessed on 24 August 2021).
91. Dobrovolskaia, M.A.; Neun, B.W.; Clogston, J.D.; Ding, H.; Ljubimova, J.; McNeil, S.E. Ambiguities in applying traditional Limulus amebocyte lysate tests to quantify endotoxin in nanoparticle formulations. *Nanomedicine* **2010**, *5*, 555–562. [CrossRef]
92. Dobrovolskaia, M.A.; Neun, B.W.; Clogston, J.D.; Grossman, J.H.; McNeil, S.E. Choice of method for endotoxin detection depends on nanoformulation. *Nanomedicine* **2014**, *9*, 1847–1856. [CrossRef]
93. Neun, B.W.; Dobrovolskaia, M.A. Detection and quantitative evaluation of endotoxin contamination in nanoparticle formulations by LAL-based assays. *Methods Mol. Biol.* **2011**, *697*, 121–130.
94. Neun, B.W.; Dobrovolskaia, M.A. Detection of endotoxin in nano-formulations using limulus amoebocyte lysate (LAL) assays. *JoVE (J. Vis. Exp.)* **2019**, *143*, e58830. [CrossRef]
95. Petersen, E.J.; Henry, T.B.; Zhao, J.; MacCuspie, R.I.; Kirschling, T.L.; Dobrovolskaia, M.A.; Hackley, V.; Xing, B.; White, J.C. Identification and avoidance of potential artifacts and misinterpretations in nanomaterial ecotoxicity measurements. *Environ. Sci. Technol.* **2014**, *48*, 4226–4246. [CrossRef]
96. Gunn, G.R.; Evans, C.; Yang, E. Immunogenicity and biomarkers: Bioanalytical challenges and considerations. *Future Sci.* **2017**, *9*, 1729–1732. [CrossRef]
97. Wadhwa, M.; Thorpe, R. Harmonization and standardization of immunogenicity assessment of biotherapeutic products. *Bioanalysis* **2019**, *11*, 1593–1604. [CrossRef]
98. Forster, R.J.; Bertoncello, P.; Keyes, T.E. Electrogenerated chemiluminescence. *Annu. Rev. Anal. Chem.* **2009**, *2*, 359–385. [CrossRef] [PubMed]
99. Wolter, T.; Richter, A. Assays for controlling host-cell impurities in biopharmaceuticals. *BioProcess Int.* **2005**, *3*, 40–46.
100. Neun, B.W.; Cedrone, E.; Dobrovolskaia, M.A. NCL Method STE 5.2: Analysis of Complement Activation by Single-Plex EIA or Multiplex ELISA. Available online: https://ncl.cancer.gov/sites/default/files/protocols/NCL_Method_ITA-5.2.pdf (accessed on 24 August 2021).
101. Potter, T.M.; Cedrone, E.; Neun, B.W.; Dobrovolskaia, M.A. NCL Method-10: Preparation of Human Whole Blood and Peripheral Blood Mononuclear Cell Cultures to Analyze Nanoparticle Potential to Induce Cytokines In Vitro. Available online: https://ncl.cancer.gov/sites/default/files/NCL_Method_ITA-10.pdf (accessed on 24 August 2021).
102. Dill, K. Sensitive analyte detection and quantitation using the threshold immunoassay system. *ACS Publ.* **1996**, *9*, 89–102.
103. Mufarrege, E.F.; Haile, L.A.; Etcheverrigaray, M.; Verthelyi, D.I. Multiplexed gene expression as a characterization of bioactivity for interferon beta (IFN-β) biosimilar candidates: Impact of innate immune response modulating impurities (IIRMIs). *AAPS J.* **2019**, *21*, 26. [CrossRef] [PubMed]

MDPI
St. Alban-Anlage 66
4052 Basel
Switzerland
Tel. +41 61 683 77 34
Fax +41 61 302 89 18
www.mdpi.com

Molecules Editorial Office
E-mail: molecules@mdpi.com
www.mdpi.com/journal/molecules

www.ingramcontent.com/pod-product-compliance
Lightning Source LLC
Chambersburg PA
CBHW061605210926
51850CB00009B/2214
9783036581668